Methods in Molecular Biology

Series Editor
John M. Walker
School of Life and Medical Sciences
University of Hertfordshire
Hatfield, Hertfordshire, UK

For further volumes:
http://www.springer.com/series/7651

For over 35 years, biological scientists have come to rely on the research protocols and methodologies in the critically acclaimed *Methods in Molecular Biology* series. The series was the first to introduce the step-by-step protocols approach that has become the standard in all biomedical protocol publishing. Each protocol is provided in readily-reproducible step-by-step fashion, opening with an introductory overview, a list of the materials and reagents needed to complete the experiment, and followed by a detailed procedure that is supported with a helpful notes section offering tips and tricks of the trade as well as troubleshooting advice. These hallmark features were introduced by series editor Dr. John Walker and constitute the key ingredient in each and every volume of the *Methods in Molecular Biology* series. Tested and trusted, comprehensive and reliable, all protocols from the series are indexed in PubMed.

Stem Cell Transcriptional Networks

Methods and Protocols

Second Edition

Edited by

Benjamin L. Kidder

Wayne State University School of Medicine, Karmanos Cancer Institute, Detroit, MI, USA

Humana Press

Editor
Benjamin L. Kidder
Wayne State University School of Medicine
Karmanos Cancer Institute
Detroit, MI, USA

ISSN 1064-3745 ISSN 1940-6029 (electronic)
Methods in Molecular Biology
ISBN 978-1-0716-0300-0 ISBN 978-1-0716-0301-7 (eBook)
https://doi.org/10.1007/978-1-0716-0301-7

This Humana imprint is published by the registered company Springer Science+Business Media, LLC, part of Springer Nature.
The registered company address is: 233 Spring Street, New York, NY 10013, U.S.A.

Preface

Advancements in stem cell research have shaped our understanding of genetics, epigenetics, developmental biology, and cancer biology, and stem cell research has sparked immense interest due to the potential of stem cells to treat human diseases. However, understanding the underlying transcriptional programs of stem cells that promote self-renewal- vs. differentiation has remained elusive. Next-generation sequencing technologies have armed stem cell biologists with powerful tools to survey single-cell transcriptomes, epigenomes, 3D chromatin landscapes, and transcription factor networks in stem cells. Use of these emerging technologies has shed light on stem cell transcriptional networks that define primitive stem cell and lineage-specific developmental landscapes. Moreover, recent advancements in reprogramming and transdifferentiation have allowed stem cell biologists to develop solutions for regenerative medicine purposes. The aim of this second edition is to provide an experimental and computational resource for biologists to interrogate stem cell transcriptional networks.

Detroit, MI, USA *Benjamin L. Kidder*

Contents

Contributors

Ash A. Alizadeh • *Division of Oncology, Department of Medicine, Stanford Cancer Institute, Stanford University, Stanford, CA, USA; Institute for Stem Cell Biology and Regenerative Medicine, Stanford University, Stanford, CA, USA; Center for Cancer Systems Biology, Stanford University, Stanford, CA, USA; Division of Hematology, Department of Medicine, Stanford Cancer Institute, Stanford University, Stanford, CA, USA*

Bandar Almutairy • *Department of Pharmaceutical Sciences, Eugene Applebaum College of Pharmacy and Health Sciences, Wayne State University, Detroit, MI, USA*

Nuno Luís Barbosa-Morais • *Instituto de Medicina Molecular João Lobo Antunes, Faculdade de Medicina, Universidade de Lisboa, Lisboa, Portugal*

Paula Beati • *Instituto de Investigaciones en Ingeniería Genética y Biología Molecular, Consejo Nacional de Investigaciones Científicas y Técnicas, Buenos Aires, Argentina*

Michael J. Betti • *University of Alabama in Huntsville, Huntsville, AL, USA*

Zhuoyue Bi • *Department of Pharmaceutical Sciences, Eugene Applebaum College of Pharmacy and Health Sciences, Wayne State University, Detroit, MI, USA; School of Health Sciences, Wuhan University, Wuhan, Hubei, China; Hubei Provincial Key Laboratory of Applied Toxicology, Hubei Provincial Center for Disease Control and Prevention, Wuhan, Hubei, China*

Ganna Bilousova • *Department of Dermatology, University of Colorado School of Medicine, Anschutz Medical Campus, Aurora, CO, USA; Charles C. Gates Center for Regenerative Medicine, University of Colorado School of Medicine, Anschutz Medical Campus, Aurora, CO, USA*

Laurel A. Brandsmeier • *HudsonAlpha Institute for Biotechnology, Huntsville, AL, USA*

Qingshan Chang • *Department of Pharmaceutical Sciences, Eugene Applebaum College of Pharmacy and Health Sciences, Wayne State University, Detroit, MI, USA*

Fei Chen • *Department of Pharmaceutical Sciences, Eugene Applebaum College of Pharmacy and Health Sciences, Wayne State University, Detroit, MI, USA*

Lin Cheng • *State Key Laboratory of Medical Genomics, Shanghai Institute of Hematology, Rui Jin Hospital, Shanghai Jiao Tong University School of Medicine, Shanghai, China*

Răzvan V. Chereji • *Division of Developmental Biology, Eunice Kennedy Shriver National Institute for Child Health and Human Development, National Institutes of Health, Bethesda, MD, USA*

Yao Fu • *Department of Pharmaceutical Sciences, Eugene Applebaum College of Pharmacy and Health Sciences, Wayne State University, Detroit, MI, USA*

Lin Gao • *School of Computer Science and Technology, Xidian University, Xi'an, Shaanxi, China*

Klaus Hansen • *The Hansen Group, Biotech Research and Innovation Centre (BRIC), Faculty of Health and Medical Sciences, University of Copenhagen, Copenhagen, Denmark*

Gangqing Hu • *Department of Microbiology, Immunology, and Cell Biology, West Virginia University School of Medicine, Morgantown, WV, USA*

Kangxin Jin • *State Key Laboratory of Ophthalmology, Zhongshan Ophthalmic Center, Sun Yat-sen University, Guangzhou, China*

WENFEI JIN • *Department of Biology, Southern University of Science and Technology, Shenzhen, Guangdong, China*
BENJAMIN L. KIDDER • *Karmanos Cancer Institute, Wayne State University School of Medicine, Detroit, MI, USA*
IGOR KOGUT • *Department of Dermatology, University of Colorado School of Medicine, Anschutz Medical Campus, Aurora, CO, USA; Charles C. Gates Center for Regenerative Medicine, University of Colorado School of Medicine, Anschutz Medical Campus, Aurora, CO, USA*
MADS LERDRUP • *Department of Cellular and Molecular Medicine, Faculty of Health and Medical Sciences, Center for Chromosome Stability, University of Copenhagen, Copenhagen, Denmark; The Hansen Group, Biotech Research and Innovation Centre (BRIC), Faculty of Health and Medical Sciences, University of Copenhagen, Copenhagen, Denmark*
WEI LIN • *Hunan Provincial Key Lab of Emergency and Critical Care, Hunan People's Hospital, Changsha, Hunan, China; Molecular Medicine Division, Translational Genomics Research Institute, Phoenix, AZ, USA*
CHIH LONG LIU • *Division of Oncology, Department of Medicine, Stanford Cancer Institute, Stanford University, Stanford, CA, USA; Institute for Stem Cell Biology and Regenerative Medicine, Stanford University, Stanford, CA, USA*
YONGJU LU • *Department of Pharmaceutical Sciences, Eugene Applebaum College of Pharmacy and Health Sciences, Wayne State University, Detroit, MI, USA*
MARK MACKIEWICZ • *HudsonAlpha Institute for Biotechnology, Huntsville, AL, USA*
SHENNEA S. MCGARVEY • *Department of Dermatology, University of Colorado School of Medicine, Anschutz Medical Campus, Aurora, CO, USA; Charles C. Gates Center for Regenerative Medicine, University of Colorado School of Medicine, Anschutz Medical Campus, Aurora, CO, USA*
PATRICK S. MCGRATH • *Department of Dermatology, University of Colorado School of Medicine, Anschutz Medical Campus, Aurora, CO, USA; Charles C. Gates Center for Regenerative Medicine, University of Colorado School of Medicine, Anschutz Medical Campus, Aurora, CO, USA*
SARAH K. MEADOWS • *HudsonAlpha Institute for Biotechnology, Huntsville, AL, USA*
ERIC M. MENDENHALL • *HudsonAlpha Institute for Biotechnology, Huntsville, AL, USA; University of Alabama in Huntsville, Huntsville, AL, USA*
RICHARD M. MYERS • *HudsonAlpha Institute for Biotechnology, Huntsville, AL, USA*
AMY S. NESMITH • *HudsonAlpha Institute for Biotechnology, Huntsville, AL, USA*
KIMBERLY M. NEWBERRY • *HudsonAlpha Institute for Biotechnology, Huntsville, AL, USA*
AARON M. NEWMAN • *Institute for Stem Cell Biology and Regenerative Medicine, Stanford University, Stanford, CA, USA; Department of Biomedical Data Science, Stanford University, Stanford, CA, USA*
E. CHRISTOPHER PARTRIDGE • *HudsonAlpha Institute for Biotechnology, Huntsville, AL, USA*
MIN QIAN • *Department of Biostatistics, Mailman School of Public Health, Columbia University, New York, NY, USA*
M' KYA ANIQUE RICE • *Department of Pharmaceutical Sciences, Eugene Applebaum College of Pharmacy and Health Sciences, Wayne State University, Detroit, MI, USA*
PEIFENG RUAN • *Department of Biostatistics, George Washington University, Washington, DC, USA*
NUNO SARAIVA-AGOSTINHO • *Instituto de Medicina Molecular João Lobo Antunes, Faculdade de Medicina, Universidade de Lisboa, Lisboa, Portugal*

CHLOÉ B. STEEN • *Division of Oncology, Department of Medicine, Stanford Cancer Institute, Stanford University, Stanford, CA, USA*
ANDREW E. TESCHENDORFF • *CAS Key Lab of Computational Biology, Shanghai Institute for Biological Sciences, Chinese Academy of Sciences, Beijing, China; Statistical Cancer Genomics, UCL Cancer Institute, University College London, London, UK*
CHITRA THUKAR • *Department of Pharmaceutical Sciences, Eugene Applebaum College of Pharmacy and Health Sciences, Wayne State University, Detroit, MI, USA*
PRIYA WADGAONKAR • *Department of Pharmaceutical Sciences, Eugene Applebaum College of Pharmacy and Health Sciences, Wayne State University, Detroit, MI, USA*
SHUANG WANG • *Department of Biostatistics, Mailman School of Public Health, Columbia University, New York, NY, USA*
YA WANG • *Department of Biostatistics, Mailman School of Public Health, Columbia University, New York, NY, USA*
CHUIJIN WEI • *State Key Laboratory of Medical Genomics, Shanghai Institute of Hematology, Rui Jin Hospital, Shanghai Jiao Tong University School of Medicine, Shanghai, China*
MENGQING XIANG • *State Key Laboratory of Ophthalmology, Zhongshan Ophthalmic Center, Sun Yat-sen University, Guangzhou, China*
DONGCHANG XIAO • *State Key Laboratory of Ophthalmology, Zhongshan Ophthalmic Center, Sun Yat-sen University, Guangzhou, China*
SHUMIN XIONG • *State Key Laboratory of Medical Genomics, Shanghai Institute of Hematology, Rui Jin Hospital, Shanghai Jiao Tong University School of Medicine, Shanghai, China*
LIPING XU • *Department of Pharmaceutical Sciences, Eugene Applebaum College of Pharmacy and Health Sciences, Wayne State University, Detroit, MI, USA*
YUSEN YE • *School of Computer Science and Technology, Xidian University, Xi'an, Shaanxi, China*
GUANGCHUANG YU • *Department of Bioinformatics, School of Basic Medical Sciences, Southern Medical University, Guangzhou, China*
QIAN ZHANG • *Department of Pharmaceutical Sciences, Eugene Applebaum College of Pharmacy and Health Sciences, Wayne State University, Detroit, MI, USA*
SHIHUA ZHANG • *NCMIS, CEMS, RCSDS, Academy of Mathematics and Systems Science, Chinese Academy of Sciences, Beijing, China; School of Mathematical Sciences, University of Chinese Academy of Sciences, Beijing, China; Center for Excellence in Animal Evolution and Genetics, Chinese Academy of Sciences, Kunming, China*
WENXIAN ZHANG • *Department of Pharmaceutical Sciences, Eugene Applebaum College of Pharmacy and Health Sciences, Wayne State University, Detroit, MI, USA*
JIANMEI ZHONG • *Department of Medical Genetics, School of Basic Medical Sciences, Southern Medical University, Guangzhou, Guangdong, China*
BO ZHOU • *Department of Biology, Southern University of Science and Technology, Shenzhen, Guangdong, China*
MIN ZOU • *Herbert Irving Comprehensive Cancer Center, Columbia University Medical Center, New York, NY, USA*

Part I

Next-Generation Sequencing Technologies and Data Analysis

Chapter 1

Epitope Tagging ChIP-Seq of DNA Binding Proteins Using CETCh-Seq

Sarah K. Meadows, Laurel A. Brandsmeier, Kimberly M. Newberry, Michael J. Betti, Amy S. Nesmith, Mark Mackiewicz, E. Christopher Partridge, Eric M. Mendenhall, and Richard M. Myers

Abstract

Chromatin immunoprecipitation followed by next-generation DNA sequencing (ChIP-seq) has been used to identify transcription factor (TF) binding proteins throughout the genome. Unfortunately, this approach traditionally requires commercially available, ChIP-seq grade antibodies that frequently fail to generate acceptable datasets. To obtain data for the many TFs for which there is no appropriate antibody, we recently developed a new method for performing ChIP-seq by epitope tagging endogenous TFs using CRISPR/Cas9 genome editing technology (CETCh-seq). Here, we describe our general protocol of CETCh-seq for both adherent and nonadherent cell lines using a commercially available FLAG antibody.

Key words CRISPR, Epitope tagging, Gene regulation, Genomics, Transcription factors, ChIP-seq, CETCh-seq

1 Introduction

As a part of the ENCODE (Encyclopedia of DNA Elements) Consortium, our lab has used ChIP-seq to identify the localization of endogenous TFs in a number of human cancer cell lines and mouse embryonic stem cells, with the goal of eventually identifying the genomic binding sites of all DNA-binding proteins important in gene regulation. One of the impediments to the ChIP-seq approach is that commercial antibodies are often unavailable for factors of interest, and despite thousands of experimental attempts, only a few hundred TFs have thus far been characterized successfully [1]. To get around this difficulty, we developed a CRISPR/Cas9 system that tags the carboxyl-terminus of endogenous TFs with three consecutive FLAG epitope sequences that can be recognized by one well-characterized, commercially available antibody (CETCH-seq) [2]. This has enabled us to scale up and significantly

Benjamin L. Kidder (ed.), *Stem Cell Transcriptional Networks: Methods and Protocols*, Methods in Molecular Biology, vol. 2117, https://doi.org/10.1007/978-1-0716-0301-7_1,

improve both the consistency and accuracy of our experiments, while rescuing the characterization of TFs that have no ChIP-seq grade antibodies.

To add the FLAG epitope tag, one or two gRNA plasmids are needed that encode the Cas9 nuclease and the specific guide sequence to direct it where to make a double stranded break in the DNA close to the stop codon. In addition, a donor plasmid is required that has the desired insertion sequence (3×FLAG epitope tag, P2A linker, and Neomycin resistance gene) flanked by regions homologous to either side of the double stranded break to complete the homology directed repair. Once these plasmids are designed and constructed, they are transfected into the cell line through nucleofection to begin cleaving the DNA and integrating the epitope tag through homology directed repair. Cells containing the correct insertion are then selected for by growing them in medium with G418. After selection is complete, the cells are cross-linked and harvested. At this point, a ChIP-seq experiment can be performed, using a standard FLAG antibody to immunoprecipitate the tagged protein along with its bound DNA. The cross-links are reversed and the eluted DNA can be sequenced to determine the genomic regions that were bound by the tagged protein.

This protocol describes our method from CRISPR design through immunoprecipitation. We also include protocols for some validation methods accepted by the ENCODE Consortium that confirm the integration of the tag and the expression of the tagged protein.

2 Materials

2.1 CRISPR Design

1. Genome database, such as UCSC genome browser (https://genome.ucsc.edu/).
2. CRISPR design software or Internet resource, such as Benchling (https://benchling.com/).
3. Microsoft Excel, Google sheets, or other spreadsheet software.
4. Microsoft Word, Google docs, or other document software.

2.2 Donor Plasmid Construction

1. FLAG donor vector stock (Addgene #63934, pFETCh_Donor).
2. BsaI-HFv2 (New England Biolabs, R3733S).
3. BbsI-HF (New England Biolabs, R3539S).
4. Cutsmart Buffer (New England Biolabs).
5. Qiaquick PCR Purification kit (Qiagen, 28104).
6. Qubit dsDNA HS Assay Kit (Thermo Fisher, Q32854).

7. Gibson Assembly Master Mix (New England Biolabs, E2611L).
8. gBlocks for homology arms (IDT).
9. Electroporator.
10. 0.1 cm cuvettes.
11. MegaX DH10B T1^R electrocompetent *E. coli* cells (Life Technologies, C6400-03).
12. Recovery medium.
13. 50 μg/mL Kanamycin LB plates (Teknova, L1025).
14. One*Taq* 2× Master Mix with Standard Buffer (New England Biolabs, M0482L).
15. Seq_HA1 forward primer "ACGCCTGTGAAACCGTACTA".
16. Seq_4 reverse primer "GACGGCCAGTGAATTGGAG".
17. 1 kb Plus DNA ladder (New England Biolabs, N3200L).
18. 2% agarose gel.
19. 14 mL round bottom tubes.
20. Kanamycin stock 50 mg/mL.
21. 60% glycerol, sterilized.
22. QIAprep Spin Miniprep Kit (Qiagen, 27106).
23. Qubit dsDNA BR Assay Kit (Thermo Fisher, Q32853).
24. Seq_P2A_v2 reverse primer "TAGCTCCGCTTCCTGAAACC".
25. Seq_HA2 forward primer "GGCCGCTTTTCTGGATTCAT".
26. Seq_5 reverse primer "AACTGTTGGGAAGGGCGATC".
27. Vacuum concentrator.

2.3 gRNA Plasmid Construction

1. T4 DNA Ligase Reaction Buffer, 10× (New England Biolabs, B0202S).
2. T4 Polynucleotide Kinase (PNK) (New England Biolabs, M0201L).
3. gRNA plasmid vector stock (Addgene #42230, pX330-U6-Chimeric_BB-CBh-hSpCas9).
4. 10× FastDigest buffer.
5. DTT (10 mM).
6. ATP (10 mM).
7. FastDigest BpiI (Thermo Fisher, FD1014).
8. T7 DNA ligase (New England Biolabs, M0318L).
9. Plasmid-Safe 10× Reaction Buffer.

10. Plasmid-Safe ATP-Dependent DNase (Epicentre, E3101K).
11. One Shot TOP10 Chemically Competent *E. coli* cells (Life Technologies, C404003).
12. SOC medium.
13. 100 μg/mL Carbenicillin plates (Teknova, L1906).
14. One*Taq* 2× Master Mix with Standard Buffer (New England Biolabs, M0482L).
15. U6_Mod_Seq forward primer "ACTATCATATGCTTACCG TAAC".
16. 100 bp ladder (New England Biolabs, N3231L).
17. 2% agarose gel.
18. 14 mL round bottom tubes.
19. Carbenicillin stock 100 mg/mL.
20. 60% glycerol, sterilized.
21. QIAprep Spin Miniprep Kit (Qiagen, 27106).
22. Qubit dsDNA BR Assay Kit (Thermo Fisher, Q32853).
23. gRNA_seq_v2 primer "AGGGCCTATTTCCCATGATTCC".
24. Vacuum concentrator.

2.4 Cell Growth and Passage

1. Growth medium: complete medium for cell line of interest, no antibiotic.
2. Selection medium: complete medium with 300 or 400 μg/mL Geneticin™ (or a low and a high dose appropriate for cell line of interest) (G418, 50 mg/mL stock; Invitrogen, 10131-035).
3. Dissociation reagent: Trypsin-EDTA 0.25% (1×, phenol-red).
4. Cell culture flasks, 25, 75, and 150 cm^2.
5. Cell culture dishes, 150 mm.
6. Phosphate buffered saline (1× PBS) without calcium and magnesium.

2.5 Transfection

1. Amaxa™ Cell Line Nucleofector™ Kit V (Lonza, VVCA-1003) with cuvettes.
2. Eppendorf tubes, 1.5-mL.
3. Amaxa™ Nucleofector™ II transfection machine.

2.6 Cross-Linking

1. Formaldehyde, 37%.
2. Glycine, 2.5 M (20×).
3. Ice-cold phosphate buffered saline (1× PBS) without calcium or magnesium.
4. Ice-cold Farnham Lysis Buffer with protease inhibitors.
5. Cell scrapers.

6. Dry ice or liquid nitrogen.
7. Conical centrifuge tubes, 15 and 50 mL.

2.7 ChIP

1. Farnham Lysis Buffer: 5 mM PIPES pH 8.0, 85 mM KCl, 0.5% NP-40 (filtered 0.2–0.45 μm filter unit), Protease Inhibitor Cocktail (Sigma, 11697498001) 4 °C. [Add protease inhibitor tablet just before use, good for 24 h when refrigerated.]
2. RIPA buffer: 1× PBS, 1% NP-40, 0.5% sodium deoxycholate, 0.1% SDS (filtered 0.2–0.45 μm filter unit), Protease Inhibitor Cocktail (Sigma, 11697498001), 4 °C. [Add protease inhibitor tablet just before use, good for 24 h when refrigerated.]
3. Needles, 20 G.
4. Syringes, 3 mL.
5. Bioruptor Pico sonication device (Diagenode, B01060010).
6. Bioruptor Pico Microtube with cap, 1.5 mL (Diagenode, C30001610).
7. Dynamag-2 Magnet (Invitrogen, 12321D).
8. Sigma F1804 Monoclonal Anti-FLAG M2 antibody produced in mouse, 1 mg/mL.
9. Dynabeads M-280 Sheep Anti-Mouse IgG (Invitrogen 11201D).
10. PBS/BSA wash buffer: 1× PBS, 5 mg/mL BSA (fraction V), (filtered 0.2–0.45 μm filter unit), Protease Inhibitor Cocktail (Sigma, 11697498001), 4 °C. [Add protease inhibitor tablet just before use, good for 24 h when refrigerated.]
11. LiCl IP Wash Buffer: 100 mM Tris pH 7.5, 500 mM LiCl, 1% NP-40, 1% sodium deoxycholate (filtered 0.2–0.45 μm filter unit) 4 °C.
12. TE: 10 mM Tris–HCl pH 7.5, 0.1 mM Na_2EDTA 4 °C.
13. IP Elution Buffer: 1% SDS, 0.1 M $NaHCO_3$ (filtered 0.2–0.45 μm filter unit).
14. QIAquick PCR Purification Kit (Qiagen, 28106).
15. Protease Inhibitor Cocktail (Sigma, 11697498001).

2.8 PCR

1. DNeasy Blood and Tissue kit (Qiagen, 69506).
2. RNase A (100 mg/mL) (Qiagen, 19101).
3. Phusion High-Fidelity PCR Master Mix with HF Buffer (Thermo Fisher, F531L).
4. P2A linker reverse primer "TAGCTCCGCTTCCTGAAA CC".
5. Neomycin resistance gene forward primer "ACCGCTTCCT CGTGCTTTAC".

6. 100 bp ladder (New England Biolabs, N3231L).
7. 2% agarose gel.

2.9 Western Blot

1. Thermo Scientific NE-PER Nuclear and Cytoplasmic Extraction Reagents (Thermo Scientific, 78835).
2. Spectrophotometer.
3. Halt Protease Inhibitor Cocktail (100×) (Thermo Fisher, 78429).
4. Molecular-grade water.
5. NuPage LDS Sample Buffer (4×) (Thermo Fisher, NP0007).
6. NuPage Sample Reducing Agent (10×) (Thermo Fisher, NP0009).
7. Eppendorf tubes, 1.5 mL/2 mL.
8. Dry bath incubator.
9. NuPage 4–12% Bis-Tris Protein Gels, 1.5 mm, 10-well (Thermo Fisher, NP0335BOX).
10. NuPage MOPS SDS Running Buffer (20×) (Thermo Fisher, NP0001).
11. WesternSure Pre-stained Chemiluminescent Protein Ladder (LI-COR, 92698000).
12. Mini Gel Tank (Thermo Fisher, A25977).
13. PowerEase 90 W system (115 VAC) (Thermo Fisher, PS0090).
14. iBlot 2 NC Regular or Mini Stacks (Thermo Fisher, IB23001 or Thermo Fisher, IB23002).
15. iBlot 2 Gel Transfer Device (Thermo Fisher, IB21001).
16. TBS Tween (20×).
17. Milk, nonfat powdered.
18. Sterile conical tubes, 50 mL.
19. Plastic wrap.
20. Laboratory rocker.
21. Bovine serum albumin, 30% solution (BSA).
22. Monoclonal anti-FLAG M2-Peroxidase (HRP) antibody produced in mouse (Sigma, A8592).
23. SuperSignal West Femto Maximum Sensitivity Substrate kit (Thermo Fisher, 34094).
24. Imaging machine (with chemiluminescence functionality).

3 Methods

3.1 CRISPR Design

1. Choose a stop codon of the specific transcription factor or gene to be tagged if there is more than one stop codon (*see* **Note 1**). In Benchling, paste the DNA sequence of the stop codon, including 25 bp upstream and 40 bp downstream, from the genome database (*see* **Note 2**). Alternatively, put in the genomic coordinates into Benchling, and it will fetch the sequence for you from the reference.
2. To design the gRNAs, click on the "Design and Analyze Guides" button. Choose the following settings in the Guide Parameters box and click Finish: single guide, 20 bp guide length, appropriate genome assembly that matches your imported sequence, that is, hg19 *Homo sapiens* genome, and NGG PAM site (*see* **Note 3**).
3. Click the "+" button next to the target region to create the guides. The target region should be 1–68.
4. Click on the blue words "No region set" to set the genome region of the location to be tagged so it will be removed from the off-target score calculations. In the dialogue box, click on "Find Genome Matches" then choose the correct chromosome region to be tagged. Click the "Set genome region" button.
5. Choose the best two guide RNAs based on their distance to the stop codon, on-target score, and their off-target score (*see* **Note 4**).
6. Record the forward and reverse complement sequences of each gRNA (*see* **Note 5**). We denote the gRNA with the highest on-target score as gRNA_A and the other as gRNA_B to distinguish them in future protocols.
7. For each gRNA, examine the forward sequence. The first base should be a "G". If it is not a "G", delete up to two base pairs to make it first. If this is not possible, add a "G" to the beginning of the gRNA. For the reverse complement sequence, alter the sequence as necessary by adding a "C" to the end or deleting bases to keep it the reverse complement of the forward sequence.
8. Add the tail sequence "CACC" to the 5′ end of the forward gRNA sequence and "AAAC" to the 5′ end of the reverse complement sequence.
9. Denote the on-target and off-target scores next to the gRNAs to refer back to in case there are problems cutting and integrating the tag sequence.

10. Order both the forward and reverse complement gRNA sequences as single stranded oligos. We order our oligos in tubes with 25 nm concentration and with standard desalting.
11. Next, design gBlocks for the homology regions upstream and downstream of the cut sites for the donor plasmid. Return to the genome database and extract the DNA sequence of the stop codon including 2000 bp upstream and 2000 bp downstream.
12. We find it easiest to paste this sequence into a text document. Label the sequence with the gene name and highlight or denote the locations of the stop codon and the gRNA cut sites.
13. Create homology arm 1 (HOM1) for the donor plasmid. Copy 708 bp of sequence directly upstream of the stop codon. If one of the gRNA cut sites is in the region, check the gRNA sequence to see if it will also cut the donor plasmid homology arm. If it has a good chance of cutting the donor plasmid, alter the sequence using the redundancy of the genetic code to keep the amino acids the same but changing the bases just enough so that it will not cut. It may be easiest to alter the PAM site sequence.
14. Paste sequence in the gBlock order form on the IDT website and click on "Test complexity" (*see* **Note 6**). If the gBlock passes, add the tail sequence "tccccgacctgcagcccagct" to the beginning or 5′ end of the sequence and add "gggagcggaggagg ttccgg" to the end or 3′ end of the sequence for Gibson Assembly.
15. Test the sequence for complexity again and adjust the sequence as necessary until it passes (*see* **Note 7**). Denote the chosen gBlock sequence in the text document and record the gBlock sequence with tails for ordering later.
16. Create homology arm 2 (HOM2) for the donor plasmid. Copy 708 bp of sequence directly after the last gRNA cut site. If there are no cut sites downstream of the stop codon, copy the sequence directly after the end of the stop codon (*see* **Note 8**).
17. Paste sequence in the gBlock order form on the IDT website and click on "Test complexity" (*see* **Note 9**). If the gBlock passes, add the tail sequence "agttcttctgattcgaacatc" to the beginning or 5′ end of the sequence and add "cttggaaagtcctctc-cactg" to the end or 3′ end of the sequence for Gibson Assembly.
18. Test the sequence for complexity again and adjust the sequence as necessary until it passes. Denote the chosen gBlock sequence in the text document and record the gBlock sequence with tails for ordering later.
19. Order both HOM1 and HOM2 gBlocks from IDT.

3.2 Donor Plasmid Construction

1. Digest FLAG donor vector stock using the following protocol in a final volume of 50 μL:
   - 5 μL 10× Cutsmart Buffer.
   - 0.5 μL (or 10 units) BbsI-HF (20,000 units/mL).
   - 0.5 μL (or 10 units) BsaI-HF (20,000 units/mL).
   - 1 μg FLAG donor vector.
   - Bring to 50 μL with dH_2O.
2. To make a large batch of digested donor vector stock, make a total of eight reactions in individual PCR tubes.
3. Incubate reactions 37 °C for 1 h (hour), then 65 °C for 20 min.
4. Combine all eight PCR tubes of digested vector into a 1.5 mL tube.
5. Load 10 μL of digest onto a 2% agarose gel. Run gel with 1 kb Plus DNA ladder for 30 min. There should be two bands around 4.5 and 1 kb.
6. If gel looks correct, clean up the digests using the Qiaquick PCR kit and splitting it evenly over four columns. Elute each column with 50 μL of molecular water.
7. Combine elutions into one tube and measure the DNA con centration using the dsDNA HS Assay Kit with the Qubit fluorometer. Adjust the concentration to 10 ng/μL with water and aliquot 100 μL each into 1.5 mL tubes. Store at −20 °C until ready to use.
8. Next, assemble the donor plasmid using Gibson Assembly. Resuspend the gBlocks in 20 μL of TE buffer. The gBlocks should have 500 ng of DNA and will be 25 ng/μL after resuspension. Make a dilution of each gBlock to 10 ng/μL in PCR strip tubes with 1 μL of gBlock and 1.5 μL of water.
9. Assemble each donor plasmid in PCR tubes on ice using the following protocol in a final volume of 10 μL. Use the New England Biolabs ligation calculator, or similar calculator, to determine how much of each homology arm to add.
   - 5 μL Gibson Assembly 2× Master Mix.
   - Homology arm 1 (3:1 insert to vector ratio).
   - Homology arm 2 (3:1 insert to vector ratio).
   - 2.5 μL Digested FLAG donor vector (25 ng).
   - Up to 10 μL dH_2O.
10. Incubate reactions at 50 °C for 1 h in the thermocycler. Gibson assemblies can be transformed directly without purification.
11. Transform the donor plasmids into cells using electroporation. Place recovery medium and one kanamycin LB agar plate for each reaction in 37 °C incubator to warm them up.

12. Place one cuvette per reaction on ice.
13. In strip tubes on ice, add 1 μL of each assembled donor plasmid Gibson assembly reaction to 23 μL of water.
14. Thaw DH10B electrocompetent cells (~100 μL per tube) on ice just until thawed. Briefly centrifuge at low speed for about one second to bring cells to bottom of tube.
15. Add 2 μL of competent cells to each diluted assembled donor and stir with pipette tip to distribute cells evenly. Do not pipet up and down to mix. If there are remaining cells after finishing all samples, refreeze at −80 °C to be used in later experiments.
16. Let sit on ice for 1 min.
17. Turn on electroporator and make sure the settings are correct for *E. coli*.
18. Wipe cuvette dry with a kimwipe and put on micropulser arm with the notch on the cuvette facing out.
19. Pipet the cell–DNA mixture into the center well of the cuvette and replace the cap taking care to avoid bubbles when pipetting (*see* **Note 10**).
20. Slide the arm with the cuvette into the device and push "pulse" once. Redo an individual electroporation with new cells, DNA, and cuvette if you hear a popping noise or it arcs.
21. Immediately add 300 μL of warmed recovery medium and transfer the whole solution to a 1.5 mL tube. Repeat electroporation steps for each sample. Close the caps tightly.
22. Incubate transformations at 37 °C for 1 h with shaking horizontally.
23. Plate out 15 μL of recovered cells mixed with 30 μL water onto warmed kanamycin LB agar plates.
24. Place plates upside-down in 37 °C incubator overnight.
25. Save leftover transformation in the refrigerator overnight in case more cells need to be plated for a particular donor the following day.
26. The following day, test at least three colonies from each plate with PCR to check for correctly assembled donor plasmids using the following protocol, in a final volume of 25 μL:
    - 12.5 μL One*Taq* 2× Master Mix.
    - 1.25 μL Seq_HA1 forward primer (10 μM).
    - 1.25 μL Seq_4 reverse primer (10 μM).
    - Colony template.
    - Up to 10 μL dH_2O.
27. Using a sterile pipette tip or inoculating loop, pick a colony to test and mix it in the PCR master mix as the DNA template. Using the same tip or loop, mix it into a 1.5 mL tube that has

80 μL of LB broth. Store these tubes in the 4 °C refrigerator until needed to inoculate overnight cultures if the colony is correct.

28. Incubate the PCR reaction in the thermocycler using the following conditions:
    - Denature at 95 °C for 5 min.
    - 25 cycles of 95 °C for 15 s, 52 °C for 20 s, and 68 °C for 2 min.
    - Extend at 68 °C for 7 min.
    - Hold at 4 °C.
29. Run 5 μL of the PCR reaction on a 2% agarose gel to test for the correct size. If both arms have 708 bp homology arms, the band size should be 2820 bp (*see* **Note 11**).
30. For each correct clone, prepare a sterile 14 mL round bottom tube with 6 mL of LB broth and 6 μL Kanamycin stock antibiotic. Inoculate each tube with 50 μL of the colony/LB broth solution from the PCR step. Grow overnight in 37 °C incubator with shaking (*see* **Note 12**).
31. Create a glycerol stock for each overnight culture by mixing 500 μL overnight culture with 500 μL 60% glycerol in a 1.5 mL cryovial and store at −80 °C.
32. Miniprep each culture with the QIAprep Spin Miniprep Kit using 3–5.5 mL of overnight culture depending on the density of the bacterial growth. Elute each column with 50 μL molecular water two times into the same 1.5 mL tube for a total of 100 μL. Measure the plasmid DNA concentration using the dsDNA BR Assay Kit with the Qubit fluorometer.
33. Send an aliquot of each plasmid prep for Sanger sequencing to verify that the assembly worked correctly and did not lose or change any of the bases. To sequence homology arm 1, use primers Seq_HA1 and Seq_P2A_v2. To sequence homology arm 2, use primers Seq_HA2 and Seq_5.
34. In order to analyze the sequences, make a reference sequence for each homology arm of the donor plasmid with about 175 bp upstream and downstream sequence of the homology arm. Make sure to delete any bases that will be eliminated in the assembly, such as the stop codon, or change ones that were altered in the gBlocks. Use Benchling, or other sequence alignment tool, to align the sequences to the reference and check to make sure they are correct (*see* **Note 13**).
35. For each correct donor plasmid, 10 μg is needed for two replicates of each transfection. If there is at least 20 μg of plasmid prep left, dry down the remaining plasmid prep in a vacuum concentrator until the concentration is approximately

1–2 μg/μL measured with the dsDNA BR Assay Kit in the Qubit fluorometer. If there is less than 20 μg, use the saved glycerol stock to grow up more bacteria and miniprep it until the desired amount is obtained, then proceed to concentrating the plasmid as before (*see* **Note 14**). The donor plasmid is now ready for transfection. Store at −20 °C until ready to use.

3.3 gRNA Plasmid Construction

1. Phosphorylate and anneal each pair of oligos in PCR strip tubes by combining the following reagents:
   - 1 μL gRNA oligo 1 (100 μM).
   - 1 μL gRNA oligo 2 (100 μM).
   - 1 μL 10× T4 DNA Ligase Buffer (*see* **Note 15**).
   - 6.5 μL dH_2O.
   - 0.5 μL T4 PNK.
2. Anneal oligos for 30 min at 37 °C, then 5 min at 95 °C. Then ramp down to 25 °C at the rate of 5 °C/min.
3. Dilute annealed oligos with molecular water to 1:250 in a new set of PCR strip tubes.
4. Digest the gRNA plasmid and ligate in the annealed oligos in PCR strip tubes by combining the following reagents:
   - 1 μL PX330 gRNA plasmid vector stock (100 ng/μL).
   - 2 μL Diluted oligo duplex from the previous step.
   - 2 μL 10× FastDigest buffer.
   - 1 μL DTT (10 mM).
   - 1 μL ATP (10 mM).
   - 1 μL FastDigest BpiI.
   - 0.5 μL T7 DNA ligase.
   - 11.5 μL dH_2O.
5. Incubate the reaction with 6 cycles of 37 °C for 5 min and 25 °C for 5 min.
6. Treat the ligation reaction with Plasmid-Safe exonuclease to prevent unwanted recombination products by combining the following reagents in a final volume of 15 μL:
   - 11 μL Ligation reaction from previous step.
   - 1.5 μL 10× Plasmid-Safe Reaction Buffer.
   - 1.5 μL ATP (10 mM).
   - 1 μL Plasmid-Safe DNase.
7. Incubate the reaction at 37 °C for 30 min.
8. To transform the assembled plasmid, dilute plasmid-safe reactions 1:4.
9. Warm SOC medium and one carbenicillin LB agar plate per reaction in 37 °C incubator.

10. Prepare a water bath or heat block filled with water to 42 °C.
11. Thaw One Shot TOP10 chemically competent cells on ice just until liquid. Each vial should contain 50 μL of competent cells.
12. Add 5 μL of competent cells to a 1.5 mL tube on ice. Add 1 μL of diluted ligation reaction to cells and stir with the pipette tip to mix. Do not pipet up and down. Discard any thawed extra cells when completed with all samples.
13. Let sit on ice for 30 min.
14. Heat shock cells in 42 °C water bath for 30 s. Immediately transfer to ice for 2 min.
15. Add 250 μL of warmed SOC medium to each tube and move to a rack at room temperature. Make sure the tubes are closed tightly.
16. Shake tubes horizontally for 1 h at 37 °C.
17. Plate 80 μL of cells on warmed carbenicillin LB agar plates and place upside down in 37 °C incubator overnight.
18. The following day, test 1 colony of each gRNA plate with PCR to see if it assembled correctly using the following protocol in a final volume of 25 μL:
    - 12.5 μL One*Taq* 2× Master Mix.
    - 1.25 μL U6_Mod_Seq forward primer (10 μM).
    - 1.25 μL gRNA reverse complement oligo (10 μM).
    - Colony template.
    - dH_2O up to 10 μL.
19. Using a sterile pipette tip or inoculating loop, pick a colony to test and mix it in the PCR master mix as the DNA template. Using the same tip or loop, mix it into a 1.5 mL tube that has 80 μL of LB broth. Store these tubes at 4 °C until needed to inoculate overnight cultures if the colony is correct.
20. Incubate the PCR reaction in the thermocycler using the following conditions:
    - Denature at 95 °C for 5 min.
    - 35 cycles of 95 °C for 15 s, 52 °C for 20 s, and 68 °C for 30 s.
    - Extend at 68 °C for 7 min.
    - Hold at 4 °C.
21. Run 20 μL of the PCR reaction on a 2% agarose gel to test for the correct size. The band size should be 101 bp if using the standard 20 bp gRNA not counting the tail sequence.
22. For each correct clone, prepare a sterile 14 mL round bottom tube with 6 mL of LB broth and 6 μL Carbenicillin stock antibiotic. Inoculate each tube with 50 μL of the colony/LB

broth solution from the PCR step. Grow overnight in 37 °C incubator with shaking.

23. Create a glycerol stock for each overnight culture by mixing 500 μL overnight culture with 500 μL 60% glycerol in a 1.5 μL cryovial and store at −80 °C.
24. Perform minipreps on each culture with the QIAprep Spin Miniprep Kit using 4.5 mL of overnight culture. Elute each column with 50 μL molecular water two times into the same 1.5 mL tube for a total of 100 μL. Measure the plasmid DNA concentration using the dsDNA BR Assay Kit with the Qubit fluorometer.
25. Send an aliquot of each plasmid prep for Sanger sequencing to verify that the assembly worked correctly. Use primer gRNA_-seq_v2 for sequencing. Just check to make sure that the returned sequence has the correct sequence of the gRNA.
26. For each correct gRNA plasmid when using two gRNAs per donor, 5 μg is needed for two replicates of each transfection (*see* **Note 16**). If there is at least 10 μg of plasmid prep left, dry down the remaining plasmid prep in a vacuum concentrator until the concentration is approximately 1–2 μg/μL measured with the dsDNA BR Assay Kit in the Qubit fluorometer. If there is less than 10 μg, use the saved glycerol stock to grow up more bacteria and miniprep it until you have the desired amount, then proceed to concentrating the plasmid as before. The gRNA plasmids are now ready for transfection. Store at −20 °C until ready to use.

3.4 Preparation of Cells for Transfection

1. Adherent cells.
   (a) Grow adherent cells per normal protocol to approximately 80% confluence (*see* **Note 17**).
   (b) Aspirate medium from flask.
   (c) Rinse with prewarmed 37 °C 1× PBS and aspirate wash, removing all liquid.
   (d) Apply approximately 1 mL prewarmed 37 °C Trypsin 0.25% for every 50 cm^2 of flask area to be trypsinized and incubate at 37 °C for 5 min.
   (e) Neutralize the reaction with 10 mL prewarmed 37 °C culture medium and swirl to coat the flask bottom.
   (f) Pipet up and down to break up clumps, and remove the cells from the flask to a 15 mL centrifuge tube.
   (g) Centrifuge the cells at 300 rcf for 5 min, and pour off the supernatant.
   (h) Resuspend the pellet in 5 mL PBS and count cells using Trypan Blue exclusion dye.

(i) Transfer 2×10^6 cells into a new 15 mL conical tube (or multiples of that, depending on the number of transfections desired (*see* **Note 18**)), add 1× PBS up to 10 mL, and centrifuge as before to wash away any remaining medium.

(j) Set aside the pelleted cells with their supernatant in a rack at 4 °C.

2. Nonadherent cells.

(a) Grow nonadherent cells per normal protocol. Do not allow the cells to grow past a density of 5×10^6 mL^{-1}. Make sure they are at more than 90% viability and that they are at a relatively early passage number before transfection.

(b) Spin them out of their culture medium at 300 rcf for 5 min. Pour off supernatant.

(c) Follow **steps 1**(**h**, through **j**).

3.5 Nucleofection

1. Label and prepare one sterile Eppendorf tube for each transfection to be performed; add 5 μg donor plasmid and 5 μg gRNA (if two gRNA, use 2.5 μg each), in a volume not to exceed 10 μL (*see* **Note 19**). Additionally, prepare a tube with 10 μL of sterile, nuclease-free water to use as a control or "no template" transfection. The timeline of death of cells in this sample will help confirm when G418 can be discontinued.
2. Label and prepare two 25 cm^2 flasks; fill one flask with 10 mL complete cell culture medium (with no antibiotic) and place in the incubator at 37 °C, 5% CO_2; set the other flask to the side in the biosafety cabinet.
3. Take each needed cuvette out of the package, label the top, and set aside in the biosafety cabinet.
4. Retrieve the cell pellet from 4 °C and aspirate as much PBS as possible from the tube. Resuspend pellet by flicking the tube by hand.
5. Add 82 μL of the proprietary Nucleofector Solution and 18 μL of Supplement from the Amaxa kit per 2×10^6 cells (or ratio appropriate for cell line of interest) and mix with the pipette tip (*see* **Note 20**). This pipetting step is critical in order to resuspend remaining small clumps of cells that can interfere with efficient transfection.
6. With a new tip, add all of the cell mixture to the waiting Eppendorf tubes with the plasmids and mix gently with the pipette tip.
7. Transfer the cell–plasmid mixture to the cuvette, put on the cuvette top, insert into the slot in the Amaxa Nucleofector II,

and run the program for the cell line of interest, selecting the "high efficiency" setting when available (*see* **Note 21**).

8. Place the cuvette back into the biosafety cabinet until all cuvettes are run.
9. Retrieve the prefilled flask with warming medium from the incubator.
10. Add 1 mL of fresh warm medium to each cuvette and pipet the entire volume up and down to break up cells (*see* **Note 22**).
11. Transfer the cells from each cuvette into its waiting flask with either a 1 mL pipette tip or the Pasteur pipette that the kit provides.
12. Resuspend the cells in the flask 5 mL pipette and remove half of the volume into the waiting, empty duplicate flask. There will be two flasks of 5.5 mL each for each nucleofection (*see* **Note 23**).
13. Incubate for 24 h at 37 °C, 5% CO_2. Adherent flasks should be laid down, and nonadherent flasks should be sitting upright for the first 24 h.

3.6 Growth and Maintenance of Cell Lines

1. Adherent cell lines.
   (a) At 24 h after transfection, replace medium with 5 mL of 300 μg/mL G418 (or low dose appropriate for cell type) in complete medium. Replace medium once more during the week with the low dose of G418 regardless of color (*see* **Note 24**).
   (b) Between days 6 and 8, when cells are more confluent, trypsinize and passage cells to two 25 cm^2 flasks and change the culture medium to include 400 μg/mL G418 (or high dose appropriate for cell line) (*see* **Notes 25** and **26**).
   (c) Between 4 days and a week later, depending on confluence, passage the cells to 15 mL medium containing high dose G418 in a 75 cm^2 flask. At this point, start replacing medium based on color and metabolism as cell growth rate increases (*see* **Note 26**). The growth of the control sample should start lagging behind the others during the second week.
   (d) Continue to passage into flasks not more than double the previous area. G418 can be discontinued after the control sample dies (*see* **Note 27**).
   (e) The final passage of the adherent lines before cross-linking should be into 150 mm dishes with 30 mL medium volume each. Make sure that the cells are in medium without G418 for at least 2 days before harvest (*see* **Note 28**).

2. Nonadherent cell lines.
   (a) 24 h after transfection, spin the cells down and replace medium with 5 mL complete medium with 300 μg/mL G418 (or low dose appropriate for cell line). At this point, lay the flask flat in the incubator.
   (b) During the first week, add fresh medium with the low dose of G418 as cells metabolize and turn the medium orange, as necessary.
   (c) At day 7, spin the cells out of their medium completely and replate into 10 mL complete medium with 400 μg/mL G418 (or high dose appropriate for cell line).
   (d) Passage the cells as needed into larger volumes, keeping the higher dose of G418 until the control sample dies. Make sure to change the medium completely with each passage. After the control sample dies, culture without antibiotic for at least 2 days before cross-linking (*see* **Note 26**).

3.7 Cross-Linking

(Standard laboratory precautions should be taken when working with and disposing of formaldehyde.)

1. Adherent cells.
   (a) When cells in the dishes are 80–90% confluent, trypsinize at least one representative dish to get an idea of the cell count per plate.
   (b) To one dish, add 810 μL of 36.5% formaldehyde directly to the 30 mL medium in the dish to achieve a 1% final concentration; swirl to mix, and incubate at room temperature (RT) for 10 min.
   (c) Stop the reaction by adding 1.55 mL of 2.5 M glycine to achieve a final concentration of 0.125 M. Swirl to mix.
   (d) Pour off the medium into a formaldehyde waste container and wash plates gently with 20 mL ice-cold 1× PBS.
   (e) Pour off the PBS and add 5 mL ice-cold Farnham Lysis Buffer with protease inhibitors.
   (f) Scrape the cells off of the plate with a cell scraper and transfer the mixture to a 15 mL conical tube on ice in aliquots for your application (our lab uses 20 million cells per tube).
   (g) Pellet the cells at 800 rcf for 5 min at 4 °C.
   (h) Remove supernatant and place pellet on ice.
   (i) Proceed to sonication step or snap-freeze pellet in LN_2 and store at −80 °C to await further processing.
2. Nonadherent cells.
   (a) Transfer desired number of cells from culture into a 50 mL conical tube (we use 2×10^7 cells).

(b) Add enough complete medium to achieve a final concentration of 5×10^5 cells/mL in the tube.

(c) Add formaldehyde 37% to obtain a final concentration of 1%, close tube and invert to mix, then incubate for 10 min at RT.

(d) Stop the reaction by adding enough 2.5 M glycine for a 0.125 M final concentration. Swirl to mix.

(e) Spin cells at 300 rcf for 5 min at 4 °C and discard supernatant into appropriate waste container.

(f) Resuspend cells in 5 mL ice cold Farnham Lysis Buffer and repeat the spin, this time at 800 rcf at 4 °C.

(g) Discard supernatant and snap freeze the pellet.

3.8 Sonication (Diagenode, Model Pico Bioruptor Sonicator)

1. Resuspend the fresh or frozen pellet (we use 2×10^7) on ice in 1 mL Farnham Lysis Buffer and pipet up and down gently (*see* **Note 29**).
2. Collect the crude nuclear prep by centrifugation at 800 rcf at 4 °C for 5 min.
3. Pour off supernatant and resuspend pellet in 200–300 μL RIPA Buffer.
4. Transfer to a 1.5 mL Bioruptor Pico Microtube with cap.
5. Process the samples in the Bioruptor Pico per the manufacturer's instructions. The number of cycles and settings necessary to sufficiently shear the chromatin must be determined empirically for each cell line (*see* **Note 30**) (*see* Fig. 1).
6. Spin the sonicated mixture at 18,400 rcf in a microcentrifuge for 15 min at 4 °C and collect the supernatant.
7. Snap freeze the sample in LN_2 and store at −80 °C, or continue with the fresh sample with the immunoprecipitation.

3.9 Immunoprecipitation

1. Coupling the FLAG antibody to magnetic beads.

   (a) Add 200 μL resuspended magnetic bead slurry to a 1.5 mL Eppendorf tube on ice containing 1 mL PBS/BSA and vortex briefly.

   (b) Place the microfuge tube on the magnet and remove the supernatant.

   (c) Resuspend the beads in 1 mL PBS/BSA.

   (d) Repeat **steps b** and **c** three times.

   (e) Add 1 mL PBS/BSA to beads.

   (f) Add 5 μg of FLAG antibody.

   (g) Gently mix on a rotator platform for at least 2 h at 4 °C.

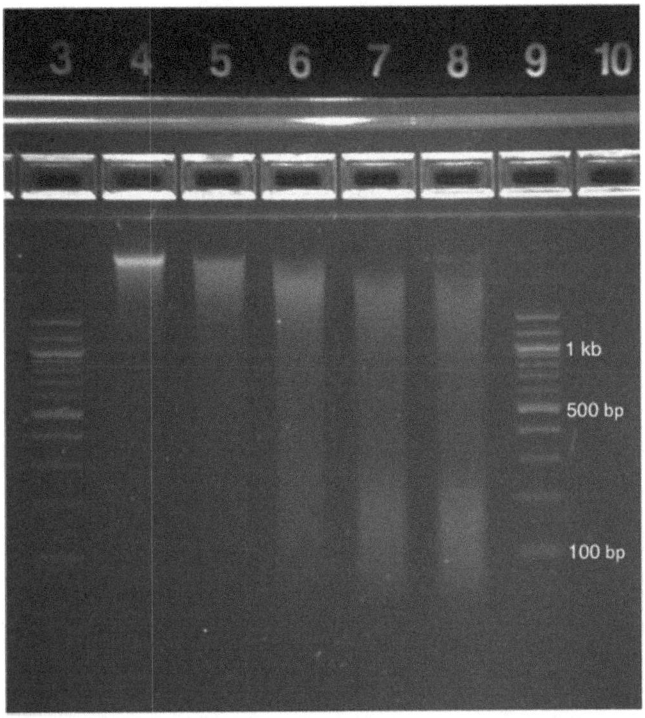

Fig. 1 Cross-linked chromatin was sheared in a Bioruptor Pico for 1, 3, 5, 7, or 15 cycles (1 cycle = 30 s on, 30 s off). The cross-links were reversed and the DNA was purified. One microgram of each condition was run on a 2% agarose gel. In this case, 5–7 cycles would be sufficiently sheared, whereas going to 15 cycles did not result in much additional shearing and would be unnecessary

(h) Wash beads three times (**steps b–c**), resuspending the beads by inverting the tubes during each wash.

(i) Resuspend in 100 μL PBS/BSA and proceed to **step 2**.

2. Add 100 μL of antibody-coupled beads from **step 1(i)** to the chromatin preparation from the sonication protocol and incubate on a rotator for 1 h at room temperature, followed by at least 1 h at 4 °C.
3. Perform all steps in the cold room at 4 °C. Collect beads containing immuno-bound chromatin by placing the microfuge tube on the magnet.
4. Remove and discard supernatant.
5. Wash beads five times with LiCl Wash Buffer, mixing 3 min for each wash on a rotator.
6. Wash bead pellet with 1 mL TE Buffer, mix 1 min on rotator, and place tubes on magnet.
7. Discard supernatant, resuspend the bead pellet in 200 μL IP Elution Buffer at room temperature, and vortex to mix.

3.10 Reversing the Cross-Links

1. Incubate in a 65 °C water bath for 1 h to elute the immuno-bound chromatin from the beads, shaking or vortexing during the incubation every 15 min.
2. Spin at 18,400 rcf in a microcentrifuge at room temperature for 3 min. Collect the supernatant, which contains the ChIP'd DNA. The tubes can be placed on the magnet to facilitate supernatant recovery.
3. Incubate the supernatant containing the ChIP'd DNA in a 65 °C water bath overnight to complete the reversal of the formaldehyde cross-links.
4. To prepare a reverse cross-link total chromatin input DNA library, incubate a sheared chromatin sample in a 65 °C water bath overnight to reverse the cross-links in that sample.
5. Add five volumes of Qiagen Buffer PB (PCR Cleanup kit) to one volume of ChIP'd DNA or Reverse Cross-link sample. Upon addition of Buffer PB, the sample should be yellow, indicating the correct pH. If the sample is not yellow, the pH should be adjusted with 3 M sodium acetate as recommended by the manufacturer (Qiagen).
6. Add about half (~600 μL) of the solution to a Qiagen PCR Cleanup column, centrifuge, and then repeat with the other half to bind the 1.2 mL sample on a Qiagen column.
7. Wash the column with 750 μL Qiagen Buffer PE, centrifuge, empty, and centrifuge the column, which contains the DNA, to dry.
8. Elute the DNA from the column with two 30 μL aliquots of warmed (~55 °C) Qiagen Buffer EB.
9. Measure DNA concentration by using a Qubit (Invitrogen) fluorometer. DNA is ready for library construction of choice.

3.11 Validation of Tag Integration

We use PCR to verify the molecular integration of the tag and Western blot to assess the presence of the tagged protein. Further analysis could include IP-mass spectrometry or IP-Western blotting.

3.11.1 PCR

1. Extract DNA from one million HepG2 FLAG tagged cells for each replicate using the Qiagen DNeasy Blood and Tissue kit, making sure to add in the optional RNase A step and eluting with 165 μL of molecular-grade water.
2. Perform PCR using the following protocol, with amplification of homology arm 1 integration validation using a forward primer upstream of the beginning of HOM1 and the reverse primer in the P2A linker. Homology arm 2 integration validation should amplify with the forward primer in the Neomycin resistance gene and the reverse primer downstream of the end

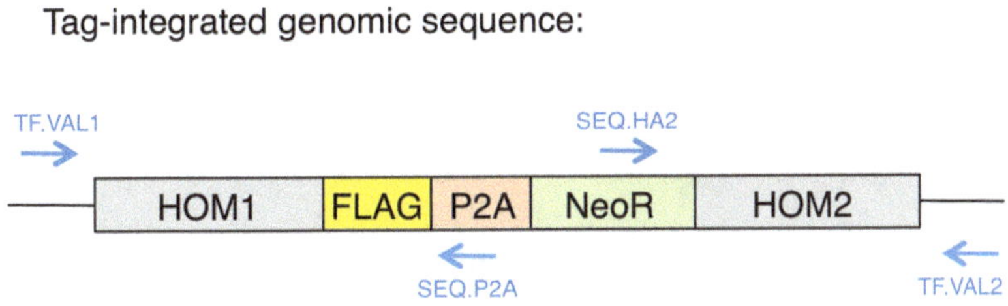

Fig. 2 Diagram of the donor plasmid with arrows marking the locations and directions of the validation PCR primers to test for correct insertion of the epitope tag

of HOM2 (*see* Fig. 2). Make sure to do a negative control PCR for each primer set with wild-type HepG2 genomic DNA to show any background bands due to mispriming that are not due to integration within the genome.

The final volume should be 50 μL.

- 2.5 μL Fwd Primer (10 μM).
- 2.5 μL Rev. Primer (10 μM).
- 25 μL Phusion High Fidelity Master Mix.
- 19 μL dH_2O.
- 1 μL ~20–30 ng DNA.

3. Run the following program:
   - Denature at 98 °C for 3 min.
   - 35 cycles of 98 °C for 15 s, 63 °C for 20 s, and 72 °C for 1 min.
   - Extend at 72 °C for 2 min.
   - Hold at 4 °C.
4. Run out 5 μL of each PCR sample on a 2% agarose gel with a 100 bp ladder to check for the correct sizes.

3.11.2 Western Blot

1. Trypsinize desired number of cells (if adherent cell line) and follow the protocol outlined in the NE-PER Nuclear and Cytoplasmic Extraction kit to obtain nuclear protein lysates. Use the largest number of cells recommended by the protocol.
2. Quantify the protein concentration of nuclear cell lysates using a spectrophotometer. We use a BioTek Synergy H4 Hybrid Reader (*see* **Note 31**).
3. Prepare a diluent solution using one 1 mL of molecular-grade water and 20 μL of Halt Protease and Phosphatase Inhibitor (20×). Vortex briefly to evenly mix.
4. Normalize each protein sample with diluent so that 50 μg of each protein sample is suspended in 22 μL of volume (per single gel) (*see* **Note 32**).

5. Add a uniform volume of NuPage LDS Sample Buffer (4×) and Reducing Agent (10×) master mix to each sample (6.67 μL of sample buffer and 2.96 μL of reducing agent per gel) (*see* **Note 33**). The final mix should contain the following:
   - 50 μg protein.
   - 22 μL diluent (volumetric equivalent of 50 μL of protein).
   - 9.63 μL sample buffer and reducing agent mastermix.
6. Using a dry bath incubator, boil the standardized samples at 90 °C for 15 min.
7. Allow the samples to cool on ice for 2–3 min.
8. Vortex briefly and spin down all samples using a microcentrifuge.
9. Insert NuPage 4–12% Bis-Tris Protein Gels into the Mini Gel Tanks, and evenly remove the gel comb isolating the wells (*see* **Note 34**).
10. Prepare a 1× NuPage MOPS Running Buffer solution by diluting 50 mL of 20× stock solution with 450 mL of deionized water.
11. Fill the front portion of the gel tank to the top fill line with 1× buffer solution, and fill the rear portion up to the second from the bottom indentation on the gel cassette.
12. Flush any remaining preservative out of the gel wells before loading samples (*see* **Note 35**).
13. Load the samples into gel(s) (25 μL per well), along with a ladder. We use the WesternSure Pre-Stained Chemiluminescent Protein Ladder (5 μL volume recommended) (*see* **Note 36**).
14. Secure the top of the gel tank, and insert the leads into a PowerEase 90 W system. Run the gel(s) for 1 h at 150 V.
15. While the gel is running, prepare 50 mL of 5% milk solution in a 50 mL conical tube (2.5 g of nonfat milk powder in 50 mL of 1× TBS Tween). Store at 4 °C for later use (*see* **Note 37**).
16. Upon completion of the gel run, remove the gel from its cassette and into a container of deionized water (*see* **Note 38**).
17. Insert an iBlot 2 NC Regular or Mini Stack into an iBlot 2 blotting system (*see* **Note 39**).
18. Place the gel(s) onto the nitrocellulose layer (toss out the separator), followed by dampened filter paper (wet in deionized water), top stack, and then absorbent pad (*see* **Note 40**).
19. Close the iBlot 2 and run using the following custom 13-min protocol:

- 20 V for 1 min.
- 23 V for 4 min.
- 25 V for 8 min.

20. Open up iBlot 2 when blot is completed, and discard absorbent pad, top stack, and filter paper (*see* **Note 41**).
21. Using plastic tweezers, remove the nitrocellulose membrane and transfer to a fresh container of deionized water to briefly wash (*see* **Note 42**).
22. Pour out the water, and immediately pour on the 50 mL of chilled 5% milk solution (prepared in **step 15**).
23. Cover the container with plastic wrap and place on a rocker (inside of a refrigerator or cold room) to block at 4 °C for a full 24 h.
24. The following day, prepare a 1:5000 anti-FLAG M2-Peroxidase antibody solution in 20 mL of 5% BSA (*see* **Note 43**).
25. After the blot has blocked for a full 24 h, pour off the milk solution (can be poured down a drain) and immediately pour on the prepared antibody solution. Rock at room temperature for 1 h.
26. Pour off the antibody solution (cannot be poured down a drain), and immediately perform four 5-min washes on the blot with 1× TBST (*see* **Note 44**).
27. Using scissors, trim excess edges off of the nitrocellulose (necessary for NC Mini Stack). If using the NC Regular Stack, cut the nitrocellulose membrane into two separate blots.
28. In a 1.5 or 2 mL Eppendorf tube, prepare a visualization buffer solution using the SuperSignal West Femto Maximum Sensitivity Substrate kit (500 μL of Stable Peroxide Buffer and 500 μL of Luminol/Enhancer Solution per blot being visualized).
29. Using plastic tweezers, lift each blot out of the TBST wash solution and dab off excess TBST from a corner using a paper towel (*see* **Note 45**).
30. Once dry, immediately place the blot in a suitable clean container and pipet 1 mL of prepared visualization buffer over the blot for 3 min (*see* **Note 46**).
31. Place each blot onto a smoothed piece of plastic wrap and fold over to cover the front surface of the blot. Roll with a blotting roller to remove any folds or air bubbles.
32. Place the blot inside of an imaging machine (we use the Thermo Scientific myECL Imager [Thermo Fisher PI62236]) and visualize the blot using the chemiluminescence setting.

3.11.3 Prepping Whole Cell Lysates for IP-Mass Spectrometry or IP-Western Blot

1. Wash fresh cell pellet by suspending the cell pellet with 1× PBS.
2. Transfer 2×10^7 cells to a 1.5 mL microcentrifuge tube and pellet by centrifugation at 500 rcf for 3 min.
3. Use a pipet to carefully remove and discard the supernatant, leaving the cell pellet as dry as possible.
4. Add 1 mL of RIPA Buffer with protease inhibitors to the cell pellet and vortex to mix well.
5. Incubate on ice for 30 min.
6. Pass the cells through a 20 G needle 20 times.
7. Centrifuge the pellet for 10 min at maximum speed (~16,000 rcf) in a microcentrifuge at 4 °C.
8. Transfer the supernantant to a clean, prechilled tube and place the tube on ice until use, or if use is not immediate, store extract at −80 °C until needed.
9. Immunoprecipitate the HepG2 whole cell lysate using the primary FLAG antibody.
10. Further analyze the material by mass spectrometry or perform an IP-Western blot.

4 Notes

1. To decide which stop codon to tag, we look at the expression level of all the transcripts. Combining the transcript levels that share the same stop codon, we choose the stop codon of the combined highest level of expression. Specific isoforms of interest can be targeted with a carboxyl-terminus tag only if these have a unique stop codon.
2. If the gene is in the reverse direction, make sure you copy the reverse complement DNA sequence so you can work with the forward sequence.
3. If there are no NGG PAM sites, we have also successfully tagged TFs using the Cpfl nuclease that has a TTTV PAM site. The Cpfl nuclease (also called Cas12a) and gRNA constructs are available to order from IDT.
4. For gRNA design, use the guidelines below to pick the two that fit within the cutoffs and are the best overall choice. Sometimes we proceed with only one gRNA if there are no more acceptable options.
   (a) Keep the gRNA cut site as close as possible to the stop codon or within the stop codon. In general, our cutoffs are within 10 bp upstream and 20 bp downstream.
   (b) Choose the gRNA with the highest on-target scores. In general, a score of 40 or above has worked well.

(c) Choose the gRNA with the highest off-target scores. In general, a score of 50 or higher will work but above 65 is preferred. Click on the off-target score to get a full list of where it might cut elsewhere in the genome and examine whether these sites are acceptable for the experiment.

(d) If choosing two gRNAs, make sure they cut within 20 bp of each other.

5. Be aware when using the copy feature in Benchling, the forward sequence provided is always based on the original sequence and not adjusted for the gRNA direction. Therefore, if you have a gRNA that cuts on the reverse strand, when copying this gRNA to your document using their copy function, the reverse complement box sequence is actually the forward gRNA sequence and the sequence box contains the reverse complement gRNA sequence. To double check which is the correct forward sequence, the list of gRNAs created on the right hand portion of the screen are all displayed in the forward gRNA sequence direction.

6. If the HOM1 gBlock does not pass, make the gBlock smaller from the end farthest from the stop codon until it passes up to ~200 bp long. If IDT cannot synthesize the sequence at all, try creating primers to amplify this region making sure the right primer is immediately adjacent to the stop codon (including the nucleotide just before the stop codon) so that you don't lose any sequence. If amplifying HOM1 with primers, add the sequence "tccccgacctgcagcccagct" to the 5′ end of the forward primer and "ccggaacctcctccgctccc" to the 5′ end of the reverse primer. If the sequence is too complex for gBlock synthesis and also fails to PCR amplify, try ordering the sequence as a gene fragment in a plasmid from IDT or Thermofisher's GeneArt gene synthesis.

7. Keeping the tail sequences the same, you can shorten the homologous gBlock sequence from the 5′ end until it passes. If the ends are the problem, you can also increase the tail sequences up to ~60 bp using the donor plasmid backbone sequence.

8. If the gBlock design for HOM2 is complex at the 5′ end, you can start the gBlock up to ~8 bp downstream of the recommended start location.

9. If the HOM2 gBlock does not pass, make the gBlock smaller from the end farthest from the stop codon until it passes up to ~200 bp long. If IDT cannot synthesize the sequence at all, try creating primers to amplify this region with the left primer starting directly after the last gRNA cut site. If amplifying HOM2 with primers, add the sequence "agttcttctgattcgaacatc" to the 5′ end of the forward primer and "tggagaggactttccaag" to the 5′ end of the reverse primer.

10. Sometimes bubbles will be inevitable in the pipet tip. In this case, you can just pipet to the first stop and not eject the last few microliters that contain the bubbles. The small amount left in the pipette tip should not affect the experiment.
11. Sometimes the sequence of the homology arm or arms will cause the bacteria to grow more slowly and there will be a range of colony sizes on the plate. In this case, pick colonies of different sizes and mark down the size you picked. If a colony tests positive with PCR, the other colonies of the same size will likely also be positive.
12. If the colony was very small, due to complexities in the homology arms, the bacteria may not grow sufficiently in 12–16 h as it should. It can either be grown for a longer time until the culture becomes cloudy or the remaining 30 μL of the colony/LB broth solution can be plated out until there are a sufficient amount of bacteria to inoculate an overnight culture. Also, since the bacteria might be spun down at different times for different donor plasmids due to the varying growth rates, the pelleted bacteria can be stored at −20 °C until the whole batch has grown, and then they can be miniprepped all at the same time.
13. If there are repeated problems obtaining a positive colony based on incorrect PCR band sizes or problems with the assembled sequence, such as deletions or insertions, after several attempts, it may be possible to add one homology arm at a time. Use a two step Gibson Assembly where you add one homology arm to a single digested donor plasmid stock. After verifying the correct sequence has been added, digest with the other restriction enzyme to add the other homology arm and proceed as before. When setting up a two step Gibson Assembly, check to see if the restriction enzymes will cut either homology arm as this may dictate the order of which arm to add first; that is, when digesting to add the second arm, the restriction site should not be present in the first arm. If it is not possible to do a two step Gibson Assembly because the restriction enzymes will cut both arms, try to reduce the size of the homology arms. This may be enough for the single step assembly to work correctly. Another option is to try the HiFi Assembly Master Mix to assemble the donor plasmid.
14. We use 20 μg as a cutoff just to be safe as you will lose some of the plasmid to the side of the tube while concentrating the DNA and to measuring the concentration. It is possible to proceed with only ~15 μg of plasmid.
15. Thoroughly vortex the 10× T4 ligation buffer if there is some precipitation in the tube to fully resuspend.

16. If only using one gRNA for the transfection, you will need 10 μg of the gRNA for two replicates of transfection. Dry down and concentrate as you would for the donor plasmid minipreps.
17. Make sure that cells are not passaged more than three times before transfection. Cells will do best if replated 2 days before harvest, and viability should be over 90% when counted using Trypan Blue. It is important to be familiar with the normal growth rate and needs of the cell line before transfecting, as this will help make determinations later about passaging. When doing more than one transfection, it is very important to make sure that the biosafety cabinet is as clean as possible and supplies should be appropriate for molecular work. Special precautions should be taken, such as using filtered pipette tips, so that no cross-contamination occurs when working with the different plasmids.
18. We find it convenient to do batches of five transfections at a time; for example, in adherent line HepG2: centrifuge 11 million cells (need ten million but include extra for an overage), resuspend in 451 μL Nucleofector reagent and 99 μL Supplement, and finally aliquot 100 μL of the cell mix into each of five waiting Eppendorf tubes with plasmids. If there are more than five nucleofections to be performed, it is wise to do them in this batch size and pull additional batches of the pelleted cells out of the refrigerator as needed, as it is detrimental for the cells to be in the transfection reagents for more than 15 min before being diluted in medium and cultured (this will vary by transfection kit and machine). Two million per transfection is twice the number of cells that the kit calls for, but we find that it works well for HepG2.
19. When preparing the plasmids, make up any shortage to 10 μL with sterile, molecular grade water.
20. While the main setup is the same with the Nucleofector reagent and Supplement, Amaxa transfection kits are individualized for different cell lines. In addition, protocols can vary depending on the transfection machine model. There may be a specific protocol for your line of interest.
21. We have found that the "high-efficiency" program is harsher and leaves more cells dead at first, but the eventual cell growth is much better than with the gentler "high-viability" program. Depending on the cell line, one manufacturer program might work better than another and this should be tested.
22. There will be a floating clump of cells after the nucleofection program is run. This can be broken up somewhat with a pipette tip during resuspension and included in the transfer of cells into the culture flask; it is not detrimental and there is no reason to try to avoid its transfer.

23. Our lab treats these as separate cell lines until harvest because we require two replicates. You may choose to combine the cell lines for your own application and passage them together as one population after the first week.
24. This protocol covers an example of a timeline for cell growth after transfection. However, it is important to note that each transfected cell line will grow at a slightly different rate depending on its nucleofection efficiency. While the nonadherent cell lines have been rather straightforward to culture, we have had to try some different culture conditions in the adherent line HepG2 to optimize growth. Cells that are low expressors of the transcription factors of interest have tended to do better with lower amounts of selection agent in the first week. We used to apply 400 μg/mL on day 1 and keep it constant throughout culture, but have since determined that 300 μg/mL of G418 during the first week helps cells edited at low-expressed genes succeed, so we have modified our standard protocol to include this step. It is advisable to do preliminary experiments to test different amounts of antibiotic with each cell type with TFs of known high and low expression to get the best idea of an efficient recipe.
25. A standard kill curve is essential to understand the dynamics in each cell line. We have noted that all cells die by day 15 in a HepG2 no-plasmid control sample. However, we keep 400 μg/mL G418 on all of the cell lines until day 18 as a precaution. After that, the lines seem to be stable, even after freezing and thawing, and medium without antibiotic can be used for culture. The earliest one should harvest the HepG2 cell line after transfection is day 20, after it has had at least 2 days without G418. K562, a nonadherent cell line, seems to be successfully selected at about day 12. G418-containing medium should be made fresh each time or stored at 4 °C and used within 1 week to ensure highest activity. Also, assume that any cultures that have been in the incubator for a week will need freshly made G418 medium.
26. Instead of counting the cells to assess viability and numbers, we observe confluence and medium color to determine whether or not to passage. There is one exception at day 6–8, at the time of the first passage, however. Even if adherent cells do not seem to be filling the 25 cm^2 flask, it is important to trypsinize, centrifuge them, and resuspend to break up the colonies, even if it means putting them back into a flask of the same area. Alternatively, if there is a very aggressive line in the first week of culture that has gotten confluent quickly, do not passage early or passage more lightly than 1:2, because during the second week there will be significantly more cell mortality. In the unlikely event that there is confluence before day

6, change medium or add fresh low-dose G418 medium up to 10 mL to keep the cells nourished for an additional 1–2 days until passage. Passaging at day 5 or earlier into a larger area and/or the higher concentration of G418 may kill the cell line.

27. G418 works best on dividing cells. If adherent cells get too crowded or, if passaged too lightly, start to grow in towers, this could inhibit proper selection because the division is slowed in both cases. During the first 2 weeks of growth, it is imperative to passage the cells at the most 1:2 so that enough cells are maintained to create quickly a conditioned medium in which the cells can efficiently communicate. If the cells are too far apart when they adhere, they will form isolated towers; this causes growth to be extremely slow or stop altogether. Towers can many times be seen with the naked eye just by looking at the underside of the flask. If this happens, trypsinize after a couple of days and replate in a flask with a smaller area, or in the same area with reduced medium so that cell communication is enhanced. The health of HepG2 seems best when a carpet-like morphology is maintained, and this is seen after day 11 or 12 in most successful transfected lines, even ones that grow very slowly at first. After that point, cells can be passaged when they reach confluence every 2 days, but under no circumstances should they remain in the same flask for more than a week without trypsinization and replating, even if growing slowly. If cells are not dividing like typical wild-type cells by the third week, it might be worth redoing the transfection, as selection will have been unreliable.

28. When we grow HepG2, one cross-linked dish gives more than enough cells for two replicates of 2×10^7 each.

29. We typically include a douncing step here, passing the lysate through a 20 G needle 20 times. We have found this increases our yield for the cell lines we use, but this step is optional and should be tested for your application.

30. Our lab uses a Bioruptor Pico to shear chromatin, but different sonicator models or even different individual instruments of the same sonicator model may vary in the settings and number of cycles needed to sufficiently shear the chromatin. A titration experiment with cross-linked chromatin should be performed for each cell line before attempting a ChIP experiment. The chromatin should be sheared until a sizable fraction of fragments appear in the 100–500 bp range on an agarose gel. In our experience, it is not necessary to continue shearing until all of the fragments are in that range, and larger fragments may still be present.

31. Use the Nuclear Extraction Reagent (NER) from the NE-PER protein extraction as a blank when quantifying the nuclear lysates.

32. Use the measured protein concentrations from **step 2** to calculate how much protein and diluent to add to achieve the desired 50 μg of protein in 22 μL of solution. While 50 μg has been ideal for the vast majority of proteins run, some with unusually high or low expression levels may require a greater or lesser mass of protein to show distinct banding.
33. When preparing the mastermix of sample buffer and reducing agent, add two additional volumes of each reagent to account for pipetting error. Prepare the mastermix in a separate 1.5 or 2 mL microcentrifuge tube. Pipet up and down to mix evenly. When adding to the protein with diluent, pipet up from the bottom several times, and let the volume of sample run down the side of the tube. This is to ensure an even distribution of sample buffer and reducing agent throughout the sample volume. We prepare enough volume of each sample for three gels, in most cases.
34. A putty knife is ideal for gently removing the gel comb. If running a single gel in a Mini Gel Tank, remove the unused cassette clamp before starting. Before inserting gels into the tank, be sure to remove the thin strip of tape running across the bottom of the gel cassette's backside.
35. A 1 mL pipette is ideal for flushing out the wells. Set to a full 1 mL and take up the same running buffer immersing the gels, expelling it into the wells. Preservative should be visibly displaced upon flushing. It is ideal to flush the wells multiple times to ensure that no residual preservative remains in any of the wells.
36. Keep the samples on ice as they are waiting to be loaded. When loading each sample, keep the pipette tip up against the back of the well and insert until the tip is near but not touching the bottom. Steadily eject sample up to the first stop, and then halfway to the second stop until a bubble forms at the end of the tip. Gently remove the tip with its bubble attached, keeping it up against the back of the gel cassette until it comes loose at the top. This is to ensure that the sample is not displaced from the well by any bubbles.
37. The 1× TBS Tween can be prepared from a 20× stock solution (50 mL of 20× solution in 450 mL of deionized water).
38. Crack the gel case open by starting at the top of one side and progressively moving towards the other side, inserting the corner of a putty knife and twisting to separate the two pieces of plastic. Let the gel stay bound to whichever side it sticks to. Using the same putty knife, cut off the teeth of the gel (small pieces at the top forming the wells), along with the portion of gel beneath the colored bottom region (marking the end of the sample run). It can also help to cut a top corner

of the gel to denote proper orientation. To move the gel into deionized water, dip the cassette half holding the gel into the water container continuously until the gel comes loose and slides off. If the gel will not fall off, lift a corner of the gel when dipping into the water to facilitate removal. However, further physical manipulation could lead to breaks in the gel.

39. Use a Regular Stack to blot two gels simultaneously or a Mini Stack to blot a single gel.
40. When placing the gel onto the nitrocellulose, be sure to push out any air bubbles using a finger. Do not touch the actual nitrocellulose layer. After applying the filter paper and each successive layer, use the Blotting Roller to push out any air bubbles.
41. Use plastic tweezers to avoid touching and/or damaging the nitrocellulose membrane.
42. We used a container with dimensions of approximately 20 cm × 10 cm × 2.5 cm.
43. Start with 20 mL of 1× TBS Tween in a 50 mL conical tube. Under a laminar flow hood (optional), remove 2 × 1.67 mL of the TBST using a 1 mL pipette. Replace with the same volume of 30% bovine serum albumin, resulting in a 5% BSA solution. Pipet up and down when adding BSA to mix evenly. To achieve an antibody concentration of 1:5000, add 4 μL of anti-FLAG M2-Peroxidase antibody to the BSA solution.
44. Perform washes on a rocker at room temperature by pouring on enough TBST so that an even wave of solution runs over the blot as it rocks from side to side. Volumes of TBST used for this step can be liberal. Pour off the TBST after each of the first three washes (can be poured down a drain), and replace with new solution.
45. Hold the blot by a corner to avoid touching any of the lanes.
46. Continuously pipet the volume of visualization buffer over the blot to ensure that the entire surface has been thoroughly treated.

Acknowledgments

Research reported in this publication was supported by the National Human Genome Research Institute of the National Institutes of Health under Award Number U54HG006998 to R.M.M. and E.M.M. The content is solely the responsibility of the authors and does not necessarily represent the official views of the National Institutes of Health. This work was also supported by funds from The HudsonAlpha Institute for Biotechnology.

References

1. Landt SG, Marinov GK, Kundaje A et al (2012) ChIP-seq guidelines and practices of the ENCODE and modENCODE consortia. Genome Res 22(9):1813–1831
2. Savic D, Partridge EC, Newberry KM et al (2015) CETCh-seq: CRISPR epitope tagging ChIP-seq of DNA-binding proteins. Genome Res 25(10):1581–1589

Chapter 2

User-Friendly and Interactive Analysis of ChIP-Seq Data Using EaSeq

Mads Lerdrup and Klaus Hansen

Abstract

ChIP-seq is a central method to gain understanding of the regulatory networks in the genome of stem cells and during differentiation. Exploration and analysis of such genome-wide data often leads to unexpected discoveries and new hypotheses. It therefore accelerates and improves the discovery phase, when scientists with biological understanding are enabled to analyze and visualize data. EaSeq (http://easeq.net) offers integrated exploration of genome-wide data in a visual, versatile, user-friendly, and interactive manner that connects abstract interpretations to the signal distribution at the underlying loci. Here we introduce the interface, data types, and acquisition, and guide the reader through two example workflows. These workflows will enable the reader to perform genome-wide analysis and visualization of transcription factor binding sites and histone marks. This includes making basic plots; finding, annotating, sorting, and filtering of peaks; using EaSeq as a genome browser; measuring ChIP-seq signal and calculating ratios; as well as data import and export.

Key words ChIP-sequencing, Next-generation sequencing, Analysis, Visualization, Exploration, Genomics, Epigenetics

1 Introduction

Chromatin immunoprecipitation followed by DNA sequencing (ChIP-seq) has been a workhorse for the identification of the transcriptional networks that drive cellular growth and differentiation and in gaining knowledge related to how histone marks are associated with the different chromatin states that maintain cellular identity [1–5]. Since the initial development of the method more than a decade ago [6–8], thousands of publications have used the method, and it is hard to imagine how the current detailed knowledge regarding transcription factor binding and histone modifications in the genome in different cellular stages could have been uncovered without the method.

ChIP-seq analysis is however often considered data-heavy and computationally intensive. The processing and analysis of data is

Benjamin L. Kidder (ed.), *Stem Cell Transcriptional Networks: Methods and Protocols*, Methods in Molecular Biology, vol. 2117, https://doi.org/10.1007/978-1-0716-0301-7_2,

therefore often done by computational experts, which are in high demand. Thus, this process tends to be time-consuming and test the patience of stakeholders. With genome-wide experiments such as those based on ChIP-seq, the exploration of the data also frequently leads to unexpected discoveries and the generation of new hypotheses. Data are therefore often reanalyzed several times, while the biological understanding grows. In many cases, the disconnection between the biological understanding and the ability to analyze and visualize data impedes this process. Therefore, user-friendly software, which allow the biologists to perform quick and interactive genome-wide exploration and analysis, can accelerate and improve the discovery phase. We developed EaSeq [9] (http://easeq.net) to enable a more integrated exploration of genome-wide data in a visual, versatile, and user-friendly manner, which allows scientists with biological understanding to get more direct contact with the data and analyses, and to test hypotheses more rapidly. A secondary aim was to tightly connect abstract interpretations in, for example, clustered heatmaps or scatter plots to the signal distribution at the underlying loci, thus the interactivity in EaSeq enables users to check whether a point in a scatter plot actually represent the assumed state at the locus. EaSeq is aimed at facilitating and accelerating the explorative and often unpredictable part of the analysis work, whereas the preceding data processing can often be standardized and pipelined efficiently with existing tools such as Galaxy [10]. Given the low predictability and the frequent variation in how data are handled during this stage, EaSeq is designed to provide a "sandbox" with wide degrees of freedom and open workflows. Due to the diversity of data and the versatility of the uses, this chapter present advices on how to use the different functionalities on example data and there is no unified procedure on how to use EaSeq. As the program contains more than 50 plot and tool types and an exceeding number of possible actions, it is also limited what can be covered in this chapter beyond the most basic operations. Yet as demonstrated in Subheading 3.2.10, **step 4**, this is sufficient to make figures similar to those in high impact publications.

2 Materials

2.1 Installation and Computer Requirements

EaSeq was developed to be used on a windows PC. A zip-file containing the program can be downloaded free of charge from http://easeq.net. To use the program, simply unzip the files, and move the file easeq.exe into a suitable folder (avoid network drives as EaSeq will need to have write permissions to its home directory, and this might be disabled for network drives due to security reasons). EaSeq was designed to have no dependencies on any external code. The only requirement is Microsoft .NET, which is a standard component of Windows installations. Recent versions of EaSeq require 64-bit windows, which are more or less standard on PCs today. It will run on most hardware, but efficient analysis of more than a handful of datasets would benefit from more than 4 GB RAM and a multicore CPU. EaSeq is designed for a monitor with a screen resolution of 1920 × 1080 or 1920 × 1200, and while all modern monitors will provide this resolution or better, some laptops might not fulfill this requirement and will benefit from an external monitor.

For PCs using MacOS or Linux, a "virtual machine" containing a Windows desktop will need to be installed first. Commercially available virtual machines for MacOS include Parallels and VMWare Fusion. Once this has been installed together with Windows, the EaSeq installation is similar to that of a Windows PC. Another solution is to have a low-latency virtual Windows desktop running on a remove server. Company and university IT-departments might offer this option for its employees, and several commercial solutions exist for running such a solution on a reasonably priced pay-per-use basis.

2.2 Interface and Basic Concepts

Once EaSeq has been installed and started, the main window will be opened, and this is subdivided into a number of panels (Fig. 1). In the following, the different parts of the main user-interface will be referred to in [brackets], and thus [Tools/Quantify] implies that the button named "Quantify" in the "Tools" section within the main window should be clicked.

In the upper part of the interface, EaSeq offers tools to work on three types of imported data as well as so-called sessions, which are single-file assemblies of all the imported data, plots, and descriptions of the data handling that has been done. Below these sections, the interface also contains two toolbars for plots and analysis tools. The lower half of the main window contains a large gray area that can be filled with plots of all types with genome-wide relationships. Once the user has generated the intended plots to understand the data, they can all be exported as high-quality images, which are ready for use in reports. In the right side of the interface, there are buttons to open and overlay four different specialized panels. Finally, the procedures that involve a tool or a plot will often lead to the opening of a new window, which is usually closed again once the procedure is initiated.

2.2.1 Data Types in EaSeq

The three types of data that EaSeq uses are (Fig. 1): (1) "Datasets," which are the imported ChIP-seq or other types of next generation sequencing libraries, such as RNA-seq. Before import, the sequenced libraries need to be mapped (*see* Subheadings 2.2.3 and 2.3.1). (2) "Regionsets," which as the name implies are sets of regions in the genome that would be interesting to study (enhancers, CpG-islands, published peak-sets, etc.). Regionsets

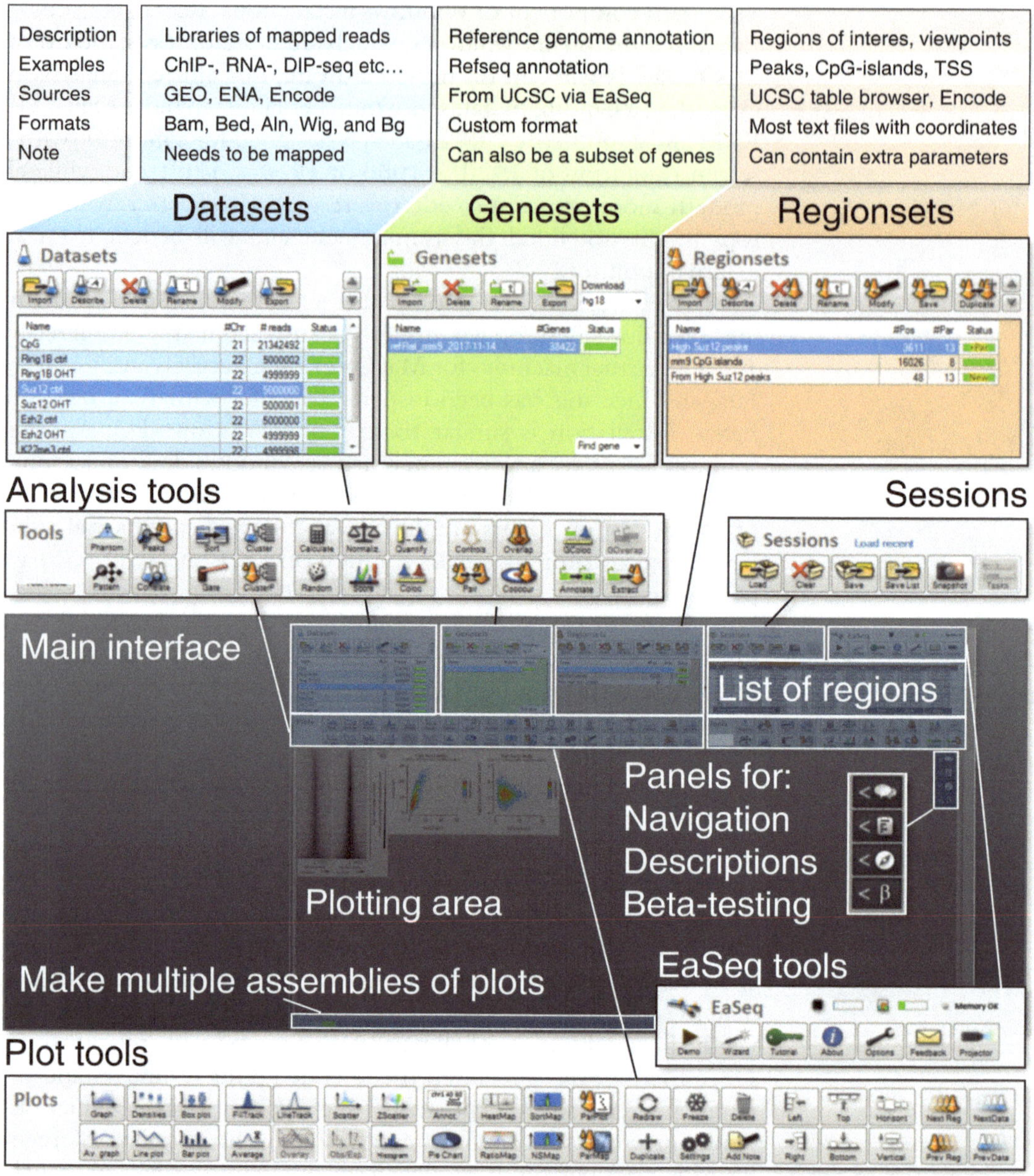

Fig. 1 Overview of the main interface in EaSeq and enlargements of the central panels and areas

could also originate from an imported analysis of RNA-seq or another type of genome-wide expression data (*see* **Note 1**). In many cases, Regionsets are used as the viewpoints for plots, but they can also be handled or analyzed in relation to other Regionsets. (3) The final type of data, "Genesets," contains the annotation of features for all genes in a reference genome (or in some cases subsets of such annotations). These data can be downloaded directly from UCSC [11, 12] within EaSeq.

To illustrate this, imagine a user, who have generated ChIP-seq samples from mouse cells grown under different conditions and mapped these to the mm10 reference genome, and would like to study differences in these ChIP-seq signals at a published set of transcription factor binding sites and transcriptional start sites (TSSes). To do so, the user would need to import the libraries mapped to mm10 as datasets (Covered in Subheading 3.1.2) and the mm10 coordinates for the transcription factor binding sites as a Regionset (Covered in Subheading 3.2.1, **step 2**). Either the mm10 coordinates for the TSSes can then be imported as a Regionset too, or the Refseq [13] annotation for mm10 genes can be downloaded from UCSC [11, 12] and stored as a Geneset within EaSeq (Covered in Subheading 3.1.4, **steps 1** and **2**). The coordinates for the TSSes within this Geneset then subsequently needs to be converted to a Regionset before further use (Covered in Subheading 3.1.4, **step 6**). EaSeq also offers a range of other ways to convert one type of data to another (*see* **Note 2**).

2.2.2 The Differences Between Datasets and Regionsets

It is common for new users to be uncertain about the exact distinction between Datasets and Regionsets. As mentioned above, libraries of mapped reads should be imported as Datasets, whereas a previously generated set of peaks from the exact same data should be imported as a Regionset. In many plot types, Regionsets would often serve as the viewpoints for genome-wide presentations of, for example, a set of transcription factor binding sets, whereas as the Datasets would serve as the depicted signal (Fig. 2). Datasets are typically millions of lines long, whereas files containing such peaks would be a few thousand lines long, and EaSeq will try to warn the user, if it detects a potential mix of the data types based on file sizes. Regionsets could also be lists of established genomic features, and such data can vary a lot in their formatting. Therefore, EaSeq includes a wizard that helps to import atypical data. Besides genomic coordinates, Regionsets can also contain additional properties that can be visualized or, for example, used for subselection. In EaSeq these properties are called Regionset Parameters.

2.2.3 Datasets: Unmapped Reads vs. Mapped Reads vs. Coverage Vectors

Next generation sequencing of ChIP-seq samples (or other types of -seq, e.g., DIP-seq, RIP-seq, RNA-seq, ATAC-seq, or CAGE-seq) leads to a high number of individual sequences, which range from 75 to 150 bp with the currently common sequencing technology. These sequences are commonly called reads, and before they can be used for most visualization and analyses, each read will need to be matched to a reference genome (currently named mm10 and hg38 for the mouse and human genome, respectively) in a process called aligning or mapping. Not all reads will be successfully matched, but those that are will contain a set of coordinates (chromosome, start, end, and strand) matching to the corresponding sequence in the

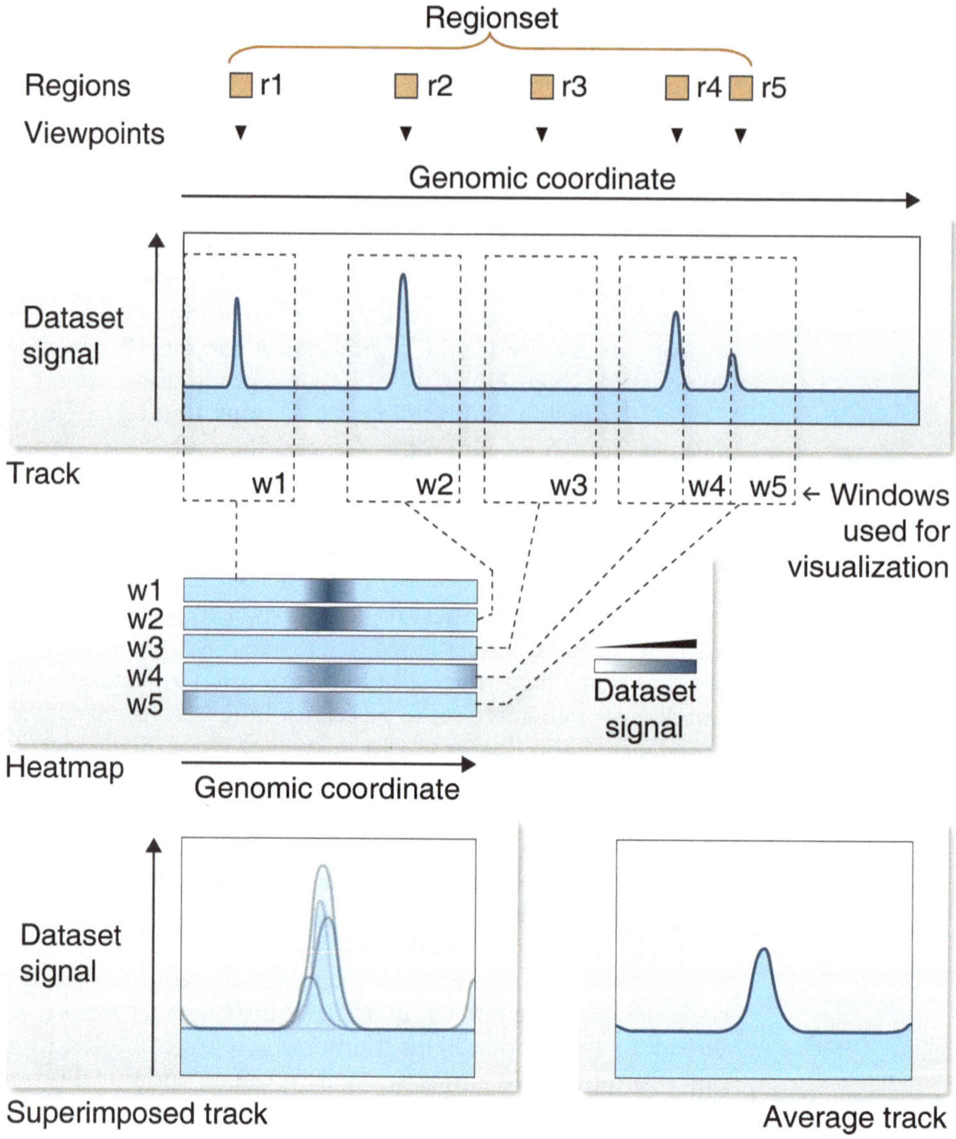

Fig. 2 Upper part: Panels illustrating the relationship between Regionsets, viewpoints, visualized windows, and dataset signal. Lower part: examples of how the five Regions will be displayed by three different plot types

reference genome, and these are often called mapped reads. Before data from a sequencing run can be imported into EaSeq, they will need to be mapped (*see* Subheading 2.3.1). During import, only the genomic coordinate is preserved, whereas other information (e.g., sequence or quality metrics) is discarded to save memory and storage.

Sometimes data have been processed further to give a coverage vector, which is a long list of values that reflect the aggregated number of reads along different positions within the chromosomes. Therefore, the information about each individual read is lost, once data from a library is converted to coverage vectors, and we recommend importing and using files containing individually mapped reads as they provide higher processing speed, better resolution and a more efficient normalization than coverage vectors do. Files with coverage vectors are, however, needed, for example, when uploading custom tracks to UCSC (REF) or when depositing data on GEO (REF), and they can be generated in EaSeq as described in Subheading 3.1.6, **step 3**.

2.2.4 Formats

Datasets containing individual reads can be imported from bed-, bam-, and aln-files, or most types of text files with columns containing the genomic coordinates (Fig. 1). Coverage-based Datasets can be imported from wig-files and bedgraph/bg-files. Often data that would be imported as a Regionset are also stored in the .bed format, and the use of the same format for mapped reads imported as Datasets and, for example, peaks imported as Regionsets, is of course an understandable source for confusion. Regionsets can however also be imported from a rich variety of file formats, and it is possible to import almost any type of text files where the information is separated into fields by tabs, spaces, semicolons, and so on as long as the file contains information regarding the genomic coordinates (covered in Subheading 3.2.1, **step 2**). The format requirements for Genesets are a bit more stringent, as they will need to contain information for several gene features. If needed the best way to learn about the formatting requirements, is to download and export a Geneset within EaSeq and mimic the formatting in a custom-made file. It is important that all imported data adheres to the same conventions on chromosome names (*see* **Note 3**).

2.3 Getting Data

2.3.1 Acquiring and Processing Libraries for Import as Datasets

After sequencing, the typical processing procedure for the data includes the following steps (Fig. 3): (1) Basecalling, which is the process of obtaining a DNA sequence from the more raw data in the sequencer. This is often an integrated part of the processing done automatically by the sequencing machine itself or staff in the sequencing facility. This results in unmapped reads in the fastq format, which also contain information on the sequencing quality for each base. It is highly advisable to store fastq-files for submission into repositories such as GEO [14] during manuscript preparation. (2) Assessment of the overall sequencing quality using FastQC [15] and possible contaminants using FastQ Screen [16]. (3) Trimming of adaptor sequences or sequence cycles with low sequencing quality, for example, using Trimmomatic, Trimgalore, or cutadapt [17–19]. (4) Mapping (aka alignment) to a reference genome. Typical input for this process is one or more fastq-files, which have been trimmed, and the output can be in many formats, most commonly bam- or sam-files. It is important to assure that the coordinates from all data analyzed together are from the same reference genome

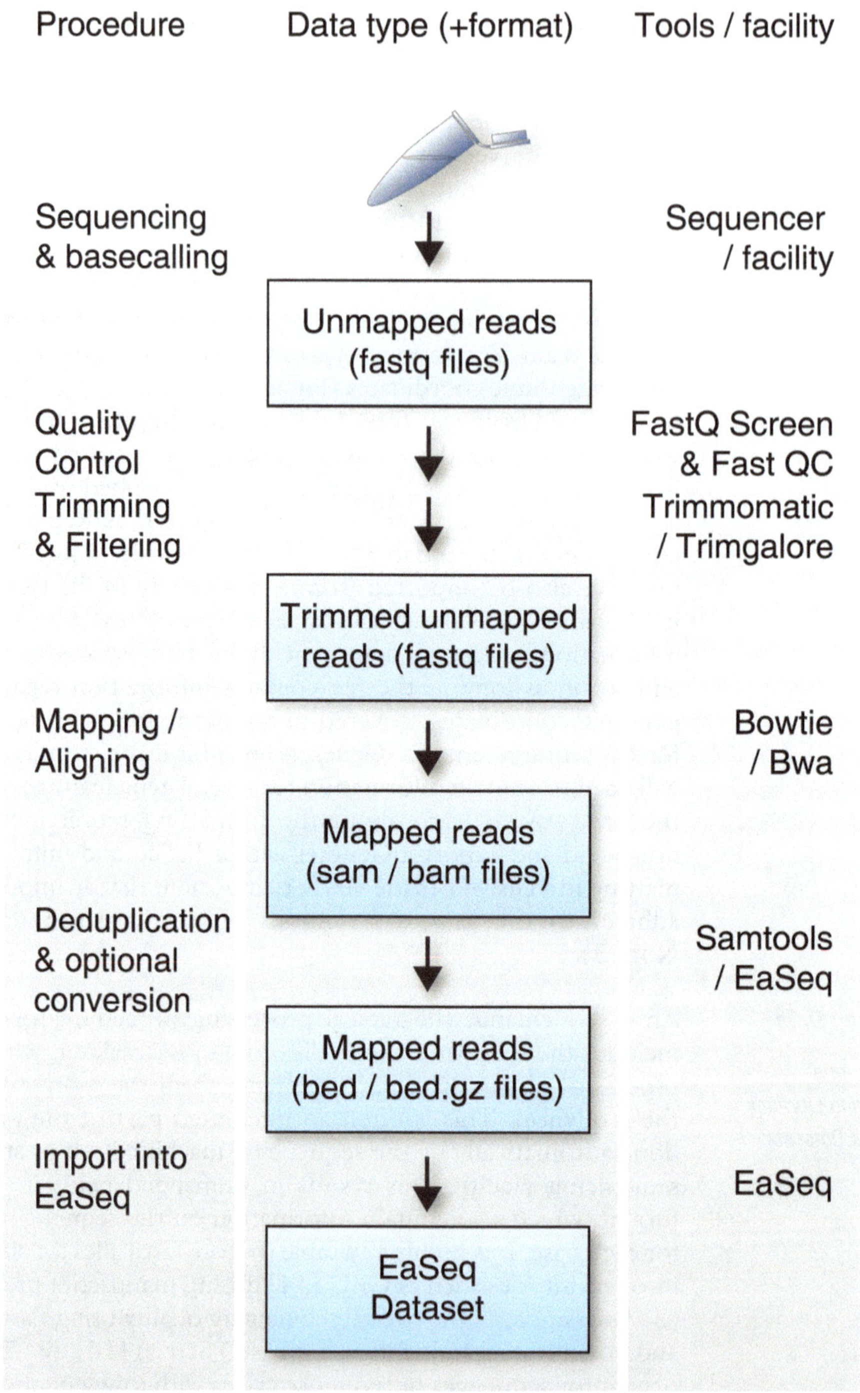

Fig. 3 Overview of the acquisition and processing of libraries prior to import into EaSeq

(*see* **Note 4**). (5) For import into EaSeq, the files need to be converted to either bam-files or bed-files (the latter is often gzip-compressed and can be stripped of information that is no longer needed and is therefore more compact). A useful tool for such conversions is samtools [20]. Some processing pipelines include one or more filtering steps to remove low quality reads and overly amplified PCR products, which are undesirable to have in the data. The latter part is also done per default when importing data into EaSeq, and this step is sometimes called deduplication. The procedures described above in **steps 1–5** above often also needs to be applied to published data that have been deposited at repositories such as GEO [14] or ENA (https://www.ebi.ac.uk/) in an unmapped format such as fastq-files. These procedures typically require a bit of preceding experience with tools that are developed for Linux/UNIX, so it will often be helpful to get assistance from the sequencing or bioinformatics core facility on setting up the pipelines needed for this. Users, who are less comfortable with the command-line based UNIX environments, will often find that the popular and user-friendly GUI-based software Galaxy is an attractive alternative [10].

2.3.2 Where to Get Data?

As mentioned above, GEO [14] (https://www.ncbi.nlm.nih.gov/geo) and ENA (https://www.ebi.ac.uk/) are excellent sources for libraries of unmapped reads, and most publications contain a GEO accession number that can guide readers to the location of the data that were used for the study. The GEO depositions also often contain coverage vectors mapped to a particular reference genome, and while wig and bedgraph/bg files can be imported directly in EaSeq, bigwig and bigbed files will first need to be converted by an external tool such as bigWigToWig (https://www.encodeproject.org/software/bigwigtowig/) before import. Consortia such as Roadmap Epigenetics [21] (http://www.roadmapepigenomics.org/) and Encode [22] (https://www.encodeproject.org/) offers resources with attractive collations of both mapped and unmapped libraries, as well as peaks, which have been identified from the data as well as other genomic regions that can be imported as Region-sets. A popular and comprehensive resource is also the Table Browser from UCSC [12] (https://genome.ucsc.edu/cgi-bin/hgTables), which can provide coordinates for a wide range of interesting genomic structures in a variety of reference genomes.

3 Methods

In the following sections, we will cover two example workflows, which are composed to illustrate the analyses and visualization of genome-wide transcription factors binding sites (Subheading 3.1) and histone marks (Subheading 3.2). The data used in the examples are parts of central and published stem cell studies [23, 24], and we have generated zip-files containing mapped ChIP-seq libraries and other relevant data for these examples to help the reader getting started on the example workflows in a quick and easy manner. The first example (Subheading 3.1) will cover peak-finding, annotation of peaks, saving data, descriptions, making a few plots, and using EaSeq as a genome browser. The second example (Subheading 3.2) will cover data import, heatmaps, inspection of peaks, quantitation of ChIP-seq signal, ratio calculations, peak sorting, filtering, and a couple of additional plot types. The two example workflows can be handled independently of each other, whereas the individual steps within each workflow often requires actions that are done in the preceding steps. Therefore, it is recommendable to perform the steps in each workflow in a sequential manner.

3.1 Example 1: Peak-Finding and -Annotation from Transcription Factor ChIP-Seq Data

The aim of this example is to provide the reader with some basic experience on how to handle ChIP-seq data for transcription factors, including data import and export, peak-finding, annotation of peaks, and some basic visualization. A zip-file containing the data needed for this example workflow can be downloaded at http://easeq.net/wf1.zip. When the zip-file is decompressed, four .bed-files become available. These files contain mapped reads corresponding to three transcription factors (CTCF, Oct4, and Nanog) fused to GFP and enriched using an anti-GFP antibody as well as a control sample with GFP alone. These data were originally a part of the data presented in a pioneering ChIP-seq paper [24], and the sequencing reads were mapped to mm9 by us and down-sampled to reduce the file size and import time.

3.1.1 Optional Disabling of Hints

Whereas pop-up hints might be a useful way to learn the functionality of new software, they might be too much of a distraction during a guided procedure as this workflow. It is therefore advisable to disable the hints that show up by default when running a new EaSeq installation.

1. This can be done by clicking on the white wrench icon in the popups and unchecking the "Show hints" box.
2. Hints can be enabled later by opening the "Options" panel and click on the tab called "Panels."

3.1.2 Import of ChIP-Seq Data

1. To import the mapped data, click on [Datasets/Import] and select all four .bed-files: "CTCF.bed", "Oct4.bed", "Nanog.bed", and "GFP.bed".
2. When the import wizard shows up, it is able to recognize the files, so the default settings will suffice for these files. Click "OK" in the import form to start the process.
3. Keep an eye on the list in the "Datasets" panel as the imported datasets will appear here. Once the small progress indicator becomes green, then the Datasets are available for subsequent work.

3.1.3 Peak-Finding

Peak-finding is the process of systematic identification of areas in the genome where the enrichment of a sample is strong, and this is almost always a requirement to, for example, identify transcription factor binding sites or areas with high levels of a histone modification. This procedure results in a set of "Peaks," which essentially are a set of genomic regions that can be used as viewpoints in plots or for a diverse range of other tasks. In EaSeq, this is done by dividing the genome into "windows" of default 100 bp and then identifying those with significantly higher levels of signal. This process is repeated four times to avoid false negatives where peaks are divided between two windows. In EaSeq, this procedure requires two Datasets, a sample and a negative control, where the latter is used to model the unspecific background distribution in the library. Peak-finding typically takes 5–10 min and will result in a new Regionset, which can be used as viewpoints for subsequent genome-wide analysis.

1. To start the peak-finding, click on the [Tools/Peaks] icon, and in the window that appears.
2. Select "Oct4" as the "Sample Dataset" and "GFP" as the "Negative Control Dataset."
3. *Optional*: Shorten the name of the resulting peakset, since long names tend to reduce the overview.

   Three settings adjust the stringency of the peak finding, "*p*-value," "False Discovery Rate," and "Log2 fold diff." If one would like to detect areas enriched in broadly distributed ChIP-seq signals, such as the histone marks H3K27ac or H3K27me3, then it is recommendable to increase window sizes and merge distances from the default to larger sizes, such as 500 bp, 2000 bp, or more (Fig. 4).
4. For this example, use the default settings and start the peak-finding by clicking "Find Peaks."

   After a while, the plots in the window will start to monitor the currently analyzed (right side) and cumulatively analyzed (left side) parts of the genome. The plots show the level of positive signal and negative control signal in all windows (top) and the signals that were found to be enriched in the initial parts of the analysis (bottom) and considered positive. By studying the distribution of signal within these populations, it is possible to assess if the positive windows are rich in the signal from the negative control or not. If this is the case, it might be a reason to increase stringency of the peak-calling.
5. It is not required to keep the window open during the procedure, so you are welcome to close it by clicking "Exit."
6. *Optional* tasks: Repeat the peak-finding for CTCF and Nanog as well.

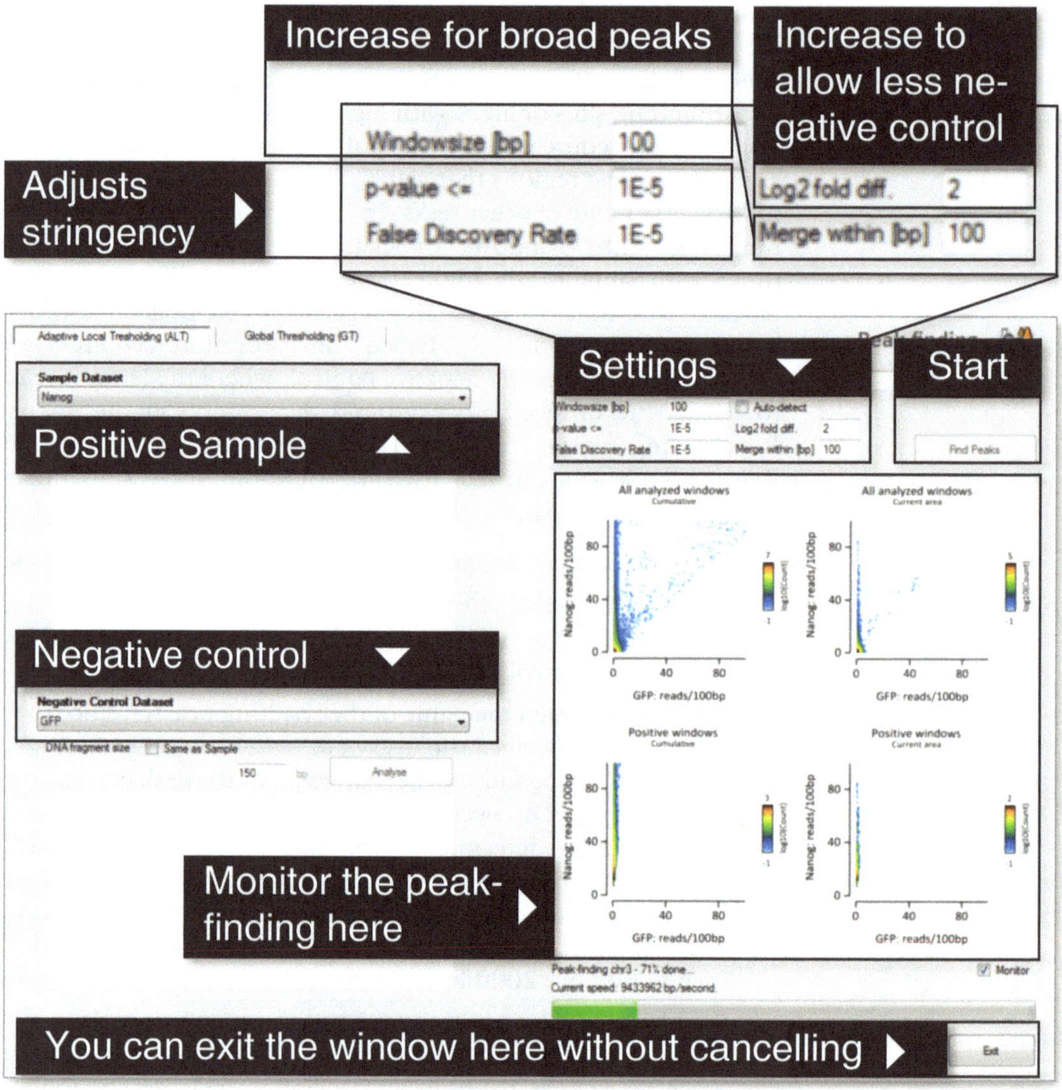

Fig. 4 Example of the peak-finding window and annotation of its contents

7. *Optional* task: Monitor the progress of the peak-calling by clicking on "Tasks" in the main menu below "Sessions."

Peak-finding will lead to the generation of a new Regionset, which contains all the coordinates and quality scores of the peaks. This can already be seen during the peak-calling, and it will appear in the list of Regionsets in the Regionset panel. Once the peak-finding completes, the Regionset will be marked as "New" in the Regionset list, and you can inspect the coordinates as well as additional parameters, such as false discovery rates (FDR) values and read counts of each peak by selecting the Regionset and observe and browse the list that appears in the panel to the right side of the Regionsets.

3.1.4 Annotation of Peaks

Once a set of peaks is generated, an often-sought procedure is to learn the identity of neighboring genes. To annotate the peaks (or any other Regionset) with gene names, accession numbers and distances, EaSeq will require a file with the coordinates of the genes in the used reference genome. EaSeq offers a quick and simple integration to acquire the Refseq [13] annotations directly from UCSC [11, 12]. Please remember to cite these sources when using the data they provide.

1. To download gene annotations for mm9, select "mm9" in the drop-down box below the Genesets panel.
2. In the import wizard, click "ok" to the default settings, and EaSeq will import all the currently annotated mouse genes in the Refseq database and the mm9 coordinates of their features, and store this as a Geneset.
3. To run the actual annotation of the peaks, click [Tools/Annotate].
4. In the window that shows up, choose the "Start" of the genes as the position to use for the annotation. This can be altered, and a typical variation would be to select "Starts + Ends", if one would like the nearest gene to be annotated to each peak regardless of which part of the gene that is closest.
5. The annotation will add a number of new parameters to the existing Regionset (in this case the Oct4 peaks), including gene names, accession numbers, and distances. Inspect these more closely in the viewer in the right side of the Regionset list.
6. *Optional*: A commonly used procedure is to visualize ChIP-seq signal at gene features such as TSS or the entire gene bodies. The methodology to do this will be covered in steps described in Subheading 3.2, but such a visualization will require that a Regionset corresponding to the gene feature of interest (e.g., TSSes) is generated. Now, with the mm9 annotation file downloaded and imported above in **steps 1** and **2** above, it is simple to make such a Regionset from the mm9 Geneset. Just click on [Tools/Extract], change the end position to "start," set the start offset to −500 and the end offset to 500, and click ok. This will result in a set of regions corresponding to the ±500 bp surrounding all the annotated TSSes. If kept unchanged, the default settings will result in a Regionset where each region corresponds to the entire range from the transcription start (TSS) to the end sites of each annotated transcript.

3.1.5 Saving Sessions

So far, the procedure has required a considerable amount of handling. To limit the risk of data loss and to accelerate later access to data and analyses, then save a session containing all the data and work. Such sessions are generally also more space efficient and loaded more quickly than the input data.

1. Save a session by clicking [Sessions/Save].
2. *Optional*: After the session is saved, try to close and reopen EaSeq, and then reimport the data in the session by clicking [Session/Load].

3.1.6 Exporting Regionsets and Datasets

Often it might be relevant to export a peak-set or another Regionset. In this case, do this for the newly generated and annotated Oct4 peaks as follows:

1. Select the Regionset containing the peaks and thereafter [Regionsets/Export]. This outputs in a tab-separated text file containing several columns of peak coordinates, quality scores, and now also the identity of the gene with its TSS closest to each peak.
2. *Optional*: import and inspect these data in Excel or R [25].

   An often-used procedure is to inspect and share data in the UCSC genome browser. For a better visualization, it is advisable to generate a coverage vector (see above) that can be uploaded to the UCSC genome browser [11].
3. In this case, export the Nanog ChIP-seq library (Dataset) as a "wig-file", by clicking [Datasets/Export] and use the default settings.
4. *Optional*: Upload the compressed Nanog.wig.gz file as a custom track in the UCSC genome browser.

3.1.7 Autogenerated Descriptions

When exporting a coverage vector from the Nanog Dataset, another file named "Nanog.wig.description.txt" is stored in the same folder as the wig-file. This is an autogenerated description of the wig-file, including all handling that was done in EaSeq.

1. Try to open "Nanog.wig.description.txt" in the folder with the wig-file using a text editor.

   These files can be exported along with all types of data exported from EaSeq, and they will be coimported if the exported set of data is reimported later (provided that the description file and the data are kept in their original state). The autogenerated descriptions are saved as a part of session files and can also be accessed and studied from within EaSeq.
2. To inspect the autogenerated descriptions within EaSeq, open the "Description panel" by clicking on the notepad icon in the right side of the screen.

3.1.8 Making Tracks

A key functionality in EaSeq is the abundance of visualization tools to study the imported data in an interactive manner. For a start, this example workflow will demonstrate how to make a set of tracks showing the one-dimensional distribution of ChIP-seq signal along the genome.

1. To make one or more tracks, click [Plots/Filltrack].
2. In the window that appears, select the desired combination (s) of Regionset(s) (horizontal) and Datasets(s) (vertical).

   The default viewing style of the FillTracks is to show the cumulative selected ChIP-seq signal (Dataset) at the center of all regions in the Regionset, but this can be changed to show single or subsets of loci (see later). If multiple combinations of Datasets and Regionset were selected in **step 2**, multiple plots will appear, and during this step their order can be changed upfront by dragging them in the panel in right side. All plots in EaSeq have a number of icons, which appears when the mouse in moved over the plot (Illustrated in Fig. 4c in [9]). The most important of these is the "Plot Settings," which gives the user many options to customize each plot (*see* Subheading 3.2.8 for instructions on how to do this).
3. An important step is to export the plots made in EaSeq. A bitmap image (tif, jpeg, etc.) can be exported by clicking on [Sessions/Snapshot]. To export a pdf file, click on the icon in the right side of the user interface containing a beta sign to open the panel with recently added tools. In this panel click on "Export snapshot as pdf."

3.1.9 Using EaSeq to Browse the Genome

Once an EaSeq session contains Dataset(s), Regionset(s), and one or more track(s), EaSeq can be used as a genome browser similar to the UCSC Genome Browser [11], but often quicker as all computation is handled locally.

1. To start navigating the genome in EaSeq, try to type the symbol of a gene in the drop box below Genesets. This will cause the track(s) to change viewpoints to the selected gene.
2. To alter the view, (for example zooming out), open "Navigation panel" by clicking on the compass icon in the right side of the screen.
3. Once the Navigation panel is open, several controls including a red knob appears. The red knob is a quick way to zoom in (drag up), zoom out (drag down), pan left (drag left), pan right (drag right), or any combination hereof.

   A practical way to inspect the ChIP-seq signal at a set of peaks (see above) is to click on one of the peaks appearing in the list of regions on the right side of the Regionsets. This procedure optionally combined with the use of heatmaps (described in Subheading 3.2.2) can generally be applied to iteratively improving the quality of identified peak-sets (Fig. 5).
4. Try to select the Regionset containing the Oct4 peaks.
5. Click on one of the peaks.
6. Then zoom out by dragging the red knob down in the Navigation panel.

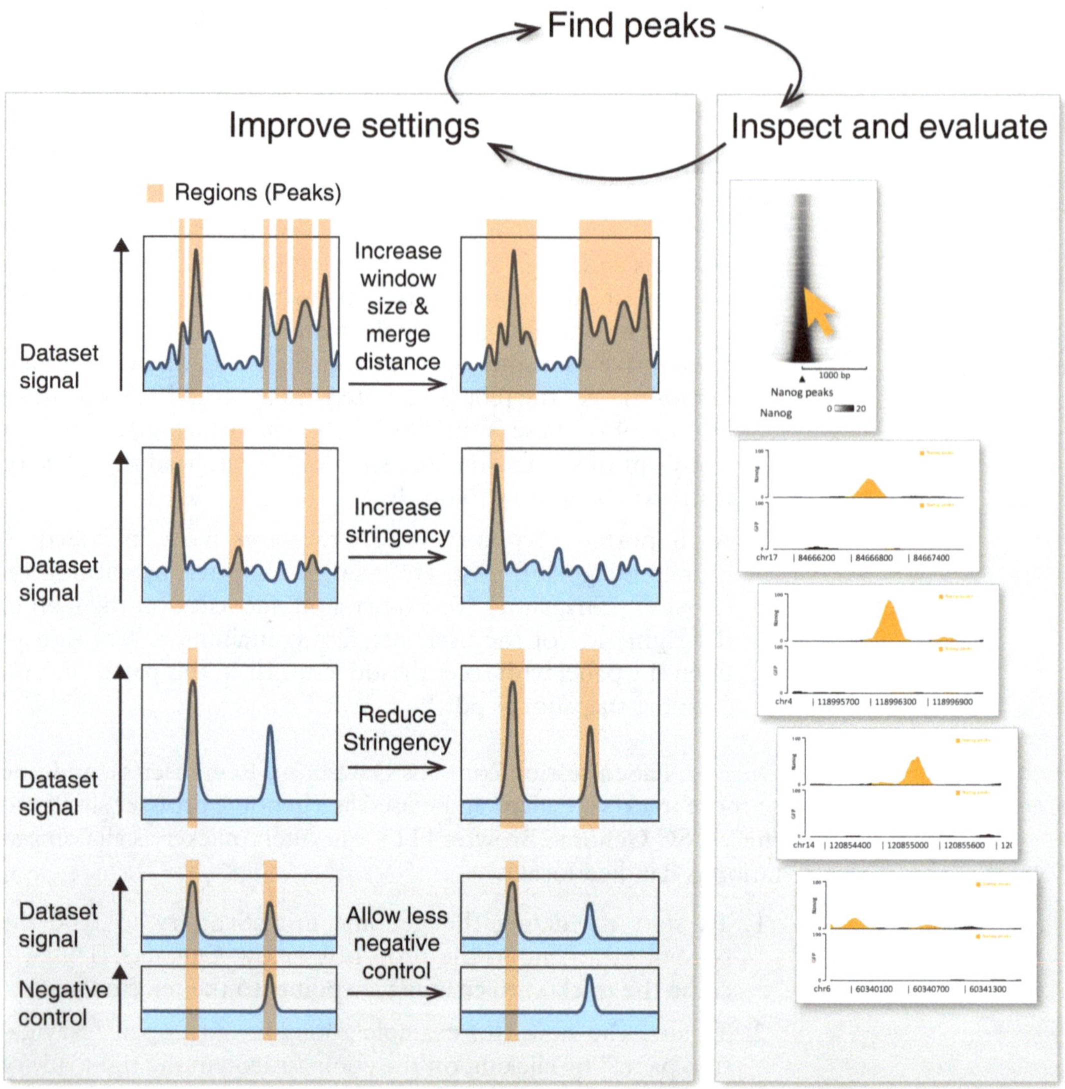

Fig. 5 Upper part: Overview of the process of iteratively improving peak-sets through rapid inspection of the peaks. Lower part, some guidance on how to improve peak-calling in cases of suboptimal output peaks

3.2 Example 2: Quantifying, Calculating Ratios, and Visualizing Histone Mark ChIP-Seq Data

The aim of this workflow example, is to provide basic experience on how to handle ChIP-seq data for histone marks, including how to make plots such as heatmaps, how to inspect the signal at a predefined set of peaks (such as that generated in Subheading 3.1.3), how to quantify signal and calculate ratios, as well as how to sort and filter peaks based on this. A zip file containing the data needed for this example workflow can be downloaded at http://easeq.net/wf2.zip. When the zip-file is decompressed, a text-file with coordinates for CpG-islands and an EaSeq session file becomes available. The session file contains several already imported ChIP-seq datasets of a histone mark (H3K27me3) and chromatin modifiers from the group of Polycomb proteins (Ring1B, Suz12, and Ezh2). These are enriched under two different conditions, an inducible knockout of the Ring1B polycomb protein (OHT) and control samples. These data were originally a part of central work in the polycomb field making use of ChIP-seq [23], and the sequencing reads were mapped to mm9 by us and down-sampled to reduce the file size and import time. To save time, the session file also contains a Regionset of Suz12 peaks that were identified previously (*see* Fig. 8 in [9]) based on an independent set of data in a procedure that was roughly as described in (Subheading 3.1.3). These peaks will be used as viewpoints for the ChIP-seq signal and as our primary regions of interests for this example.

3.2.1 Import of ChIP-Seq Data and Regions

1. Unpack the zip-file and import the session "Blackledge compact.eas" by clicking [Sessions/Load].
2. *Optional*: Data formats tend to vary a lot, but the import Wizard in EaSeq, offers a lot of flexibility that allows import of diverse formats. To get some experience with import of an interesting but more challenging set of viewpoints from an external source, then import "mm9 CpG islands.gz" from the zipped data as a Regionset by clicking [Regionset/Import]. The default view of the file browser is to look for text files, so it will need to be instructed to look for all files or .gz-files. These CpG-islands were originally downloaded from the UCSC table browser [12] with contains a rich collection of relevant genomic loci for a large set of reference genomes. In this case, the import will lead EaSeq to ask whether it should skip the first line or not. This occurs because the first line of the file contains a hashtag (#), which often also is used for comments with content that should be ignored by software. In this case, it is instead used in the name of header, but EaSeq does not know this, and needs instructions on how to handle it. Therefore, instruct EaSeq that it should *not* skip the line. If the import goes well, then the wizard should identify that chromosome names, start coordinates, and end coordinates are found in columns 2, 3, and 4, respectively, and will suggest that the data in columns 2–4 should not be imported as parameters. In this case, just use the default suggestion and click OK. If EaSeq did not recognize the information structure properly, then it will need instructions from the user regarding in which columns the different coordinates can be found.

3.2.2 Make and Use Heatmaps for Genome Browsing and Inspection of Peaks

A quick way to get an overview of the signal in the imported datasets, relative to the Suz12 peaks, is to make heatmaps, which also can be used to explore the signal at the peaks in more depth.

1. To make one or more heatmaps, click on [Plots/Heatmaps].

2. This will open a window where it is possible to select the desired combination(s) of Regionset(s) (horizontal) and Datasets(s) (vertical).
3. To use the heatmaps as a starting point to explore ChIP-seq signal, then also make a set of tracks, by clicking [Plots/Filltrack].
4. Again, in the window that appears, select the desired combination(s) of Regionset(s) (horizontal) and Datasets(s) (vertical).
5. Click on a single point in the heatmap. The tracks will now update to show that particular locus. This is a quick and good strategy to assess if peak-finding settings were appropriate or it should be repeated with more or less stringency (Fig. 5).
6. *Optional*: It is possible to combine this with manual navigation within the genome by opening the "Navigation panel" (click on the compass icon in the right side of the screen), and, for example, drag the red knob to zoom in (drag up), zoom out (drag down), pan left (drag left), pan right (drag right). In this way, the surroundings of many peaks can readily be inspected at any magnification to assess how strong the signal is at the peak relative to interesting genomic features or the general background.
7. To select multiple loci in the heatmap, and get the signal visualized in the tracks, press down the mouse button inside a heatmap, and keep it pressed down while dragging it over a subset of the peaks. Once it is released, the tracks will update to show the superimposed signal at all of the selected peaks. This superimposed track state is a useful way to get an overview of the signal distribution in the selected subpopulation of the peaks.
8. *Optional*: Select all peaks in the Regionset by clicking the small icon next to the selected area.
9. This is also a good occasion to consider the information provided in the heatmap. Heatmaps shows the signal from all loci within a Regionset, with the loci stacked vertically (Fig. 2). The horizontal center in the heatmap is by default also the center of each visualized locus, and the default view then shows the 10 kbp upstream loci to the left and the 10 kbp downstream loci to the right. Both the visualized window as well as the vertical order of the heatmaps can be adjusted (*see* Subheading 3.2.5 and **Note 5** for instructions on how to change this).

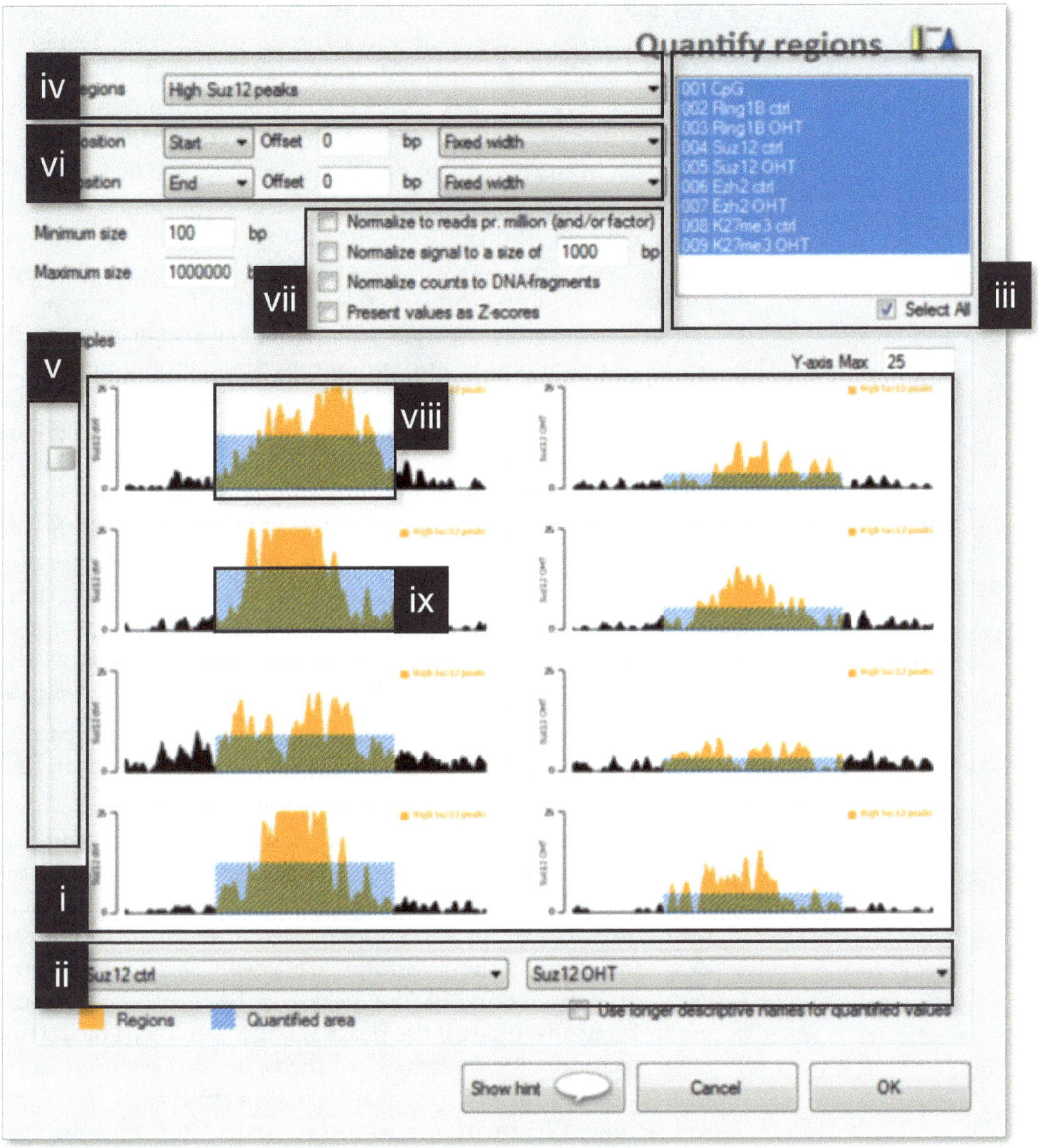

Fig. 6 (i–ix) Example of the Quantify window and annotation of its contents

3.2.3 Quantify ChIP-Seq at a Set of Peaks or Other Regions

In the following, we will quantify the signal of all imported ChIP-seq datasets at the midpoint and surrounding 1 kbp of all Suz12 peaks.

1. To quantify ChIP-seq signal, click [Tools/Quantify].

   This will open a new window (Fig. 6), which also offers an intuitive way to assess the signal distribution at peaks, and to evaluate if the applied settings will capture the signal appropriately. The eight tracks in the center of the window (Fig. 6i) are visualizations of the ChIP-seq signal from the four first regions (vertical) in the Regionset that will be used for the quantification. The two columns of tracks shows two example Datasets, and changing those in the drop-down menu below the track examples (Fig. 6ii) has no effect on the final outcome. The datasets that are used for the actual quantitation can be selected in the list in the upper right corner (Fig. 6iii), and by default all datasets are selected. The Regionset to be used can be changed in the uppermost drop-down menu (Fig. 6iv), and the slider in the left side (Fig. 6v) will allow a quick inspection of a large number of regions from the selected Regionset. The window also contains several options to instruct specifically which area relative to each region in the Regionset that should be used as the basis for the quantitation (Fig. 6vi). Finally, the Quantify tool provides a range of options for how to normalize the quantified values (Fig. 6vii), and it will by default provide FPKM values (*see* **Note 6**). When most of these options are changed the visualization in the tracks (Fig. 6i) will change accordingly. The orange area (Fig. 6viii) show each region (and possibly neighboring regions as well), whereas the blue transparent area (Fig. 6ix) gives an indication of the extent of each area that is used for the quantitation as well as how the normalization will affect the amplitude of the quantified signal.

2. *Optional.* Explore the functions and settings in the Quantify tool by changing the selection in the drop boxes or by changing the offset values—what do you see?

3. Once you have explored sufficiently, use the default settings for the quantitation, with start as start position, end as end position, and set both offsets to 0 (If many settings were changed in **step 2** in Subheading 3.2.3 above, then it might be simpler to close and reopen the window). This will lead to the signal within the boundary of the peaks to be quantified, which might be an advantage if the peaks vary a lot in size. The default normalization will take the size of the peaks into account. If the start, end, and offset settings are left as default, then the signal will be quantified within a window starting 500 bp upstream and ending 500 bp downstream of the peak centers, which in many cases also might be a useful option as the size of the quantified regions is kept constant.

   The quantified values will be added to the Regionset as new "Parameters" (*see* Subheading 2.2.2), and the list of Regionsets will display that this Regionset has been updated.

4. *Optional:* Inspect these values by clicking on the updated Regionset in the table showing all the Regionsets, and then in the list of regions in the right side, new columns of data will appear (one for each Dataset that was included in the quantification).

3.2.4 Calculate Ratios of Quantified ChIP-Seq Signal at Peaks

The next steps will demonstrate how to calculate the ratio between Suz12 in the conditional knockout (OHT) and the control samples. Although this is a useful way to assess overall changes at a set of regions and to find those with the most pronounced changes, it is advisable do this with caution (*see* **Note 7**).

1. To calculate the log2 fold difference between the quantified Suz12 OTH and Suz12 control values, first open the Calculate tool by clicking [Tools/Calculate].
2. Then in the appearing window, select the "Division" radiobutton, check the "Logarithm" flag, from the drop-down boxes select the newly quantified values in the Suz12 OHT parameter as "x" and Suz12 ctrl parameter as "y," and click "OK."

As for the quantified values in Subheading 3.2.3, the ratios will now appear as a new "Parameter" (*see* Subheading 2.2.2), and the list of Regionsets will display that the Regionset has been updated.

3.2.5 Sort Peaks According to Quantified ChIP-Seq Ratios

To change the order of appearance of regions in certain plot types, such as heatmaps, they will need to be sorted first. To do this:

1. Open the Sort tool by clicking on [Tools/Sort].
2. In the appearing window, select the newly calculated ratios.

Any heatmap that shows the sorted Regionset will now automatically update to show the new and updated order, in this case the Suz12OTH/ctrl ratio.

3.2.6 Making 1D- and 2D-Histograms

A useful way to get an overview of a population of quantified values is to make a histogram of these. If the relationship between two sets of quantified values or, for example, imported Parameters in a Regionset is in question, then a 2D-histogram is ideal. 2D-histograms are rather similar to scatter plots, but they use color to show the density of regions with certain sets of values. They segment the plotted area into "bins" for this counting, similarly to what a 1D-histogram does, and the binning can be coarse or fine depending on data density, resulting in very different appearances.

1. To make a (1D) histogram of the quantified Suz12 OHT values, click [Plots/Histogram].
2. In the window that appears, select the desired combination (s) of Regionset(s) (horizontal) and Parameter(s) (vertical). If multiple combinations of Regionsets and Regionset Parameters are clicked, multiple plots will appear, and their order can be changed upfront by dragging them in the right side panel.

3. To make a 2D-histogram of the quantified Suz12 OHT and Suz12 ctrl values, select the Regionset containing the values that should be visualized and click [Plots/Scatter].
4. In the window that appears, select the desired combination (s) of Regionset Parameter(s) (both horizontal and vertical). If multiple combinations are clicked, multiple plots will appear, and their order can be changed upfront by dragging them in the right side panel.

3.2.7 Interactively Exploring Data in 2D-Histograms

Once the new plots showing the Suz12 OHT and Suz12 Ctrl levels has appeared. These plots are interactive and allow you to select one or more regions by clicking on the area in the plot. Once a region is selected it will be highlighted in other plots displaying the same Regionset, and tracks will automatically update to show the locus surrounding the selected regions(s).

1. To select a single region in the 2D-histogram, click on the dot in the plot.
2. To select a subset of multiple regions from the Regionset displayed by the 2D-histogram, press down the mouse button inside the 2D-histogram, and keep it pressed down while dragging it over the regions that you would like to select.

   Plots can be moved, resized, and duplicated to build up an exportable (*see* Subheading 3.1.8, **step 3**) panel, which supports the scientific message.
3. To duplicate a plot, click on the small icon with a plus appearing inside the plot when the mouse pointer moves over it.
4. To move a plot, press the appearing small icon with four arrows in the upper left corner of the plot and hold down the mouse button while moving the plot.
5. To resize a plot, press the appearing small icon with the square and arrow in the lower right corner of the plot and hold down the mouse button while resizing the plot.
6. *Optional*: If multiple plots are selected, then they can be moved and resized in the same way. To duplicate all plots, click on the [Plots/Duplicate] icon in the main menu. Duplicated plots will appear in a larger unoccupied area.
7. *Optional*: Make some tracks by clicking on [Plots/Tracks], and select one or more regions in the 2D-histogram to show the signal at this/these regions.

3.2.8 Adjusting Plot Settings and a Quick Way to Explore Subsets of Regions

Each plot type has multiple variations, and the possibilities to adjust settings and vary the plots are too comprehensive to cover in detail here. Inspiration and an overview of the possibilities can be found here: http://easeq.net/plots.pdf. In the steps below, we will only briefly cover how a few of these settings are accessed and changed.

1. To enter the settings for the 2D-histogram generated in Subheading 3.2.6, **steps 3** and **4**, move the mouse over the plot, and click on the small icon with the gears that appear in the upper right corner of the plot.

   This will open a rather busy panel with many settings that can be changed, and when the cursor exits this panel, the changes will be applied and the plot will update automatically. Settings that are not relevant to the selected plot type are greyed out and inactive. The best way to learn about their functions is to duplicate a plot, change a setting, and study the outcome. Often, the effect can be deduced rather intuitively by comparing the plots.
2. For the 2D-histogram, we will change the default log scale on the x- and y- axes from log (default) to linear. First duplicate the plot to compare the difference by clicking on the small icon with a plus appearing inside the plot when the mouse pointer moves over it. Then open the Plot Settings as described in Subheading 3.2.8, **step 1** above, and uncheck the "Log scale" checkboxes beneath the *X*-axis and *Y*-axis headers.
3. *Optional*: Click on some data within the 2D-histogram with linear scales on the axes, select an area, and observe the corresponding regions in the other plot with the log scales.
4. *Optional*: Make another 2D-histogram of Ring1B OHT vs. Ring1B ctrl. This can be done either as in Subheading 3.2.6 **steps 3** and **4** or by duplicating the histogram as described in Subheading 3.2.7 **step 3** open the Plot Settings as in **step 1**, Subheading 3.2.8, above and change the Parameters in the drop-down menus named "Parameter" beneath the *X*-axis and *Y*-axis headers to that of the Ring1B OHT and Ring1B ctrl values quantified in Subheading 3.2.3.

3.2.9 Making Subpopulations of Regions from Plots and by Using the Gate Tool

If you are interested in understanding or exploring a certain subset of regions further, it is a good idea to make a new Regionset consisting of this subset. This can be done either by:

1. Select an area with regions of interest in the 2D-histogram made in Subheading 3.2.6, **steps 3** and **4** showing the Suz12 OHT and Suz12 ctrl values.
2. Click on the small icon showing a pair of scissor. A new Regionset should appear in the Regionset panel.

Alternatively, it can be done in this manner:

3. Open the Gate tool by clicking [Tools/Gate]. This will open a new window, where the criteria that should be used can be defined.
4. First, make sure that the selected Regionset in the uppermost drop-down menu is the Suz12 peaks, then to use the ratio calculated in Subheading 3.2.4 select the Parameter named "log2('Suz12 OHT' divided by 'Suz12 ctrl')", and to get the subset of regions with a ratio higher than 0 select the ">" operator in the drop-down menu next to the selected Parameter and type 0 in the textbox. This should result in a small Regionset consisting of a few regions (48 if done similar to our example).

3.2.10 Quick Ways to Make Identical Figures of Different Populations of Regions

It is often useful to investigate ChIP-seq signal at several sets of Regions—either to compare signal, for example, at different populations of regions (e.g., enhancers, CpG-islands, and TSSes) or to explore a subset of regions as covered in Subheading 3.2.9.

1. Start by selecting the heatmap(s) made in Subheading 3.2.2.
2. A faster way to duplicate multiple plots, rather than clicking on the small duplicate icon in each of plots, is to click on [Plots/Duplicate].
3. Then try to click on either [Plots/Next Reg.] or [Plots/Prev. Reg.] and observe what happens.

   The selected heatmaps should show the next or previous Regionset on the lists. The same would happen to any other selected plot type, but for 2D-histograms, for example, the Regionset would need to have identical Parameters for the plots to show similar information.
4. *Optional:* With the above steps done, you might be ready for a larger and more demanding exercise. Access the paper by Blackledge et al. [23] and remake panels d–f, and h from Fig. 4 of this paper using the data available in the session, the approaches described above, and some creativity.

4 Notes

1. An often-sought type of data integration, is to do combined studies of ChIP-seq and RNA-seq to reveal the relationship between histone modifications and transcriptional readout. RNA-seq data can be imported directly as datasets and visualized in, for example, heatmaps or tracks as for ChIP-seq data. However, to identify differentially expressed genes, we recommend using dedicated tools such as Deseq2 [27] possibly within the Galaxy environment [10], and then import the coordinates for the genes together with the statistics as a

Regionset into EaSeq for further analysis and visualization. Then this regionset can be used as a basis to explore the ChIP-seq data similar as how the Suz12 peaks were used in the above examples.

2. The different types of data in EaSeq; Datasets, Regionsets, and Genesets; can often be used to provide another type of data. Conversion from one data type to another or making a derived set of the same type can be done by several tools, including (sorted according to usage frequency): [Tools/Extract] (Geneset → Regionset), [Tools/Peaks] (Dataset → Regionset), [Tools/Gate] (Regionset → Regionset), [Tools/Controls] (Regionset → Regionset), [Regionsets/Modify] (Regionset → Regionset), [Datasets/Modify] (Dataset → Dataset), and [Tools/Pattern] (Dataset → Regionset).
3. To provide as much freedom as possible, EaSeq does not adhere to a specific chromosome nomenclature. The names within different types of data, which should be compared, do however need to be the identical. While it is most common to use chr3 when referring to chromosome 3, shorter names are sometimes used to save space, and data where chromosome 3 is referred to as "3" are not uncommon. If for example an imported Regionset uses a different naming than the already imported Datasets, then EaSeq opens a wizard that will allow you to check, if the chromosome names have sufficient correspondence and possibly allow the chromosomes to be renamed.
4. EaSeq does not keep track of which reference genome that has been used to derive the different Datasets and Regionsets. It will therefore not be able to point out if two incompatible reference genomes are used, and the user has to keep track of this.
5. When newly generated, heatmaps and tracks will show the center of the visualized regions. For visualization of peaks and TSSes this is sensible, but for gene bodies this is not the case as the gene body sizes vary a lot, and the middle of the gene body is rarely an interesting viewpoint (*see* Subheading 3.1.4, **step 6** to generate a Regionset of gene bodies and Subheading 3.2.2 to make an example heatmap). Therefore, for such plot types it makes sense to enter the plot settings (Subheading 3.2.8, **step 1**), click on the checkbox named "Rel." to use relative window sizes, and change the "Window size" to, for example, 200 bp. With this setting the size of the visualized window will be adapted for each region so that it shows an area corresponding to 200% of the region size, so for a 30 kbp long gene this would be 60 kbp. In this manner, regions with highly dissimilar sizes can actually be fit into the same heatmap or track.

6. EaSeq contains several options to normalize data. Per default, tracks and heatmaps of datasets (which are not from coverage vectors, *see* Subheading 2.2.3) are denominated in FPKMs, which stands for Fragments per Kilobasepairs per Million Reads. This can be disabled in the plot settings if needed. The same applies for quantified values, but in the Quantify Tool (Covered in Subheading 3.2.3) the individual components of this normalization can be enabled or disabled on a custom basis. For quantitations that are intended for count-based statistical analysis such as in Deseq2 [27], the normalization should be disabled to provide the raw read counts. A frequently used normalization type in ChIP-seq is the use of spike-in DNA as an internal control [26]. Constants from such spike-in normalization can be entered and used to adjust Dataset values in the menu controlled by [Datasets/Describe].
7. A few words of caution. (a) ChIP-seq is inherently unsuited for detecting massive and global changes in the abundance of the studied target, although this can be alleviated using spike-in of, for example, drosophila DNA as an internal standard for global normalization [26]. (b) ChIP-seq is subject to linear and non-linear biases, even within different replicates—we have an example in Fig. 4c here [9]. (c) Stochastic variation likely contribute to quantified differences, so it is advisable to only use regions with a high density of reads to limit this variation. (d) If one set of data is used for both peak-calling and to calculate the ratio of a signal, there is bound to be a bias as the stochastic variation will both contribute to the definition of the regions and the values used for calculating the ratios. This is the case even for two technical replicates, where the one used for peak-calling would have a higher amount of signal in the peaks than the other. Some good ways to limit these effects is to use independent replicates for the peak-finding and the quantitation (in this case in Subheading 3.2 the peaks were established earlier, *see* Fig. 8 in ref. [9]), and to use ChIP-seq for studies of local changes, preferably by comparing the effects at one set of regions to those at another. Finally, it is always strengthening the scientific value of the data to have multiple biological replicates, potentially using one set of data for finding interesting relationships, and the rest of the replicates for validating these observations.

References

1. Atlasi Y, Stunnenberg HG (2017) The interplay of epigenetic marks during stem cell differentiation and development. Nat Rev Genet 18:643. https://doi.org/10.1038/nrg.2017.57
2. Barski A, Zhao K (2009) Genomic location analysis by ChIP-Seq. J Cell Biochem 107 (1):11–18. https://doi.org/10.1002/jcb.22077
3. Berdasco M, Esteller M (2010) Aberrant epigenetic landscape in cancer: how cellular identity goes awry. Dev Cell 19(5):698–711. https://doi.org/10.1016/j.devcel.2010.10.005
4. Holmberg J, Perlmann T (2012) Maintaining differentiated cellular identity. Nat Rev Genet 13:429. https://doi.org/10.1038/nrg3209
5. Ouyang Z, Zhou Q, Wong WH (2009) ChIP-seq of transcription factors predicts absolute and differential gene expression in embryonic stem cells. Proc Natl Acad Sci 106 (51):21521–21526. https://doi.org/10.1073/pnas.0904863106
6. Barski A, Cuddapah S, Cui K, Roh TY, Schones DE, Wang Z, Wei G, Chepelev I, Zhao K (2007) High-resolution profiling of histone methylations in the human genome. Cell 129 (4):823–837. https://doi.org/10.1016/j.cell.2007.05.009
7. Johnson DS, Mortazavi A, Myers RM, Wold B (2007) Genome-wide mapping of in vivo protein-DNA interactions. Science 316 (5830):1497–1502. https://doi.org/10.1126/science.1141319
8. Mikkelsen TS, Ku M, Jaffe DB, Issac B, Lieberman E, Giannoukos G, Alvarez P, Brockman W, Kim TK, Koche RP, Lee W, Mendenhall E, O'Donovan A, Presser A, Russ C, Xie X, Meissner A, Wernig M, Jaenisch R, Nusbaum C, Lander ES, Bernstein BE (2007) Genome-wide maps of chromatin state in pluripotent and lineage-committed cells. Nature 448(7153):553–560. https://doi.org/10.1038/nature06008
9. Lerdrup M, Johansen JV, Agrawal-Singh S, Hansen K (2016) An interactive environment for agile analysis and visualization of ChIP-sequencing data. Nat Struct Mol Biol 23 (4):349–357. https://doi.org/10.1038/nsmb.3180
10. Afgan E, Baker D, Batut B, van den Beek M, Bouvier D, Čech M, Chilton J, Clements D, Coraor N, Grüning BA, Guerler A, Hillman-Jackson J, Hiltemann S, Jalili V, Rasche H, Soranzo N, Goecks J, Taylor J, Nekrutenko A, Blankenberg D (2018) The Galaxy platform for accessible, reproducible and collaborative biomedical analyses: 2018 update. Nucleic Acids Res 46 (W1):W537-W544. doi:https://doi.org/10.1093/nar/gky379
11. Kent WJ, Sugnet CW, Furey TS, Roskin KM, Pringle TH, Zahler AM, Haussler D (2002) The human genome browser at UCSC. Genome Res 12(6):996–1006. https://doi.org/10.1101/gr.229102
12. Karolchik D, Hinrichs AS, Furey TS, Roskin KM, Sugnet CW, Haussler D, Kent WJ (2004) The UCSC table browser data retrieval tool. Nucleic Acids Res 32(Database issue): D493–D496. https://doi.org/10.1093/nar/gkh103
13. O'Leary NA, Wright MW, Brister JR, Ciufo S, Haddad D, McVeigh R, Rajput B, Robbertse B, Smith-White B, Ako-Adjei D, Astashyn A, Badretdin A, Bao Y, Blinkova O, Brover V, Chetvernin V, Choi J, Cox E, Ermolaeva O, Farrell CM, Goldfarb T, Gupta T, Haft D, Hatcher E, Hlavina W, Joardar VS, Kodali VK, Li W, Maglott D, Masterson P, McGarvey KM, Murphy MR, O'Neill K, Pujar S, Rangwala SH, Rausch D, Riddick LD, Schoch C, Shkeda A, Storz SS, Sun H, Thibaud-Nissen F, Tolstoy I, Tully RE, Vatsan AR, Wallin C, Webb D, Wu W, Landrum MJ, Kimchi A, Tatusova T, DiCuccio M, Kitts P, Murphy TD, Pruitt KD (2016) Reference sequence (RefSeq) database at NCBI: current status, taxonomic expansion, and functional annotation. Nucleic Acids Res 44(D1):D733–D745. https://doi.org/10.1093/nar/gkv1189
14. Edgar R, Domrachev M, Lash AE (2002) Gene expression omnibus: NCBI gene expression and hybridization array data repository. Nucleic Acids Res 30(1):207–210. https://doi.org/10.1093/nar/30.1.207
15. SA (2010) FastQC: a quality control tool for high throughput sequence data. http://www.bioinformaticsbabrahamacuk/projects/fastqc
16. Wingett SW, Andrews S (2018) FastQ Screen: a tool for multi-genome mapping and quality control. F1000Res 7:1338–1338. https://doi.org/10.12688/f1000research.15931.2
17. Bolger AM, Lohse M, Usadel B (2014) Trimmomatic: a flexible trimmer for Illumina sequence data. Bioinformatics 30 (15):2114–2120. https://doi.org/10.1093/bioinformatics/btu170
18. Martin M (2011) Cutadapt removes adapter sequences from high-throughput sequencing

reads 2011 17 (1):3. doi:https://doi.org/10.14806/ej.17.1.200

19. Krueger F (2015) Trim galore. A wrapper tool around Cutadapt and FastQC to consistently apply quality and adapter trimming to FastQ files
20. Li H, Handsaker B, Wysoker A, Fennell T, Ruan J, Homer N, Marth G, Abecasis G, Durbin R (2009) The sequence alignment/map format and SAMtools. Bioinformatics 25 (16):2078–2079. https://doi.org/10.1093/bioinformatics/btp352
21. Bernstein BE, Stamatoyannopoulos JA, Costello JF, Ren B, Milosavljevic A, Meissner A, Kellis M, Marra MA, Beaudet AL, Ecker JR, Farnham PJ, Hirst M, Lander ES, Mikkelsen TS, Thomson JA (2010) The NIH roadmap epigenomics mapping consortium. Nat Biotechnol 28(10):1045–1048. https://doi.org/10.1038/nbt1010-1045
22. The EPC, Dunham I, Kundaje A, Aldred SF, Collins PJ, Davis CA, Doyle F, Epstein CB, Frietze S, Harrow J, Kaul R, Khatun J, Lajoie BR, Landt SG, Lee B-K, Pauli F, Rosenbloom KR, Sabo P, Safi A, Sanyal A, Shoresh N, Simon JM, Song L, Trinklein ND, Altshuler RC, Birney E, Brown JB, Cheng C, Djebali S, Dong X, Dunham I, Ernst J, Furey TS, Gerstein M, Giardine B, Greven M, Hardison RC, Harris RS, Herrero J, Hoffman MM, Iyer S, Kellis M, Khatun J, Kheradpour P, Kundaje A, Lassmann T, Li Q, Lin X, Marinov GK, Merkel A, Mortazavi A, Parker SCJ, Reddy TE, Rozowsky J, Schlesinger F, Thurman RE, Wang J, Ward LD, Whitfield TW, Wilder SP, Wu W, Xi HS, Yip KY, Zhuang J, Bernstein BE, Birney E, Dunham I, Green ED, Gunter C, Snyder M, Pazin MJ, Lowdon RF, Dillon LAL, Adams LB, Kelly CJ, Zhang J, Wexler JR, Green ED, Good PJ, Feingold EA, Bernstein BE, Birney E, Crawford GE, Dekker J, Elnitski L, Farnham PJ, Gerstein M, Giddings MC, Gingeras TR, Green ED, Guigó R, Hardison RC, Hubbard TJ, Kellis M, Kent WJ, Lieb JD, Margulies EH, Myers RM, Snyder M, Stamatoyannopoulos JA, Tenenbaum SA, Weng Z, White KP, Wold B, Khatun J, Yu Y, Wrobel J, Risk BA, Gunawardena HP, Kuiper HC, Maier CW, Xie L, Chen X, Giddings MC, Bernstein BE, Epstein CB, Shoresh N, Ernst J, Kheradpour P, Mikkelsen TS, Gillespie S, Goren A, Ram O, Zhang X, Wang L, Issner R, Coyne MJ, Durham T, Ku M, Truong T, Ward LD, Altshuler RC, Eaton ML, Kellis M, Djebali S, Davis CA, Merkel A, Dobin A, Lassmann T, Mortazavi A, Tanzer A, Lagarde J, Lin W, Schlesinger F, Xue C, Marinov GK, Khatun J, Williams BA, Zaleski C, Rozowsky J, Röder M, Kokocinski F, Abdelhamid RF, Alioto T, Antoshechkin I, Baer MT, Batut P, Bell I, Bell K, Chakrabortty S, Chen X, Chrast J, Curado J, Derrien T, Drenkow J, Dumais E, Dumais J, Duttagupta R, Fastuca M, Fejes-Toth K, Ferreira P, Foissac S, Fullwood MJ, Gao H, Gonzalez D, Gordon A, Gunawardena HP, Howald C, Jha S, Johnson R, Kapranov P, King B, Kingswood C, Li G, Luo OJ, Park E, Preall JB, Presaud K, Ribeca P, Risk BA, Robyr D, Ruan X, Sammeth M, Sandhu KS, Schaeffer L, See L-H, Shahab A, Skancke J, Suzuki AM, Takahashi H, Tilgner H, Trout D, Walters N, Wang H, Wrobel J, Yu Y, Hayashizaki Y, Harrow J, Gerstein M, Hubbard TJ, Reymond A, Antonarakis SE, Hannon GJ, Giddings MC, Ruan Y, Wold B, Carninci P, Guigó R, Gingeras TR, Rosenbloom KR, Sloan CA, Learned K, Malladi VS, Wong MC, Barber GP, Cline MS, Dreszer TR, Heitner SG, Karolchik D, Kent WJ, Kirkup VM, Meyer LR, Long JC, Maddren M, Raney BJ, Furey TS, Song L, Grasfeder LL, Giresi PG, Lee B-K, Battenhouse A, Sheffield NC, Simon JM, Showers KA, Safi A, London D, Bhinge AA, Shestak C, Schaner MR, Ki Kim S, Zhang ZZ, Mieczkowski PA, Mieczkowska JO, Liu Z, McDaniell RM, Ni Y, Rashid NU, Kim MJ, Adar S, Zhang Z, Wang T, Winter D, Keefe D, Birney E, Iyer VR, Lieb JD, Crawford GE, Li G, Sandhu KS, Zheng M, Wang P, Luo OJ, Shahab A, Fullwood MJ, Ruan X, Ruan Y, Myers RM, Pauli F, Williams BA, Gertz J, Marinov GK, Reddy TE, Vielmetter J, Partridge E, Trout D, Varley KE, Gasper C, Bansal A, Pepke S, Jain P, Amrhein H, Bowling KM, Anaya M, Cross MK, King B, Muratet MA, Antoshechkin I, Newberry KM, McCue K, Nesmith AS, Fisher-Aylor KI, Pusey B, DeSalvo G, Parker SL, Balasubramanian S, Davis NS, Meadows SK, Eggleston T, Gunter C, Newberry JS, Levy SE, Absher DM, Mortazavi A, Wong WH, Wold B, Blow MJ, Visel A, Pennachio LA, Elnitski L, Margulies EH, Parker SCJ, Petrykowska HM, Abyzov A, Aken B, Barrell D, Barson G, Berry A, Bignell A, Boychenko V, Bussotti G, Chrast J, Davidson C, Derrien T, Despacio-Reyes G, Diekhans M, Ezkurdia I, Frankish A, Gilbert J, Gonzalez JM, Griffiths E, Harte R, Hendrix DA, Howald C, Hunt T, Jungreis I, Kay M, Khurana E, Kokocinski F, Leng J, Lin MF, Loveland J, Lu Z, Manthravadi D, Mariotti M, Mudge J, Mukherjee G, Notredame C, Pei B, Rodriguez JM, Saunders G, Sboner A, Searle S, Sisu C, Snow C, Steward C, Tanzer A, Tapanari E, Tress ML, van Baren MJ, Walters N, Washietl S, Wilming L, Zadissa A, Zhang Z,

Brent M, Haussler D, Kellis M, Valencia A, Gerstein M, Reymond A, Guigó R, Harrow J, Hubbard TJ, Landt SG, Frietze S, Abyzov A, Addleman N, Alexander RP, Auerbach RK, Balasubramanian S, Bettinger K, Bhardwaj N, Boyle AP, Cao AR, Cayting P, Charos A, Cheng Y, Cheng C, Eastman C, Euskirchen G, Fleming JD, Grubert F, Habegger L, Hariharan M, Harmanci A, Iyengar S, Jin VX, Karczewski KJ, Kasowski M, Lacroute P, Lam H, Lamarre-Vincent N, Leng J, Lian J, Lindahl-Allen M, Min R, Miotto B, Monahan H, Moqtaderi Z, Mu XJ, O'Geen H, Ouyang Z, Patacsil D, Pei B, Raha D, Ramirez L, Reed B, Rozowsky J, Sboner A, Shi M, Sisu C, Slifer T, Witt H, Wu L, Xu X, Yan K-K, Yang X, Yip KY, Zhang Z, Struhl K, Weissman SM, Gerstein M, Farnham PJ, Snyder M, Tenenbaum SA, Penalva LO, Doyle F, Karmakar S, Landt SG, Bhanvadia RR, Choudhury A, Domanus M, Ma L, Moran J, Patacsil D, Slifer T, Victorsen A, Yang X, Snyder M, White KP, Auer T, Centanin L, Eichenlaub M, Gruhl F, Heermann S, Hoeckendorf B, Inoue D, Kellner T, Kirchmaier S, Mueller C, Reinhardt R, Schertel L, Schneider S, Sinn R, Wittbrodt B, Wittbrodt J, Weng Z, Whitfield TW, Wang J, Collins PJ, Aldred SF, Trinklein ND, Partridge EC, Myers RM, Dekker J, Jain G, Lajoie BR, Sanyal A, Balasundaram G, Bates DL, Byron R, Canfield TK, Diegel MJ, Dunn D, Ebersol AK, Frum T, Garg K, Gist E, Hansen RS, Boatman L, Haugen E, Humbert R, Jain G, Johnson AK, Johnson EM, Kutyavin TV, Lajoie BR, Lee K, Lotakis D, Maurano MT, Neph SJ, Neri FV, Nguyen ED, Qu H, Reynolds AP, Roach V, Rynes E, Sabo P, Sanchez ME, Sandstrom RS, Sanyal A, Shafer AO, Stergachis AB, Thomas S, Thurman RE, Vernot B, Vierstra J, Vong S, Wang H, Weaver MA, Yan Y, Zhang M, Akey JM, Bender M, Dorschner MO, Groudine M, MacCoss MJ, Navas P, Stamatoyannopoulos G, Kaul R, Dekker J, Stamatoyannopoulos JA, Dunham I, Beal K, Brazma A, Flicek P, Herrero J, Johnson N, Keefe D, Lukk M, Luscombe NM, Sobral D, Vaquerizas JM, Wilder SP, Batzoglou S, Sidow A, Hussami N, Kyriazopoulou-Panagiotopoulou S, Libbrecht MW, Schaub MA, Kundaje A, Hardison RC, Miller W, Giardine B, Harris RS, Wu W, Bickel PJ, Banfai B, Boley NP, Brown JB, Huang H, Li Q, Li JJ, Noble WS, Bilmes JA, Buske OJ, Hoffman MM, Sahu AD, Kharchenko PV, Park PJ, Baker D, Taylor J, Weng Z, Iyer S, Dong X, Greven M, Lin X, Wang J, Xi HS, Zhuang J, Gerstein M, Alexander RP, Balasubramanian S, Cheng C, Harmanci A, Lochovsky L, Min R, Mu XJ, Rozowsky J, Yan K-K, Yip KY, Birney E (2012) An integrated encyclopedia of DNA elements in the human genome. Nature 489:57. doi:https://doi.org/10.1038/nature11247. https://www.nature.com/articles/nature11247#supplementary-information

23. Blackledge NP, Farcas AM, Kondo T, King HW, McGouran JF, Hanssen LL, Ito S, Cooper S, Kondo K, Koseki Y, Ishikura T, Long HK, Sheahan TW, Brockdorff N, Kessler BM, Koseki H, Klose RJ (2014) Variant PRC1 complex-dependent H2A ubiquitylation drives PRC2 recruitment and polycomb domain formation. Cell 157(6):1445–1459. https://doi.org/10.1016/j.cell.2014.05.004
24. Chen X, Xu H, Yuan P, Fang F, Huss M, Vega VB, Wong E, Orlov YL, Zhang W, Jiang J (2008) Integration of external signaling pathways with the core transcriptional network in embryonic stem cells. Cell 133(6):1106–1117
25. R Development Core Team R (2011) R: a language and environment for statistical computing. R Foundation for Statistical Computing, Vienna
26. Bonhoure N, Bounova G, Bernasconi D, Praz V, Lammers F, Canella D, Willis IM, Herr W, Hernandez N, Delorenzi M, Cycli XC (2014) Quantifying ChIP-seq data: a spiking method providing an internal reference for sample-to-sample normalization. Genome Res 24(7):1157–1168. https://doi.org/10.1101/gr.168260.113
27. Love MI, Huber W, Anders S (2014) Moderated estimation of fold change and dispersion for RNA-seq data with DESeq2. Genome Biol 15(12):550. https://doi.org/10.1186/s13059-014-0550-8

Chapter 3

Evaluation of 3D Chromatin Interactions Using Hi-C

Gangqing Hu

Abstract

The invention of Hi-C has greatly facilitated 3D genome research through an unbiased probing of 3D chromatin interactions. It produces enormous amount of sequencing data that capture multiscale chromatin conformation structures. In the last decade, numerous computational methods have been developed to analyze Hi-C data and predict A/B compartments, topologically associating domains (TADs), and significant chromatin contacts. This chapter introduced the iHiC package that provides several utilities to facilitate Hi-C data analysis with public software and demonstrated its application to a Hi-C dataset generated for mouse embryonic stem (ES) cells.

Key words Hi-C, Bioinformatics, A/B compartments, TAD, Chromatin contacts

1 Introduction

DNA sequence within a nucleus is packed into a hierarchy of chromatin structures with histone proteins [1]. Chromatin conformation can be captured through chromosome conformation capture technique based on proximity ligation, such as Hi-C and its derivatives [2, 3], and ligation-free methods such as GAM [4], SPRITE [5], and Trac-looping [6]. Hi-C captures genome-wide chromatin interactions through ligation of chromatin DNA sequences located in space-proximity and cut by restriction enzymes, and applies high-throughput sequencing to measure interaction frequencies [2]. A typical Hi-C experiment would yield hundreds of millions of or even billions of pair-end tags (PETs) [7, 8], which impose computational challenges for data analysis.

Analysis of Hi-C contact matrix, which records the number of PETs linking two genomic regions, uncovers a hierarchy structure of chromatin organization. At a range up to millions of base pairs (bps), the chromatin is segregated into alternating A and B compartments, corresponding to transcription active and transcription repressive regions, respectively [2]. One compartment region may

Benjamin L. Kidder (ed.), *Stem Cell Transcriptional Networks: Methods and Protocols*, Methods in Molecular Biology, vol. 2117, https://doi.org/10.1007/978-1-0716-0301-7_3,

consist of one or more topologically associating domains (TADs); chromatin regions from one TAD preferentially interact with regions from the same TAD [9–11]. Within a TAD, interacting chromatin region may form structural loops anchored by convergent CTCT motifs [8] or regulatory loops for promoter enhancer interaction [6]. The structural loops could be readily identified by in situ Hi-C. The prediction of regulatory loops requires a much higher sequencing depth but could be addressed by alternative methods that target chromatin regions with specific features [6].

Bioinformatic tools developed for Hi-C data analysis include (but not limited to) data preprocessing such as PET alignment and data normalization, identification of A/B compartments, TADs, and significant contacts, and data visualization [12, 13]. Multiple or even dozens of tools have been developed for each of the topics [14, 15]. This chapter will introduce the iHiC package that provides a handful of utilities to facilitate Hi-C data analysis with public software, including CscoreTool for compartments [16], TopDom for TADs [17], Fit-Hi-C for significant contacts [18], and WashU Epigenome Browser for data visualization [19]. It includes a step-by-step guidance on the application of the package to a Hi-C dataset generated for mouse ES cells E14 [20].

2 Materials

2.1 Public Hi-C Data

In-situ Hi-C sequencing data for mouse ES cell line E14 [20] is downloaded from SRA under sample accession number SRX1338543 (https://www.ncbi.nlm.nih.gov/sra?term=SRX1338543).

2.2 Public Software

Sratoolkit: https://github.com/ncbi/sra-tools.

Bowtie2 [21]: https://github.com/BenLangmead/bowtie2.

Samtools [22]: https://github.com/samtools/samtools.

Bedtools [23]: https://github.com/arq5x/bedtools2.

CscoreTool [16]: https://github.com/scoutzxb/CscoreTool.

TopDom [17]: http://zhoulab.usc.edu/TopDom/ (*see* **Note 1**).

Fit-Hi-C [18]: https://github.com/ay-lab/fithic.

HTSlib for "bgzip" and "tabix": http://www.htslib.org/download.

R: https://www.r-project.org/.

2.3 Online Web Servers

WashU Epigenome Browser: https://epigenomegateway.wustl.edu.

UCSC Genome Browser: https://genome.ucsc.edu/.

Cyverse: https://www.cyverse.org.

2.4 The iHiC Package

The iHiC package includes a set of utilities (mostly written in C++) to facilitate Hi-C data analysis: it calculates statistics of PETs alignments, generates input files for several downstream software, and postprocesses their outputs for data visualization.

To install the package:

1. *"git clone https://github.com/hugangqing/iHiC.git"*: Download the package from github.
2. *"sh iHiC_install"*: compile the C++ codes with g++. Executive files will be saved to a subfolder named "bin".
3. Add the "bin" folder to system PATH or copy the executive files to "/usr/bin/" or "/usr/local/bin".

The package includes the following utilities:

(a) iHiC_BEDPE_Statistics: Calculate statistics of PET alignments from a simplified version of BEDPE file (*see* **Note 2**).

(b) iHiC_PETOrientations_Stat: Calculate % of PETs in different combinations of strand orientations as a function of genomic distance.

(c) iHiC_ContactFreq_Distance: Calculate contact probability as a function genomic distance.

(d) iHiC_BEDPE2III: Generate an intrachromosomal interaction index (III) file that records the number of PETs for interacting genomic bins from the same chromosomes.

(e) iHiC_III2WashU: Convert an III file into a "longrange" format (http://wiki.wubrowse.org/Long-range) accepted by the WashU browser.

(f) iHiC_BEDPE2HiCSummary: Convert a BEDPE file into a "HiCsummary" format (http://homer.ucsd.edu/homer/interactions/HiCtagDirectory.html) accepted by CscoreTool.

(g) iHiC_Cscore_Adj: Flip the sign of "C-score" chromosome-by- chromosome to ensure genomic regions with positive scores corresponding to A compartment status.

(h) iHiC_III2MTX4TopDom: Convert an III file into interaction matrices accepted by TopDom.

(i) iHiC_TADBoundary_TopDom: Extract TAD boundaries from TopDom predictions.

(j) iHiC_III2FitHiCInputs: Convert an III file into interaction matrices accepted by Fit-Hi-C.

(k) iHiC_FitHiC2WashU: Convert the significant contacts predicted by Fit-Hi-C into a "longrange" format accepted by the WashU Epigenome Browser.

(l) TopDom_wrapper.R: an Rscript wrapper to run TopDom.

3 Methods

3.1 Sequencing Data Extraction

The "fastq-dump" utility from the sratoolkit is a convenient way to locate and download sequencing data from NCBI SRA simply based on accession numbers; the SRA explorer (https://ewels.github.io/sra-explorer/) is also an useful tool to explore data in SRA. The accession numbers for the SRA sample SRX1338543 are SRR2658622, SRR2658623, SRR2658624, SRR2658625, and SRR2658626.

1. *"mkdir your_folder"*: Create a folder for the Hi-C analysis demonstrated in this chapter. All commands listed in the chapter are executed in this folder under Linux/Unix environment (ubuntu LTS) unless stated otherwise.
2. Download sratoolkit binaries from https://www.ncbi.nlm.nih.gov/sra/docs/toolkitsoft/ and copy them to /user/local/bin/ or /user/bin/.
3. *"fastq-dump --split-files SRR2658622"*: Extract fastq files from SRR2658622 and repeat for other accession numbers (*see* **Note 3**).

3.2 Sequence Alignment

1. Download Bowtie index from http://bowtie-bio.sourceforge.net/bowtie2/index.shtml.
2. *"bowtie2 -p 24 --local -x path-to-bowtie-index/index-prefix -1 SRR2658622_1.fastq -2 SRR2658622_2.fastq | samtools view -b>SRR2658622.bam"*: Execute PETs alignment by "bowtie2" and save the alignment results in bam file (*see* **Note 4**).
3. *"samtools view -bf 1 -F 12 SRR2658622.bam | bamToBed -i stdin -bedpe | cut -f 1-6,8,9,10>SRR2658622.bedpe"*: Extract PETs where both ends map from the BAM file and convert the output to a simplified BEDPE format (*see* **Note 2**).
4. Repeat **steps 2** and **3** for all other accession numbers.
5. *"cat SRR265862*.bedpe > SRX1338543.bedpe"*: Concatenate the BEDPE file for each accession to generate a pooled BEDPE file.

3.3 Quality Inspection

1. *"iHiC_BEDPE_Statistics SRX1338543.bedpe 10"*, where 10 is the minimal mapping quality (MAPQ) to extract uniquely mapped PETs: Calculate statistics for mapped PETs of a Hi-C library and present the output in a hierarchic order (Fig. 1a) (*see* **Note 5**).
2. *"iHiC_PETOrientations_Stat SRX1338543.bedpe 10"*, where 10 is the minimal MAPQ for uniquely mapped PETs: Calculate the percentages of PETs in different combinations of strand orientations and the standard deviation of the percentages as a function of genomic distance (Fig. 1b, c) (*see* **Note 6**).

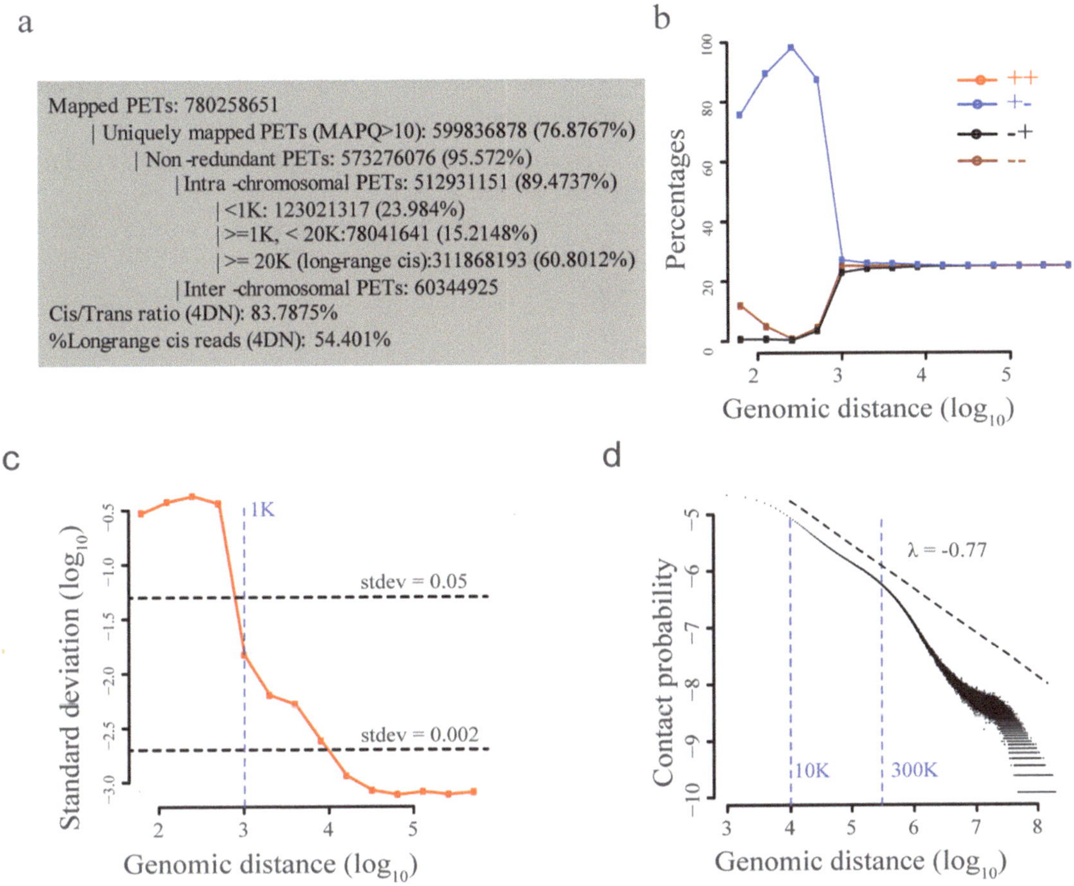

Fig. 1 Quality inspection on a Hi-C library of mouse ES E14 (SRX1338543). (**a**) Statistics on the PET alignments from Bowtie2. (**b**) Proportion of different combinations of strand orientations as a function of genomic distance between PET ends. (**c**) Standard deviation of the proportions in (**b**) as a function of genomic distance. The convergence occurs after 1 K bp (dashed blue line). (**d**) Contact probability as a function of genomic distance intervals. Genomics ranges correspond to TAD indicated by blue dash lines

3. *"iHiC_ContactFreq_Distance SRX1338543.bedpe 10 > SRX1338543_contactprob.txt"*, where 10 is the minimal MAPQ for uniquely mapped PETs: Calculate contact probability as a function of genomic distance interval (Fig. 1d) (*see* **Note** 7).

3.4 Visualization of Interaction Matrix

Tools and packages available to visualize interaction matrix of Hi-C data include the WashU Epigenome Browser [19], The 3D Genome Browser [24], Juicebox [25], HiCPlotter [26], and HiGlass [27]. The chapter uses the WashU Epigenome Browser as an example:

1. *"iHiC_BEDPE2III SRX1338543.bedpe 20000 20000 2000000 10 SRX1338543.iii"*, where the second, third, fourth, and fifth parameters stand for bin size, minimal distance, maximal

distance, and minimal MAPQ for uniquely mapped PETs, respectively: Calculate numbers of PETs for genomic bin-pairs from the same chromosome and generate an intra-chromosomal interaction index (III) file (*see* **Note 8**).

2. *"iHiC_III2WashU SRX1338543.iii 20000 SRX1338543.longrange"*, where 20,000 is bin size: Convert the III file into a "longrange" format, which should be sorted, compressed, and indexed (next three steps) before supply to the WashU Epigenome Browser.
3. *"sort -k1,1 -k2,2n SRX1338543.longrange > SRX1338543.longrange.sort"*: Sort the "longrange" file.
4. *"bgzip -f SRX1338543.longrange.sort"*: Compress the sorted file.
5. *"tabix -p bed -f SRX1338543.longrange.sort.gz"*: Index the compressed file. The output filename is ending with ".tbi".
6. Follow the instruction described in http://wiki.wubrowse.org/Track_hosting to upload the compressed file and its index to Cyverse (https://www.cyverse.org/), and obtain an URL to be supplied to the WashU Browser: briefly, (a) register an account with Cyverse and login; (b) Choose "MY SERVICES" under the user's home and launch the "Discovery Environment" app; (c) Choose the "Data" app, and click the "upload" menu to upload files. (d) Click the file to be submitted to the WashU Browser and edit the "Info-Type" into "bedgz." (e) Click the "Send to: Genome Browser" to obtain the URL (*see* **Note 9**).
7. Open the WashU Epigenome Browser (https://epigenomegateway.wustl.edu), select "mouse/mm10", go to menu "Tracks/Custom Tracks", choose "longrange" as "Track type," paste the URL obtained from **step 6**, and click "submit." A screenshot of the WashU Browser for the interaction matrix for a genomic region in chromosome 6 is shown in Fig. 2.

3.5 A/B Compartments Generation

Conventional A/B compartment prediction is based on principal component analysis (PCA) applied to a correlation matrix that records the correlation of contact profiles between genomic bins [2]. The PC1 value defines compartment status with positive and negative values corresponding to A and B compartments, respectively. Examples of A/B compartment calculation based on PCA are described elsewhere [28]. An alternative method is based on maximum-likelihood estimation by calculating the chance (or "C-score") of a genomic region being in an A compartment region from long range Hi-C interacting PETs [16].

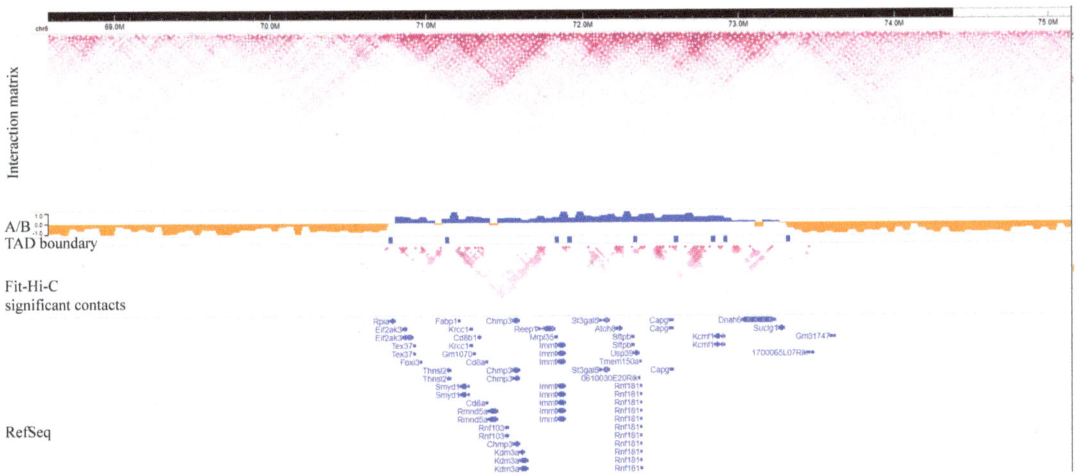

Fig. 2 A screenshot of the WashU Epigenome Browser for the interaction matrix (resolution = 20 K bps), "C-score" by CScoreTool for A/B compartments (blue for A and yellow for B), TAD boundaries by TopDom, and significant contacts by Fit-Hi-C for a genomic region of 6.6 million bps in chromosome 6 of mouse ES E14

1. Download the CscoreTool package from https://github.com/scoutzxb/CscoreTool.
2. *"chmod +x CscoreTool1.1"*: Make the file CscoreTool1.1 executable.
3. *"g++ generateEqualLengthBed.cpp -O3 -o generateEqualLengthBed"*: Generate an executive file used to partition the genome into bins of equal size. **Steps 3** and **4** are done in the CscoreTool folder.
4. *"./generateEqualLengthBed mm10_chrsizes.txt mm10_20k.bed 20000"*, where 20,000 is the bin size: Partition the genome into bins of equal size and save in "mm10_20k.bed". The "mm10_chrsizes.txt" file records the size of each chromosome, available from CscoreTool package or the "resource" folder from the iHiC package.
5. Copy the executive file CscoreTool1.1 and "mm10_20k.bed" to the Hi-C data analysis folder.
6. *"iHiC_BEDPE2HiCSummary SRX1338543.bedpe SRX1338543.hicsum"*: Convert the BEDPE file into a "HiC-summary" file (http://homer.ucsd.edu/homer/interactions/HiCtagDirectory.html) [29].
7. *"./CscoreTool1.1 mm10_20k.bed SRX1338543.hicsum SRX1338543 24 1000000"*, where "SRX1338543" is the prefix for output files, 24 is the number of threads, and 1,000,000 is the minimal distance of PET-ends used to estimate compartment status: Calculate compartment scores (namely, the "C-scores"), which are linearly transformed into values in range −1 and 1. The resultant file "SRX1338543_cscore.bedgraph" stores the "C-scores" for all chromosomal bins (*see* **Note 10**).

8. *"iHiC_Cscore_Adj SRX1338543_cscore.bedgraph mm10_CpG_ucsc.txt"*: to ensure that a positive value in "C-score" corresponds to A compartment status. The program calculates the average "C-score" for CpG islands of a chromosome and flips the sign of "C-score" if the average is negative for the chromosome (*see* **Note 11**); the adjusted "C-score" is saved in a file with name ending with "bedgraph.adj". File "mm10_CpG_ucsc.txt" is for CpG annotation, downloaded from UCSC and also available from the "resource" folder of the iHiC package.
9. Follow the steps in Subheading 3.4 to sort (**step 3**), compress (**step 4**), make index (**step 5**) for the adjusted "C-score" file, upload to Cyverse, and obtain a URL for the WashU Browser (**step 6**).
10. Follow the instruction in Subheading 3.4 (**step 7**) to supply the URL to the WashU Browser (choose "bedGraph" as "Track type"). A screenshot of the WashU Browser for the "C-scores" of compartments is shown in Fig. 2 for a genomic region in chromosome 6.

3.6 TAD Prediction

TAD prediction is an active research subject with dozens of tools published so far. A comprehensive evaluation of the methods was recently done by considering five criteria: robustness to window size and normalization method, cost efficiency in terms of number of reads for concordant prediction, prediction reproducibility by other methods, biology relevance to prior knowledge on TAD features, and computing time [15]. TopDom is the only tool that satisfies all the five criteria among others [15].

1. *"iHiC_III2MTX4TopDom SRX1338543.iii 20000 mm10_chrsizes.txt SRX1338543"*, where 20,000 is the bin size and "SRX1338543" is a prefix for output files: Convert the III file into an interaction matrix file accepted by TopDom for each chromosome specified in "mm10_chrsizes.txt". The filenames of outputs end with ".mtx4topdom".
2. Download "TopDom_v0.0.2.R" from https://github.com/HenrikBengtsson/TopDom/.
3. *"Rscript TopDom_wrapper.R Path_for_TopDom_v0.0.2.R SRX1338543.chr6.mtx4topdom"*: Run TopDom through an Rscript wrapper (shown here "chr6" as an example). TopDom predicts TADs one chromosome a time. The outputs include a "binSignal" file, which is used to extract TAD boundaries in the next step.
4. *"iHiC_TADBoundary_TopDom SRX1338543.chr6.mtx4topdom.binSignal 0.000001 SRX1338543.chr6.tadboundary.bedgraph"*: Extract and merge genomic bins that are labeled as "−1" in the "local.ext" column and is less than 0.000001 in the "p-value" column from the "binSignal" file for TAD boundaries.

5. Follow the instruction in Subheading 3.4 (**steps 3**–7) to upload the "bedgraph" file (sorted and indexed) in the previous step to Cyverse, and obtain a URL to be supplied to the WashU Browser (choosing "bedGraph" as "Track type").
6. Follow the instruction in Subheading 3.4 (**step** 7) to supply the URL to the WashU Browser (choose "bedGraph" as "Track type"). The TAD boundaries for a genomic region of 6.6 million bps in chromosome 6 are shown in Fig. 2.

3.7 Identification of Chromatin Interaction

A Hi-C interaction matrix captures both short-range and long-range significant contacts as well as those reflecting random polymer looping events. The last few years have witnessed the development of numerous tools for the identification of significant chromatin interactions. The chapter demonstrates the use of the Fit-Hi-C method, which calculates the statistical significance of chromatin interactions from a background model considering both random polymer effects and other biases in Hi-C data [18].

1. Download and install Fit-Hi-C from https://github.com/ay-lab/fithic. Take note on the path of "fithic.py", located in the "fithic" folder of the package.
2. *"iHiC_III2FitHiCInputs SRX1338543.iii SRX1338543 mm10_chrsizes.txt 20000 20000 2000000"*, where the last three parameters are bin size, minimal distance, and maximal distance: Generate "fragments file" (ending with "frag.gz") and "interactions file" (ending with cnts.gz") for the required "-f" and "-i" arguments of Fit-Hi-C, respectively. The two files will be generated for each chromosome as specified in "mm10_chrsizes.txt".
3. *"python path_to_fit-hic-c.py -f SRX1338543.chr6.frag.gz -i SRX1338543.chr6.cnts.gz -r 20000 -L 20000 -U 2000000 -o fithic_output -l SRX1338543.chr6"*: Run Fit-Hi-C to calculate the significances of chromatin interactions. Fit-Hi-C makes the prediction one chromosome a time (shown here "chr6" as an example). The output file ending with "significances.txt.gz" records the significance of the interaction for bin-pairs.
4. *"cd fithic_output"*: Locate the "fithic_output" folder.
5. *"gunzip SRX1338543.chr6.spline_pass1.res20000.significances.txt.gz"*: Decompress the file.
7. *"iHiC_FitHiC2WashU SRX1338543.chr6.spline_pass1.res20000.significances.txt 20000 0.05"*, where 20,000 is the bin size, and 0.05 is a threshold on q-value: Extract bin-pairs with q-value <0.05 from the Fit-Hi-C predictions and convert to a "longrange" format accepted by the WashU Browser for visualization.

8. Follow the instruction in Subheading 3.4 (**steps 3–7**) to obtain a URL for the "longrange" file generated from the previous step to be supplied to the WashU Epigenome Browser.
9. Follow the instruction in Subheading 3.4 (**step 7**) to supply the URL to the WashU Browser (choose "longrange" as "Track type"). The significant contacts from Fit-Hi-C for a genomic region in chromosome 6 in shown in Fig. 2.

4 Notes

1. In case that the website is not available, https://github.com/HenrikBengtsson/TopDom maintains a copy of the original TopDom R script ("TopDom_v0.0.2.R") and its documentation.
2. The BEDPE format is described in https://bedtools.readthedocs.io/en/latest/content/general-usage.html. The simplified version keeps columns corresponding to "chrom1," "start1," "end1," "chrom2," "start2," "end2," "score," "strand1," and "strand 2" to save space storage. The simplified version is called as BEDPE throughout the chapter unless stated otherwise.
3. The option "--split-files" denotes pair-end sequencing data; "fastq-dump" will generate two fastq files, one ending with "_1.fastq" for read 1 and the other ending with "_2.fastq" for read 2.
4. The Hi-C PETs are generally chimeric with one end originated from one genomic region and the other end from another region, which may be separated by tens of thousands of basepairs (bps) or even from a different chromosome. The sequence aligner Bowtie2 is flexible to capture PETs for both situations. The output SAM format demands huge space for storage. The "samtools view -b" command converts the SAM format into a binary BAM format. The "--local" option allows for an automatic trimming at read ends to maximize alignment score for "Bowtie2"; it helps to rescue PETs that are mixed with adapter sequences. The vertical bar (|) is a pipe symbol that redirects the output of one program (here "bowtie2") to the input of another (here "samtools") without creating any temporary file. The ">" symbol saves the output of "cut" to a local file.
5. Often, only PETs uniquely mapped are included for Hi-C data analysis. MAPQ measures the uniqueness of the mapping: a higher value indicates more unique, and a score higher than 10 is frequently used to exclude PETs mapped to multiple positions. PETs mapping to the same positions are likely caused by PCR amplification and therefore are redundant. The

redundancy may depend on both library complexity and sequencing depth and therefore a high redundancy does not necessarily indicate a library failure, especially in situations where cell number is limited. For this reason, I recommended to base the quality inspection on the uniquely mapped PETS after removing redundancy. The "*Cis/Trans* ratio" in Fig. 1a is defined by the number of long-range *cis* PETs ($\geq$20K bps) divided by the numbers of interchromosomal PETs and long-range *cis* PETs. According to the 4D Nucleome Consortium (4DN), a *Cis/Trans* ratio higher than 40% is required [13]. The percentage of "long-range *cis* reads" in Fig. 1a is defined by the number of long-range *cis* PETs over the number of uniquely mapped and nonredundant PETs. A number of 40% or higher indicates a good library.

6. A PET originated from the same DNA fragment should have one end map to the "+" strand and the other end to the "−" strand. For SRX1338543, the PETs less than 1000 bps are dominated by the "+−" combination, interpreted as a result of self-ligations (Fig. 1b). The standard deviation of the percentages are smaller than 0.05 after 1K bps, and <0.002 after 20K bps (Fig. 1c); according to 4DN, the convergence is "good" if <0.05 and "very good" if <0.002.
7. The curve follows a power-law relationship in the form of $P(s) \sim s^{\lambda}$ for genomic interval corresponding to TAD (10K–300K), where s and λ refer to distance and slope respectively. The slop λ is around −0.76 for mitotic samples [30]; it is around −0.77 for SRX1338543 (Fig. 1d).
8. The intrachromosomal interaction index file (∗.iii) is in the following format: chromosome, genomic position of bin1, genomic position of bin2, number of PETs.
9. The tracks submitted to the WashU Browser need to be hosted under an HTTP server, which should be accessible to the Browser. While institute's Firewall will normally block the access, services provided by CyVerse can be used to host the tracks. Be advised that both the compressed file and its index file should be uploaded.
10. Compared to PCA-based methods, CscoreTool requires less time and memory. In addition, unlike the PC1 values, the "C-score" is directly comparable between different samples [16]. CscoreTools could efficiently estimate A/B compartments at high resolution (e.g., 10K bps) with only 35 million PETs (human genome) and reproduce the results estimated from 3.5 billion PETs. The "bedgraph" file is in the following format: chromosome, start, end, and score.

11. Genomic bins from the same chromosome share compartment status if they exhibit same sign in "C-score." Whether positive values in "C-score" denote A compartment or not depends on the data initialization during the likelihood estimation. As the initialization is a randomized process, genomic bins with negative values may turn out to come from A compartment regions. Because CpG islands are enriched in A compartments, the average "C-scores" at CpG islands can be used to tell whether or not a positive score denotes A compartment status.

Acknowledgments

The author thanks Dr. Xiaobin Zheng from Carnegie Institution for Science and Mr. Aniello Infante in the lab for helpful comments. The work was partially supported by National Institute of General Medical Sciences Grant 5U54GM104942-04.

References

1. Dekker J, Belmont AS, Guttman M, Leshyk VO, Lis JT, Lomvardas S, Mirny LA, O'Shea CC, Park PJ, Ren B, Politz JCR, Shendure J, Zhong S, Network DN (2017) The 4D nucleome project. Nature 549 (7671):219–226. https://doi.org/10.1038/nature23884
2. Lieberman-Aiden E, van Berkum NL, Williams L, Imakaev M, Ragoczy T, Telling A, Amit I, Lajoie BR, Sabo PJ, Dorschner MO, Sandstrom R, Bernstein B, Bender MA, Groudine M, Gnirke A, Stamatoyannopoulos J, Mirny LA, Lander ES, Dekker J (2009) Comprehensive mapping of long-range interactions reveals folding principles of the human genome. Science 326 (5950):289–293. https://doi.org/10.1126/science.1181369
3. Denker A, de Laat W (2016) The second decade of 3C technologies: detailed insights into nuclear organization. Genes Dev 30 (12):1357–1382. https://doi.org/10.1101/gad.281964.116
4. Beagrie RA, Scialdone A, Schueler M, Kraemer DC, Chotalia M, Xie SQ, Barbieri M, de Santiago I, Lavitas LM, Branco MR, Fraser J, Dostie J, Game L, Dillon N, Edwards PA, Nicodemi M, Pombo A (2017) Complex multi-enhancer contacts captured by genome architecture mapping. Nature 543 (7646):519–524. https://doi.org/10.1038/nature21411
5. Quinodoz SA, Ollikainen N, Tabak B, Palla A, Schmidt JM, Detmar E, Lai MM, Shishkin AA, Bhat P, Takei Y, Trinh V, Aznauryan E, Russell P, Cheng C, Jovanovic M, Chow A, Cai L, McDonel P, Garber M, Guttman M (2018) Higher-Order Inter-chromosomal Hubs Shape 3D Genome Organization in the Nucleus. Cell 174(3):744–757. e724. https://doi.org/10.1016/j.cell.2018.05.024
6. Lai B, Tang Q, Jin W, Hu G, Wangsa D, Cui K, Stanton BZ, Ren G, Ding Y, Zhao M, Liu S, Song J, Ried T, Zhao K (2018) Trac-looping measures genome structure and chromatin accessibility. Nat Methods 15(9):741–747. https://doi.org/10.1038/s41592-018-0107-y
7. Bonev B, Mendelson Cohen N, Szabo Q, Fritsch L, Papadopoulos GL, Lubling Y, Xu X, Lv X, Hugnot JP, Tanay A, Cavalli G (2017) Multiscale 3D Genome Rewiring during Mouse Neural Development. Cell 171 (3):557–572. e524. https://doi.org/10.1016/j.cell.2017.09.043
8. Rao SS, Huntley MH, Durand NC, Stamenova EK, Bochkov ID, Robinson JT, Sanborn AL, Machol I, Omer AD, Lander ES, Aiden EL (2014) A 3D map of the human genome at kilobase resolution reveals principles of chromatin looping. Cell 159(7):1665–1680. https://doi.org/10.1016/j.cell.2014.11.021
9. Dixon JR, Selvaraj S, Yue F, Kim A, Li Y, Shen Y, Hu M, Liu JS, Ren B (2012) Topological domains in mammalian genomes identified

by analysis of chromatin interactions. Nature 485(7398):376–380. https://doi.org/10.1038/nature11082

10. Nora EP, Lajoie BR, Schulz EG, Giorgetti L, Okamoto I, Servant N, Piolot T, van Berkum NL, Meisig J, Sedat J, Gribnau J, Barillot E, Bluthgen N, Dekker J, Heard E (2012) Spatial partitioning of the regulatory landscape of the X-inactivation centre. Nature 485 (7398):381–385. https://doi.org/10.1038/nature11049
11. Sexton T, Yaffe E, Kenigsberg E, Bantignies F, Leblanc B, Hoichman M, Parrinello H, Tanay A, Cavalli G (2012) Three-dimensional folding and functional organization principles of the Drosophila genome. Cell 148 (3):458–472. https://doi.org/10.1016/j.cell.2012.01.010
12. Ay F, Noble WS (2015) Analysis methods for studying the 3D architecture of the genome. Genome Biol 16:183. https://doi.org/10.1186/s13059-015-0745-7
13. Lajoie BR, Dekker J, Kaplan N (2015) The Hitchhiker's guide to Hi-C analysis: practical guidelines. Methods 72:65–75. https://doi.org/10.1016/j.ymeth.2014.10.031
14. Forcato M, Nicoletti C, Pal K, Livi CM, Ferrari F, Bicciato S (2017) Comparison of computational methods for Hi-C data analysis. Nat Methods 14(7):679–685. https://doi.org/10.1038/nmeth.4325
15. Zufferey M, Tavernari D, Oricchio E, Ciriello G (2018) Comparison of computational methods for the identification of topologically associating domains. Genome Biol 19(1):217. https://doi.org/10.1186/s13059-018-1596-9
16. Zheng X, Zheng Y (2018) CscoreTool: fast Hi-C compartment analysis at high resolution. Bioinformatics 34(9):1568–1570. https://doi.org/10.1093/bioinformatics/btx802
17. Shin H, Shi Y, Dai C, Tjong H, Gong K, Alber F, Zhou XJ (2016) TopDom: an efficient and deterministic method for identifying topological domains in genomes. Nucleic Acids Res 44(7):e70. https://doi.org/10.1093/nar/gkv1505
18. Ay F, Bailey TL, Noble WS (2014) Statistical confidence estimation for Hi-C data reveals regulatory chromatin contacts. Genome Res 24(6):999–1011. https://doi.org/10.1101/gr.160374.113
19. Li D, Hsu S, Purushotham D, Sears RL, Wang T (2019) WashU Epigenome Browser update 2019. Nucleic Acids Res 47(W1):W158–W165. https://doi.org/10.1093/nar/gkz348
20. Yan J, Chen SA, Local A, Liu T, Qiu Y, Dorighi KM, Preissl S, Rivera CM, Wang C, Ye Z, Ge K, Hu M, Wysocka J, Ren B (2018) Histone H3 lysine 4 monomethylation modulates long-range chromatin interactions at enhancers. Cell Res 28(3):387. https://doi.org/10.1038/cr.2018.18
21. Langmead B, Salzberg SL (2012) Fast gapped-read alignment with Bowtie 2. Nat Methods 9 (4):357–359. https://doi.org/10.1038/nmeth.1923
22. Li H, Handsaker B, Wysoker A, Fennell T, Ruan J, Homer N, Marth G, Abecasis G, Durbin R, Genome Project Data Processing S (2009) The Sequence Alignment/Map format and SAMtools. Bioinformatics 25 (16):2078–2079. https://doi.org/10.1093/bioinformatics/btp352
23. Quinlan AR (2014) BEDTools: The Swiss-Army tool for genome feature analysis. Curr Protoc Bioinformatics 47:11.12.1–11.1234. https://doi.org/10.1002/0471250953.bi1112s47
24. Wang Y, Song F, Zhang B, Zhang L, Xu J, Kuang D, Li D, Choudhary MNK, Li Y, Hu M, Hardison R, Wang T, Yue F (2018) The 3D Genome Browser: a web-based browser for visualizing 3D genome organization and long-range chromatin interactions. Genome Biol 19(1):151. https://doi.org/10.1186/s13059-018-1519-9
25. Durand NC, Robinson JT, Shamim MS, Machol I, Mesirov JP, Lander ES, Aiden EL (2016) Juicebox provides a visualization system for Hi-C contact maps with unlimited zoom. Cell Syst 3(1):99–101. https://doi.org/10.1016/j.cels.2015.07.012
26. Akdemir KC, Chin L (2015) HiCPlotter integrates genomic data with interaction matrices. Genome Biol 16:198. https://doi.org/10.1186/s13059-015-0767-1
27. Kerpedjiev P, Abdennur N, Lekschas F, McCallum C, Dinkla K, Strobelt H, Luber JM, Ouellette SB, Azhir A, Kumar N, Hwang J, Lee S, Alver BH, Pfister H, Mirny LA, Park PJ, Gehlenborg N (2018) HiGlass: web-based visual exploration and analysis of genome interaction maps. Genome Biol 19 (1):125. https://doi.org/10.1186/s13059-018-1486-1
28. Miura H, Poonperm R, Takahashi S, Hiratani I (2018) Practical analysis of Hi-C data: generating A/B compartment profiles. Methods Mol Biol 1861:221–245. https://doi.org/10.1007/978-1-4939-8766-5_16
29. Heinz S, Benner C, Spann N, Bertolino E, Lin YC, Laslo P, Cheng JX, Murre C, Singh H, Glass CK (2010) Simple combinations of

lineage-determining transcription factors prime cis-regulatory elements required for macrophage and B cell identities. Mol Cell 38 (4):576–589. https://doi.org/10.1016/j.molcel.2010.05.004

30. Sanborn AL, Rao SS, Huang SC, Durand NC, Huntley MH, Jewett AI, Bochkov ID, Chinnappan D, Cutkosky A, Li J, Geeting KP, Gnirke A, Melnikov A, McKenna D, Stamenova EK, Lander ES, Aiden EL (2015) Chromatin extrusion explains key features of loop and domain formation in wild-type and engineered genomes. Proc Natl Acad Sci U S A 112 (47):E6456–E6465. https://doi.org/10.1073/pnas.1518552112

Chapter 4

MSTD for Detecting Topological Domains from 3D Genomic Maps

Yusen Ye, Lin Gao, and Shihua Zhang

Abstract

In this chapter, we introduce a generic and efficient method to identify multiscale topological domains (MSTD), including *cis*- and *trans*-interacting regions, from a variety of 3D genomic datasets. We first applied MSTD to detect promoter-anchored interaction domains (PADs) from promoter capture Hi-C datasets across 17 primary blood cell types. The boundaries of PADs are significantly enriched with one or the combination of multiple epigenetic factors. Moreover, PADs between functionally similar cell types are significantly conserved in terms of domain regions and expression states. Cell type-specific PADs involve in distinct cell type-specific activities and regulatory events by dynamic interactions within them. We also employed MSTD to define multiscale domains from typical symmetric Hi-C datasets and illustrated its distinct superiority to the state-of-the-art methods in terms of accuracy, flexibility and efficiency.

Key words 3D chromosomal architecture, Topologically domains, *Cis*- and *trans*-interacting regions, Promoter-anchored interaction domains

1 Introduction

Folding of mammalian chromosomes into the nucleus has increasingly been recognized as an important factor in gene regulation, cell fate decisions, and so on [1–3]. However, how chromosomes fold into the nucleus is still obscure. The chromosome conformation capture (3C) technique and its variants such as Hi-C, ChIA-PET, and capture Hi-C have been employed to uncover the chromatin loops and hierarchical chromatin structural domains in three-dimensional (3D) genome architecture. Specifically, recent studies have revealed direct physical interactions, such as long-range chromatin contacts between enhancer and target genes [4], actively coregulated genes [5], and Polycomb-repressed genes [6]. In addition, researchers have provided evidence for topologically associating domains (TADs) or sub-TADs [7–12]. Furthermore, interactions between TADs at a variable distance result in active and inactive compartments, which are further subdivided into six

Benjamin L. Kidder (ed.), *Stem Cell Transcriptional Networks: Methods and Protocols*, Methods in Molecular Biology, vol. 2117, https://doi.org/10.1007/978-1-0716-0301-7_4,

different subcompartments according to distinct patterns of histone modifications [4, 13]. Therefore, with the rapid accumulation of 3D genomic maps, developing efficient computational methods for detecting topological domains in chromosomal architecture is urgently needed.

Recently, a few methods have been designed to explore the hierarchical organization of chromosomal architecture from symmetric Hi-C maps. For example, Filippova et al. introduced a dynamic programing model to identify hierarchical domains by adjusting a single parameter [14]. Weinreb et al. proposed a matrix decomposition model to infer a hierarchy of nested TADs based on ideal empirical distributions [15]. Zhan et al. presented a method CaTCH to detect hierarchical trees of chromosomal domains given a certain degree of reciprocal insulation from Hi-C maps [16]. However, these methods can only detect *cis*-interacting regions from Hi-C maps.

Very recent studies have revealed that both *cis*- and *trans*-interacting DNA regions exist in the cell nucleus [17]. With the emerging higher resolution genome-wide interaction maps of chromatin such as promoter capture Hi-C maps [18] and ChIA-PET maps [19], many distal DNA fragments regulate their targets bypassing long chromatin regions have been observed. Therefore, how to develop a generic method to infer both *cis*- and *trans*-interacting regions from diverse types of 3D genomic data is still a grand challenge in computational biology.

To this end, we proposed a generic and efficient method to identify multiscale topological domains (MSTD) from both asymmetric and symmetric 3D genomic datasets with a single adjustable parameter controlling domain scales. MSTD can detect long-range interacting domains such as those between promoters and regulatory elements at variable distances, which cannot be addressed directly by existing methods. We detected promoter-anchored interacting domains (PADs) from promoter capture Hi-C datasets across 17 primary blood cell types. We observed that the boundaries of these PADs can be well specified by one or the combination of a few epigenetic factors, indicating that these factors should play a key role the formation of genomic conformation. The analysis of the affinity relationship among cell types showed that PADs between functionally similar cell types have distinctly high conservation and consistent expression levels. Furthermore, dynamic PADs might perform specific cellular functions, while common ones take into account the underlying conditions of regular cell activities and participate in cell type-specific regulatory events by dynamic interactions within each PAD. This suggests that PADs are important and basic units of genomic structure and function influencing gene regulation and cellular differentiation. We also employed MSTD to define multiscale domains from symmetric Hi-C datasets with distinctly superior accuracy and efficiency compared to existing methods. Interestingly, the conservation of TADs between cells during continuous differentiation cycles is significantly higher than those between interval cycles. Last but not least, TADs are strongly correlated with epigenetic and transcriptional features.

2 Materials

We first downloaded a comprehensive catalog of capture Hi-C datasets of 31,253 annotated promoters and 230,525 unique promoter-interacting regions (PIRs) in 17 human primary blood cell types [18]. The datasets detect 698,187 high-confidence unique promoter interactions, of which 9.6% are promoter-promoter interactions and 90.4% are promoter-PIR interactions. The 17 cell types are from different nodes of the hematopoietic tree, which can be roughly divided into two categories, including eight myeloid cell types (Megakaryocytes (MK), Erythroblasts (Ery), Neutrophils (Neu), Monocytes (Mon), Endothelial precursors (Endp), Macrophages M0 (Mac0), Macrophages M1 (Mac1), and Macrophages M2 (Mac2)) and nine lymphoid cell types (Naive B cells (nB), Total B cells (tB), Fetal thymus (FetT), Naive CD4+ T cells (nCD4), Total CD4+ T cells (tCD4), Nonactivated total CD4 + T cells (naCD4), Activated total CD4+ T cells (aCD4), Naive CD8+ T cells (nCD8), and Total CD8+ T cells (tCD8)). Interaction scores were computed for each fragment pairs between promoters and promoter interacting regions (PIRs). As described by (21), the scores were computed by the CHiCAGO pipeline for each cell types.

We also collected two sets of Hi-C maps. The first dataset includes mouse cortex, ES and human ES cell types binned at 40 kb resolution [7]. For human ES cell type, we obtained human housekeeping genes from [20] and TSS of protein coding genes of hg19 RefSeq in GENCODE database. The second dataset contains proliferating mouse embryonic stem cells (ESC), intermediate neuronal precursor cells (NPC) and post-mitotic neurons (Neurons) Hi-C datasets through different differentiation periods, binned at 50 kb resolution from [21]. We normalized these Hi-C maps as previously described by Yaffe and Tanay [22].

3 Methods

3.1 MSTD

MSTD was inspired by a fast density-based clustering method, which is designed for grouping data points [23]. Here we group the strong contact interactions (*see* **Note 1**), which form square or rectangle submatrices corresponding to domains defined as above (Fig. 1). Given a symmetric or asymmetric matrix A, where $A_{i,}$

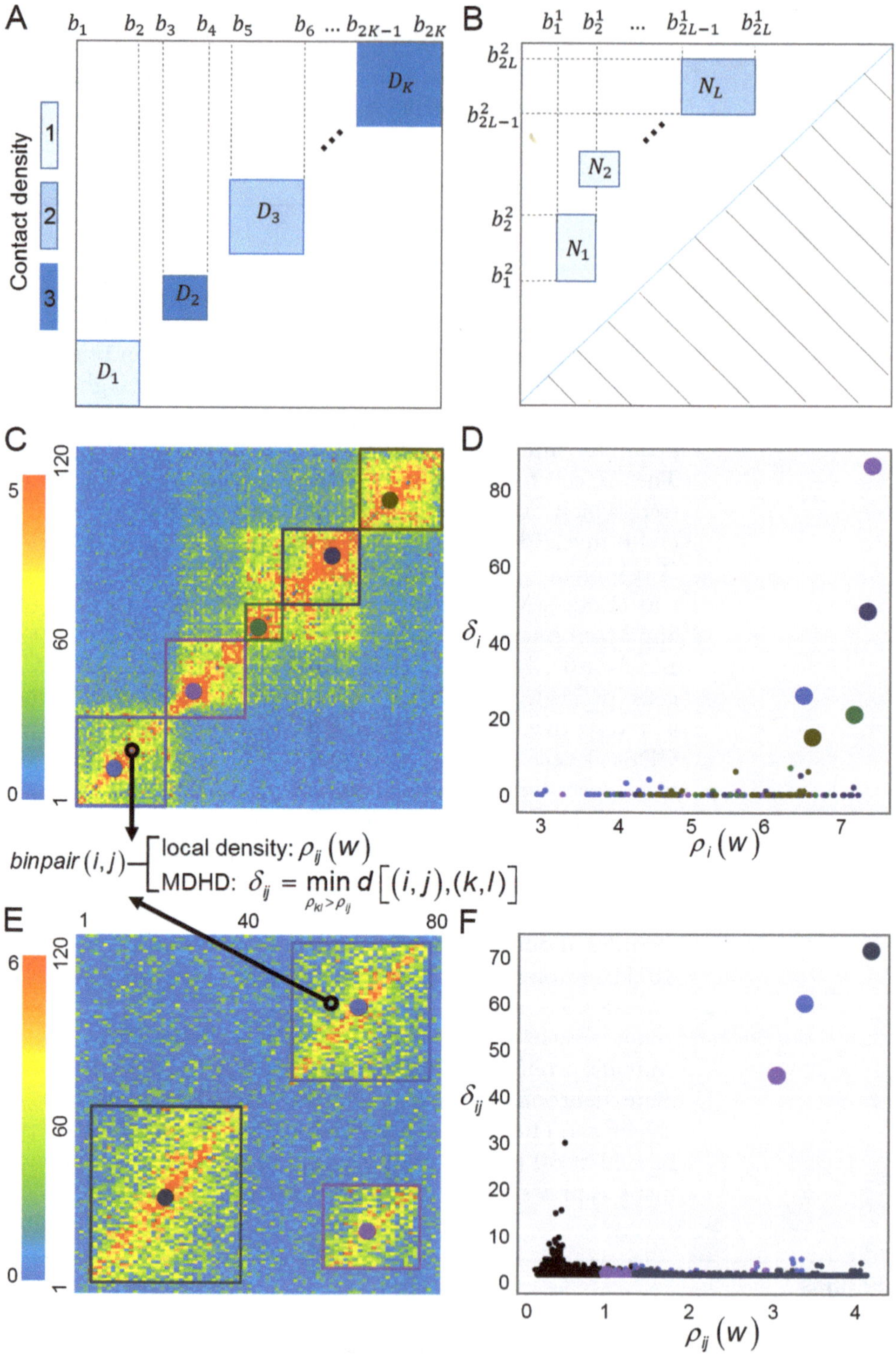

Fig. 1 Illustration of MSTD for detecting TADs and PTADs in a symmetric and asymmetric matrix respectively. (**a**, **b**) Examples of TADs $\{D_1, D_2, \cdots, D_K\}$ and PTADs$\{N_1, N_2, \cdots, N_L\}$. The color depth represents the density of domains. (**c**) MSTD computes two indexes including local density and minimum distance of higher density

$_j(1 \leq i \leq n, 1 \leq j \leq m)$ denotes the interaction frequencies between bin i and j. Firstly, for each element (i, j) (a diagonal element (i.e., $i = j$) for a symmetric matrix in Fig. 1c or a nonzero element for an asymmetric matrix in Fig. 1e), MSTD computes two indexes: (1) local density$\rho_{ij}(w)$ defined as the average of the interaction frequencies with a cutoff radius window w (*see* **Note 2**), and (2) δ_{ij} is computing by minimum distance between the element (i, j) and any other element with a higher local density (MDHD),

$$\delta_{ij} = \min_{\rho_{kl} > \rho_{ij}} d[(i,j),(k,l)], \qquad (\delta_{ij} \leq Max),$$

where $d[(i,j),(k,l)]$ is the distance between the element (i, j) and element (k, l). For the efficiency of the algorithm, we search for a nearest neighbor element of higher density of the element (i, j) in the range of Max(100 is used as the default value), which the results are not sensitive to (*see* **Note 3**). Note that local density ρ_{ij} of the clustering center element is locally or globally maximum, and its δ_{ij} values are much larger than those of its neighbors. Thus, MSTD determined clustering centers as elements for which the value δ_{ij} is anomalously large (*see* **Note 4**), and the scale of domain is control by an adjustable parameter $\delta_T(\delta_{ij} > \delta_T)$. Next, the remaining elements are assigned to the same cluster as its nearest neighbor element of higher density layer by layer. Meanwhile, individual outlier elements with high δ and low ρ (Fig. 1d) and the elements associated with these outlier elements will not be assigned to any cluster. Finally, the boundaries for each cluster are defined as the outermost elements of the same cluster in all directions. If the elements whose local density is significantly lower than their centers' ($\delta_{ij} < 0.2 \times \delta_{\text{center}}$), this element is removed from this cluster.

In summary, MSTD can identify multiscale topological domains from both asymmetric and symmetric datasets with a single adjustable parameter controlling domain scales. As the adjustable parameter increases, the number of clusters decreases and the size increases. This process furthermore controls the scale of domains, which together form the hierarchical chromatin structure.

Fig. 1 (continued) (MDHD) for each diagonal element k in the symmetric matrix. The value of heatmap represents normalized interacting scores. (**d**) The decision graph of clustering results for (**c**). Clustering centers are marked by big dots with different colors. Starting with the centers, the remaining elements are assigned to the same cluster as its nearest neighbor element of higher density layer by layer. The clustering boundaries are defined to be the outermost elements of the same cluster in the two directions. (**e**) MSTD computes two indexes similar to (**c**) for each nonzero element k in the asymmetric matrix. (**f**) The decision graph of clustering results for (**e**). For the asymmetric matrix, MSTD employs all nonzero elements and defines the clustering boundaries in four directions. Noise elements were colored by black

3.2 Installation of MSTD

1. It is recommended to install MSTD in python3.6 or above environment.
2. Installing requiring packages, including numpy, pandas, seaborn, and colormap using pip or conda.
3. Extracting the downloaded installation package to the specified directory and run the command: python setup.py install.

3.3 Using MSTD to identify TADs and PADs

1. Now you can execute the commands below to import MSTD package to your project.

```
import os
import MSTDlib.MSTD_v2 as MSTD_v2
```

2. Next you need specify the address of input file (Matrix file) and output file (Output_file). The format of the input file is triple for detecting PADs from promoter capture Hi-C maps, and the format of the file is N∗N matrix file without row and column names for detecting TADs for Hi-C maps.
3. MSTD computes two indexes for each element: (1) local density $\rho_{ij}(w)$, and (2) δ_{ij} is computing by minimum distance between the element (i, j) and any other element with a higher local density (MDHD). MSTD controls the scale of domain by an adjustable parameter $\delta_T(\delta_{ij} > \delta_T)$. For detection of promoter-anchored interacting domains (PADs), the size of PIRs is about 10 kb. Thus, the range of parameters is approximately 15–200 (150 kb/10 kb to 2Mb/10 kb), which can be interpreted as the least and the most number of PIRs contained within the PADs (Fig. 2). Similarly, for detection of topological associating domains (TADs) at a resolution of 40 kb, the range of parameter is approximately 3-50 (150 kb/40 kb to 2Mb/40 kb), which can be interpreted as the least and the most number of bins contained within the TADs (Fig. 3). Suppose you want to identify TADs (The size of TADs is approximately 1M) from Hi-C maps of 40 kb, the parameter MDHD is approximately 1M/40 kb = 25. The parameter symmetry specifies the identification of TADs (symmetry = 0) or the identification of PADs or PTADs (symmetry = 1). The parameter windows is used to compute the average contact frequency of one element (Defaults, windows = 5). According to the different resolution of the data, it can be adjusted between 5 and 10, which has little effect on the results. The parameter visualization (1/0) specifies whether the result is visual.

   You can use the following commands to perform detection of domains:

```
MSTD_v2.MSTD(Matrix_file,Output_file,MDHD=10,
symmetry=1,window=10,visualization=1)
```

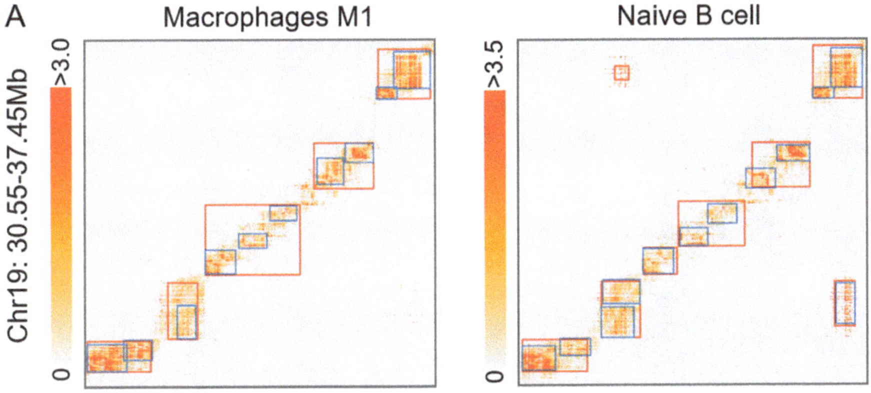

Fig. 2 Illustration of PADs detected from promoter capture Hi-C datasets. (**a**) Examples of PADs detected by MSTD with $\delta_T = 100$ (blue) and $\delta_T = 200$(red). The value of heatmaps represents CHiCAGO score

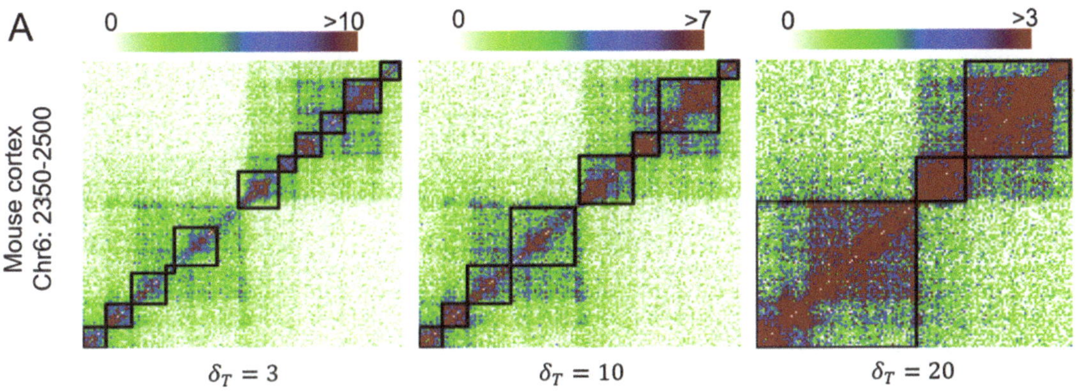

Fig. 3 Illustration of multiscale domains determined by MSTD from Hi-C datasets. (**a**) Multiscale topological domain examples detected by MSTD with different MDHD thresholds (i.e., $\delta_T = 3$, 10, 20) of chr19: 2350–2500 bins of Hi-C maps in the mouse cortex cell (40 kb resolution). The value of heatmap represents normalized interacting scores

3.4 Identifying PADs from Asymmetric Promoter Capture Hi-C Datasets

Firstly, we applied MSTD to identify multiscale PADs from promoter capture Hi-C maps across 17 human primary blood cell types (Fig. 2a) [18]. MSTD can not only find *cis*-interacting PADs within the continuous genome but also identify *trans*-interacting PADs between distal regulatory elements and target promoters (Fig. 2a). The results below are based on parameter $\delta_T = 100$ for detecting PADs, unless noted otherwise.

3.5 Epigenetic Feature Is a Good Predictor of PADs

We only defined the boundaries and centers of PADs along promoter interacting regions (PIRs) axis not promoters' axis (Subheading 2) and found that epigenetic features of PADs show strong difference on these PIRs' regions compared with random ones (Fig. 4). Specifically, for monocytes cell type, the promoter mark H3K4me3 is strongly enriched in the boundaries of PADs (even more than twice that of *intra*-PADs) and the transcribed

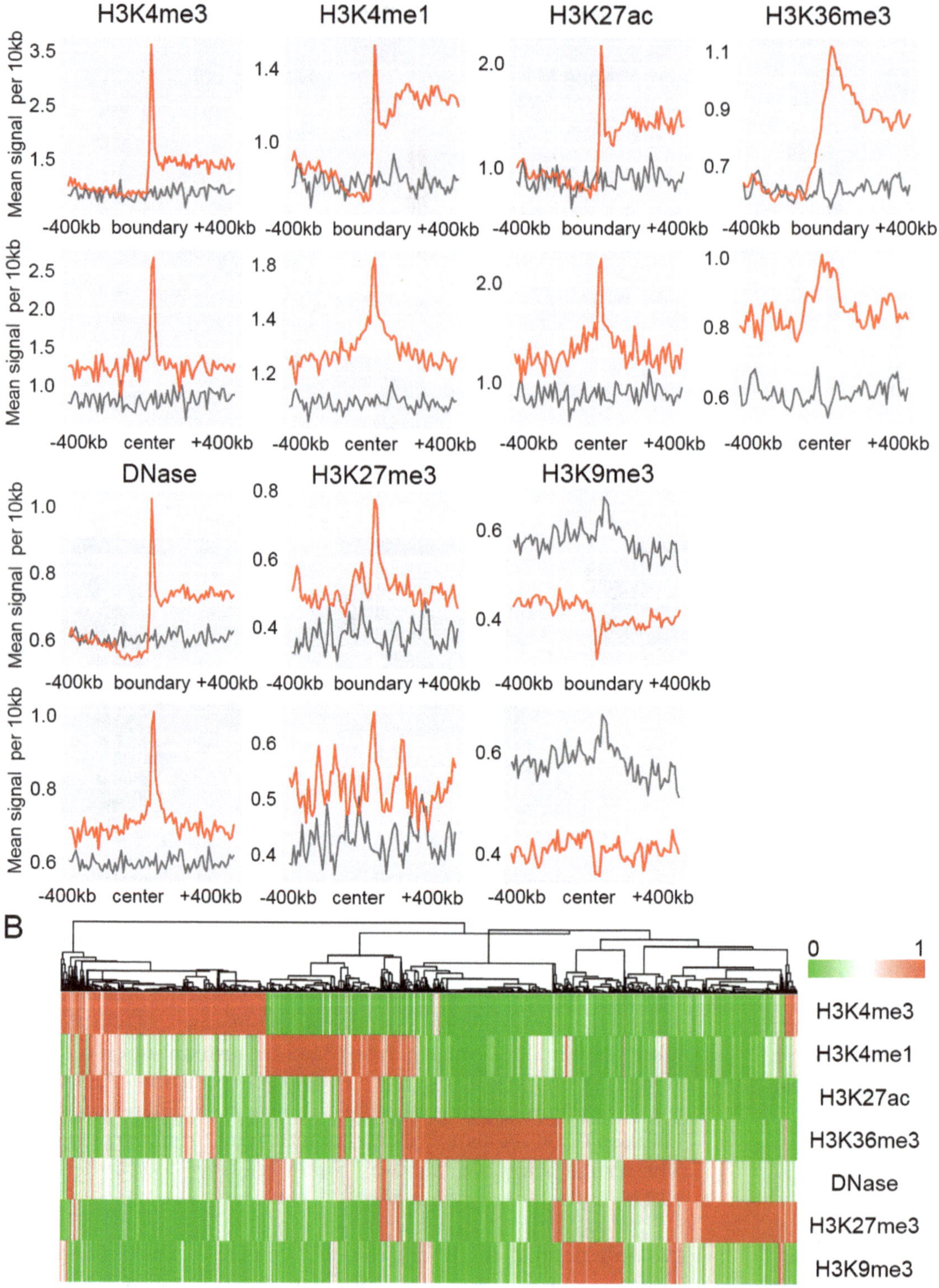

Fig. 4 PADs are relating to epigenetic factors. (**a**) Enrichment analysis of seven epigenetic factors around boundaries and centers of PADs along PIRs axis not promoters' axis in the human monocytes cell type. The observed and random signals are marked with red and gray colors, respectively. (**b**) Hierarchical clustering of boundaries of PADs according to their epigenetic factors. The value of heatmap represents normalized score of columns of different epigenetic signals (min–max normalization to [0, 1])

region mark (H3K36me3) is enriched in the regions slightly deviating from boundary regions, indicating that the boundary elements of PADs are significantly associated with promoters (Fig. 4a). Meanwhile, the enhancer marks (H3K4me1, H3K27ac) are also enriched around topological boundaries, which might be related with chromatin loops in the boundary regions (Fig. 4a). The transcribed region mark (H3K36me3) and the enhancer marks (H3K4me1, H3K27ac) demonstrate significant changes between *intra*-PADs and *inter*-PADs, suggesting PADs play key roles in active gene regulation and expression (Fig. 4a). Meanwhile, the repressive mark (H3K27me3) is enriched in topological boundaries, which can be associated with the formation of repressive domains. Interestingly, the signals of H3K9me3 in *inter*-PADs are slightly higher than those in the *intra*-PADs, and both of which are significantly lower than random ones, revealing that PADs can segregate the spread of the heterochromatin (Fig. 4a). Thus, multiple epigenetic features can identify different functional PAD boundaries. We also observed that multiple epigenetic features (except the heterochromatin mark) are enriched around the regions with high interaction intensity (PAD centers) (Fig. 4a). Furthermore, we guess that one or the combination of several epigenetic features contributes to the formation of PADs, which can be associated with different functional units (Fig. 4b). For example, H3K27ac tends to combine H3K4me3 or H3K4me1 to mark active boundaries of PADs, while H3K27me3 tends to mark repressive boundaries alone (Fig. 4b). Therefore, histone modifications and DNase signal seem to be a good predictor of PADs and different functional units may have different strategies to specify chromatin domains.

3.6 Exploring PADs' Biological Functions

Next, we identified 693 lymphoid-specific, 426 myeloid-specific and 3808 common PADs according to the occurrence features among 17 human primary blood cell types to explore their cellular functions. We further obtained 308 lymphoid-specific, 195 myeloid-specific, and 1380 common PADs, whose two interacting chromatin regions containing more than one marker genes. Among those PADs, MSTD pioneered the discovery of 180 *trans*-interacting PADs without any overlap between their two interacting chromatin regions.

For example, we found two interesting *trans*-interacting PADs within the same region of genome—one myeloid-specific PAD and a common PAD shared by 16 cell types (except aCD4) (Fig. 5). For the myeloid-specific PAD, we can see that active TSS and transcription states occupies the two interacting chromatin regions of this PAD based on the ChromHMM annotations and active markers including H3K4me3, H3K27ac, and DNase are significantly enriched in them too, which may suggest the chromatin loops within this PAD bring active promoters into the same nuclear

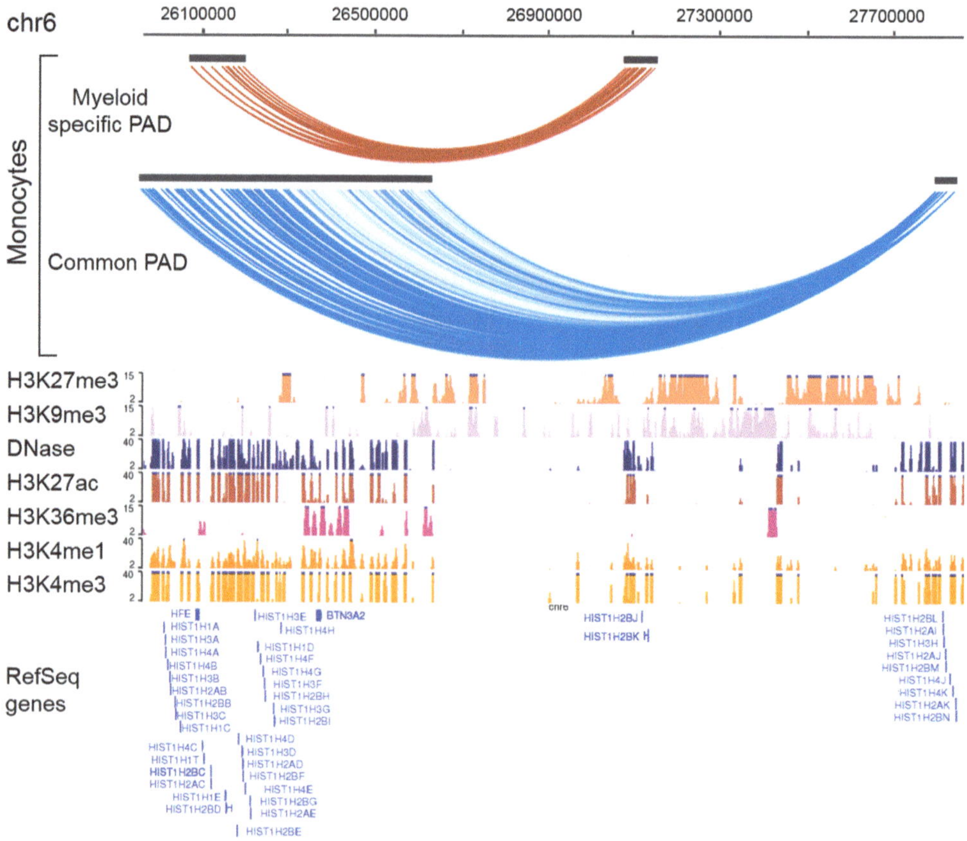

Fig. 5 Illustration of a myeloid-specific PAD and a common PAD. Both PADs appear within the same genomic region using monocytes cell type as an example (chr6:26001891-27865483). Chromatin modification signals of seven epigenetic factors are shown in the monocytes cell type. The PADs connected with red arcs and blue arcs are a myeloid-specific one and a common one respectively. The image is drawn based on WashU epigenome browser with RefSeq gene annotations. The color depth of arcs represents CHiCAGO scores of chromatin interactions

space to initiate transcription [24]. Moreover, E2F, p53 motifs, and E2F, E2F1 motifs are enriched in the upstream and downstream regions of this PAD, which contain genes related to innate immune response of myeloid tissues (upstream: HIST1H2BC, downstream: HIST1H2BJ and HIST1H2BK) [25]. This suggests that the chromatin loops among this myeloid-specific PAD play a special role in the cooperation between E2F1 and p53 to specifically activate innate immune response genes in myeloid tissues, which could promote the occurrence of apoptosis [26].

For the lymphoid and myeloid common PAD, the two interacting chromatin regions contain several genes from histone families (H1, H2A, H2B, H3, and H4), which regulates circulating iron and mediates the regulation of lymphocyte and leukocyte (such as activation, cell–cell adhesion, immune response, and antigen processing and presentation) (Fig. 5) [25]. Furthermore, we identified 529 common and 40 dynamic chromatin loops between lymphoid and myeloid tissues from high-density chromatin loops of this PAD (p-value < 0.01). Surprisingly, we found all of dynamic chromatin loops occur in lymphoid tissues and these dynamics interacting regions contains BTN3A2 gene involved in the adaptive immune response of lymphoid tissues [25]. Thus, this PAD might participate in the underlying functions of regular activities of blood cells while enforce the adaptive immune response in lymphoid tissue. It can be seen that the upstream region of the common PAD contains that of the myeloid-specific PAD, and compared with the common PAD, the myeloid-specific PAD performs specific cellular functions by dynamic interactions with the different remote-region (Fig. 5).

4 Notes

1. *Strong contact interactions.*

   MSTD group the strong contact interactions, which form square or rectangle submatrices corresponding to domains defined as above (Fig. 1). For detection of TADs, we take all the diagonal elements as strong elements. For detection of PADs or PTADs, we can set the top percentage of high frequency interactions (for example, top 25%) as strong elements.

2. *Local density for Hi-C and promoter capture Hi-C datasets.*

   We define the local density for both Hi-C and promoter capture Hi-C maps. For the symmetric Hi-C contact map A, we only search for boundaries of domains from diagonal elements for detecting TADs. For ith bin, local density is defined as,

$$\rho_i(w) = \frac{1}{(w+1)^2} \sum_{l=i}^{i+w} \sum_{m=i-w}^{i} A_{l,m}.$$

   For the asymmetric promoter capture Hi-C contact map B, considering the resolution ratio α of elements between promoters and PIRs, local density ρ_{ij} is defined as the average contact frequency within a rectangle window with length $= \alpha w$ and width $= w$,

$$\rho_{ij}(w) = \frac{1}{4\alpha w^2} \sum_{l=i-w}^{i+w-1} \sum_{o=j-\alpha w}^{j+\alpha w-1} A_{l,o}.$$

3. *Choosing max.*

   MSTD search for a nearest neighbor element of higher density of the element (i, j) in the range of *Max*, which is not sensitive to the results. For detection of promoter-anchored interacting domains (PADs), the size of PIRs is about 10 kb. Thus, *Max* should be set to a fixed value 500 ($Max = 5$ M/10 Kb). Similarly, for detection of topological associating domains (TADs) at a resolution of 50 kb, MSTD sets *Max* to a fixed value 100 ($Max = 5$ M/50 Kb).

   Since the interaction map is generally very sparse, we set a search threshold *Max* to control the largest search range for the nearest neighbor element of higher density of the element (i, j), which dramatically improves the efficiency of the algorithm. Taking TAD detection from Hi-C maps as an example, the results are not sensitive to *Max* due to three facts: Frist, MSTD determines clustering centers by an adjustable parameter $\delta_T(\delta_{ij} > \delta_T)$. The value of *Max* does not affect the determination of clustering centers with $Max > \delta_T$. Second, the δ_{ij} values of the remaining elements except clustering centers and outlier elements must be less than parameter δ_T or *Max*, That is, these elements can find a nearest neighbor element of higher density in the range of *Max*. Third, the outlier elements are with high δ ($\delta > \delta_T$) and low ρ, and the value of *Max* does not affect the determination of outliers with $Max > \delta_T$. Thus, the detection results are not sensitive to *Max*.

4. *Determining clustering centers and outliers.*

   MSTD determined clustering centers with $\delta_{ij} > \delta_T$ and ρ more than 95 percentile of density of all effective elements of 2D map on the detection of PADs and PTADs, and ρ more than 20 percentile of density of all effective elements of 1D vector on the detection of TADs. Individual outlier elements are stated by $\delta_{ij} > \delta_T$ except clustering centers.

Acknowledgments

Yusen Ye would like to thank the support of the Academy of Mathematics and Systems Science at CAS during his visit there. This work has been supported by the National Natural Science Foundation of China [No. 61873198, 61532014, 61432010, 61672407 to LG; 11661141019, 61621003, the Key Research Program of the Chinese Academy of Sciences [No. KFZD-SW-219], National Key Research and Development Program of China (2017YFC0908405) and CAS Frontier Science Research Key Project for Top Young Scientist [No. QYZDB-SSW-SYS008].

References

1. Bickmore WA, van Steensel B (2013) Genome architecture: domain organization of interphase chromosomes. Cell 152(6):1270–1284
2. Sexton T, Cavalli G (2015) The role of chromosome domains in shaping the functional genome. Cell 160(6):1049–1059
3. Pombo A, Dillon N (2015) Three-dimensional genome architecture: players and mechanisms. Nat Rev Mol Cell Biol 16(4):245–257
4. Rao SS, Huntley MH, Durand NC, Stamenova EK, Bochkov ID, Robinson JT, Sanborn AL, Machol I, Omer AD, Lander ES (2014) A 3D map of the human genome at kilobase resolution reveals principles of chromatin looping. Cell 159(7):1665–1680
5. Schoenfelder S, Sexton T, Chakalova L, Cope NF, Horton A, Andrews S, Kurukuti S, Mitchell JA, Umlauf D, Dimitrova DS (2010) Preferential associations between co-regulated genes reveal a transcriptional interactome in erythroid cells. Nat Genet 42(1):53–61
6. Bantignies F, Roure V, Comet I, Leblanc B, Schuettengruber B, Bonnet J, Tixier V, Mas A, Cavalli G (2011) Polycomb-dependent regulatory contacts between distant Hox loci in Drosophila. Cell 144(2):214–226
7. Dixon JR, Selvaraj S, Yue F, Kim A, Li Y, Shen Y, Hu M, Liu JS, Ren B (2012) Topological domains in mammalian genomes identified by analysis of chromatin interactions. Nature 485(7398):376–380
8. Nora EP, Lajoie BR, Schulz EG, Giorgetti L, Okamoto I, Servant N, Piolot T, van Berkum NL, Meisig J, Sedat J (2012) Spatial partitioning of the regulatory landscape of the X-inactivation centre. Nature 485 (7398):381–385
9. Phillips-Cremins JE, Sauria ME, Sanyal A, Gerasimova TI, Lajoie BR, Bell JS, Ong C-T, Hookway TA, Guo C, Sun Y (2013) Architectural protein subclasses shape 3D organization of genomes during lineage commitment. Cell 153(6):1281–1295
10. Dali R, Blanchette M (2017) A critical assessment of topologically associating domain prediction tools. Nucleic Acids Res 45 (6):2994–3005
11. Norton HK, Emerson DJ, Huang H, Kim J, Titus KR, Gu S, Bassett DS, Phillips-Cremins JE (2018) Detecting hierarchical genome folding with network modularity. Nat Methods 15 (2):119
12. Malik L, Patro R (2018) Rich chromatin structure prediction from Hi-C data. IEEE/ACM Trans Comput Biol Bioinform 16 (5):1448–1458
13. Lieberman-Aiden E, van Berkum NL, Williams L, Imakaev M, Ragoczy T, Telling A, Amit I, Lajoie BR, Sabo PJ, Dorschner MO, Sandstrom R, Bernstein B, Bender MA, Groudine M, Gnirke A, Stamatoyannopoulos J, Mirny LA, Lander ES, Dekker J (2009) Comprehensive mapping of long-range interactions reveals folding principles of the human genome. Science 326 (5950):289–293. https://doi.org/10.1126/science.1181369
14. Filippova D, Patro R, Duggal G, Kingsford C (2014) Identification of alternative topological domains in chromatin. Algorithm Mol Biol 9 (1):14
15. Weinreb C, Raphael BJ (2016) Identification of hierarchical chromatin domains. Bioinformatics 32(11):1601–1609
16. Zhan Y, Mariani L, Barozzi I, Schulz EG, Bluthgen N, Stadler M, Tiana G, Giorgetti L (2017) Reciprocal insulation analysis of Hi-C data shows that TADs represent a functionally but not structurally privileged scale in the hierarchical folding of chromosomes. Genome Res 27(3):479–490
17. Fraser J, Rousseau M, Shenker S, Ferraiuolo MA, Hayashizaki Y, Blanchette M, Dostie J (2009) Chromatin conformation signatures of cellular differentiation. Genome Biol 10(4): R37
18. Javierre BM, Burren OS, Wilder SP, Kreuzhuber R, Hill SM, Sewitz S, Cairns J, Wingett SW, Várnai C, Thiecke MJ (2016) Lineage-specific genome architecture links enhancers and non-coding disease variants to target gene promoters. Cell 167 (5):1369–1384. e1319
19. Zhang Y, Wong C-H, Birnbaum RY, Li G, Favaro R, Ngan CY, Lim J, Tai E, Poh HM, Wong E (2013) Chromatin connectivity maps reveal dynamic promoter–enhancer long-range associations. Nature 504(7479):306
20. Eisenberg E, Levanon EY (2013) Human housekeeping genes, revisited. Trends Genet 29(10):569–574
21. Fraser J, Ferrai C, Chiariello AM, Schueler M, Rito T, Laudanno G, Barbieri M, Moore BL, Kraemer DC, Aitken S (2015) Hierarchical folding and reorganization of chromosomes are linked to transcriptional changes in cellular differentiation. Mol Syst Biol 11(12):852
22. Yaffe E, Tanay A (2011) Probabilistic modeling of Hi-C contact maps eliminates systematic

biases to characterize global chromosomal architecture. Nat Genet 43(11):1059
23. Rodriguez A, Laio A (2014) Clustering by fast search and find of density peaks. Science 344 (6191):1492–1496
24. Sutherland H, Bickmore WA (2009) Transcription factories: gene expression in unions? Nat Rev Genet 10(7):457
25. Consortium GO (2015) Gene ontology consortium: going forward. Nucleic Acids Res 43 (D1):D1049–D1056
26. Polager S, Ginsberg D (2009) p53 and E2f: partners in life and death. Nat Rev Cancer 9 (10):738–748

Chapter 5

Creating 2D Occupancy Plots Using *plot2DO*

Paula Beati and Răzvan V. Chereji

Abstract

Chromatin organization and epigenetic marks play a critical role in stem cell pluripotency and differentiation. Chromatin digestion by micrococcal nuclease (MNase) followed by high-throughput sequencing (MNase-seq) is the most widely used genome-wide method for studying nucleosome organization, that is, the first level of DNA packaging into chromatin. Combined with chromatin immunoprecipitation (ChIP), MNase-ChIP-seq represents a high-resolution method for investigating both chromatin organization and the distribution of epigenetic marks and histone variants. The plot2DO package presented here is a flexible tool for evaluating the quality of MNase-seq and MNase-ChIP-seq data, and for visualizing the distribution of nucleosomes near the functional regions of the genome. The plot2DO package is open-source software, and it is freely available from https://github.com/rchereji/plot2DO under the MIT license.

Key words 2D occupancy plot, MNase-seq, Heat map, Paired-end sequencing, NGS data quality control

1 Introduction

Nucleosomes consist of 147 bp of DNA wrapped around a histone octamer in about 1.7 turns and represent the basic unit of DNA packaging in eukaryotes [1]. Arrays of multiple nucleosomes separated by linker DNA are further folded to produce the chromatin fiber. Access of DNA-binding factors to their target sites is much reduced when these sites are occupied by nucleosomes [2]. Thus, chromatin often needs to be remodeled in order to achieve transcription. Apart from their role in DNA packaging, nucleosomes also carry chemical modifications, known as epigenetic marks, which establish active or repressive chromatin states that are important for cell differentiation and development. Therefore, mapping nucleosomes and their epigenetic marks with high accuracy is important for understanding DNA-binding, gene regulation, and all other DNA-related processes in the cell. For example, important differences in chromatin organization and changing transcriptional circuits are observed during the differentiation of human embryonic stem cells to various derived lineages [3].

Benjamin L. Kidder (ed.), *Stem Cell Transcriptional Networks: Methods and Protocols*, Methods in Molecular Biology, vol. 2117, https://doi.org/10.1007/978-1-0716-0301-7_5,

Currently, the most commonly used method for determining nucleosome positions is MNase-seq: chromatin is digested with micrococcal nuclease (MNase) and the resulting DNA fragments are subjected to high-throughput sequencing. Unfortunately, MNase has a strong sequence preference for A/T-rich sequences [4, 5], and the nucleosomal fragments that result from an MNase-seq experiment contain biases that depend on the degree of MNase digestion [6, 7]. Thus, after a mild digestion of chromatin, a large fraction of the genome is not yet broken into mononucleosomal DNA fragments (~150 bp long) and is still contained in larger oligonucleosome fragments, which are usually not sequenced. On the other hand, after an extensive digestion of chromatin, many nucleosomes occupying A/T-rich sequences are overdigested and lost from the sample of mononucleosomal (~150 bp long) fragments [6, 7]. Therefore, MNase-seq experiments require a careful control of the level of digestion, and the variable degree of digestion must always be considered when multiple samples are compared and used to draw conclusions from the differences observed among the samples.

Here, we present plot2DO, a free package for plotting the coverage of sequencing reads of various sizes at different genomic loci. The 2D occupancy (2DO) plot is useful not only for the initial quality check of MNase-seq data, in order to assess the degree of digestion, but also for getting insights about the nucleosome organization near functional regions of the genome, as well as about the MNase digestion kinetics.

2 Materials

2.1 Operating System

Plot2DO runs both on Unix-like operating systems, such as Linux and macOS, and on Microsoft Windows. To take advantage of the speed improvements offered by multicore processors, we recommend running plot2DO on a Unix-like machine. A bash shell or a similar terminal/command-line environment is necessary in order to launch plot2DO.

2.2 Software

The following additional software packages are needed in order to run plot2DO:

1. The R statistical computing environment, downloadable from http://www.r-project.org/.
2. A number of R packages available on the Comprehensive R Archive Network (CRAN):
   - colorRamps, https://CRAN.R-project.org/package=colorRamps.

- doParallel, https://CRAN.R-project.org/package=doParallel.
- foreach, https://CRAN.R-project.org/package=foreach.
- ggplot2, https://CRAN.R-project.org/package=ggplot2.
- gridExtra, https://CRAN.R-project.org/package=gridExtra.
- optparse, https://CRAN.R-project.org/package=optparse.
- reshape2, https://CRAN.R-project.org/package=reshape2.
- yaml, https://CRAN.R-project.org/package=yaml

3. A number of Bioconductor [8] packages, specifically:
   - AnnotationHub, http://bioconductor.org/packages/AnnotationHub/.
   - biomaRt, http://bioconductor.org/packages/biomaRt/.
   - GenomicFeatures, http://bioconductor.org/packages/GenomicFeatures/.
   - GenomicRanges, http://bioconductor.org/packages/GenomicRanges/.
   - GenomeInfoDb, http://bioconductor.org/packages/GenomeInfoDb/.
   - IRanges, http://bioconductor.org/packages/IRanges/.
   - Rsamtools, http://bioconductor.org/packages/Rsamtools/.
   - rtracklayer, http://bioconductor.org/packages/rtracklayer/.
4. In order to run plot2DO on Windows or macOS, a copy of GNU wget is also required. The wget utility allows us to download files from the Internet using the HTTP or FTP protocols.

   As many of the R packages listed above are regularly updated, and new features and optimizations are constantly developed and released, we recommend using the latest stable versions of these packages and reading the documentation that is available for these packages. In Fig. 1 we provide the output of the `sessionInfo()` command, showing the package versions used in the production of this chapter.

   Because of important changes introduced in Bioconductor release 3.7, plot2DO requires a minimum version of R 3.5.0 and Bioconductor 3.7 in order to run properly.

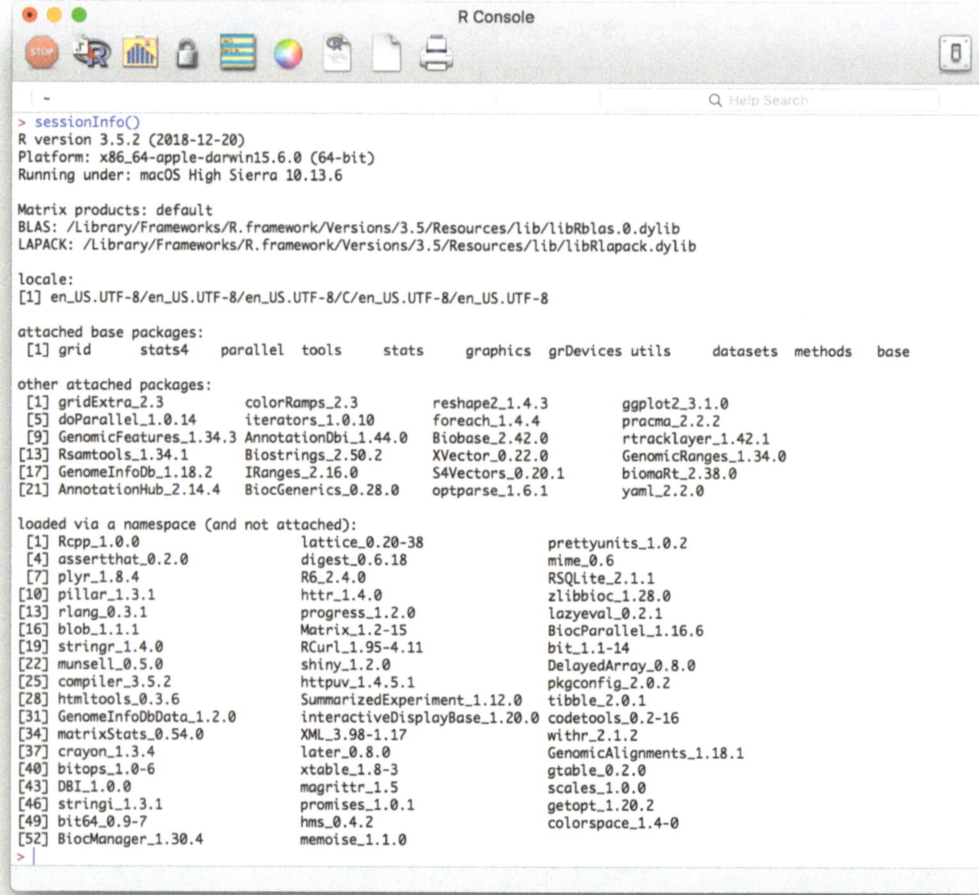

Fig. 1 The versions of R and its packages that were used to produce this chapter and to illustrate the features of plot2DO

2.3 Software Installation

In what follows, all commands that are run in the terminal are indicated by a dollar sign ($) prefix, as in the following example:

```
$ echo "Hello World"
```

To install the R statistical computing environment, download the latest precompiled binary distributions from the following website: http://cran.r-project.org.

In Windows and macOS, the GNU wget package is not installed by default, and should be downloaded and installed separately. The Windows binary executable file can be obtained from the following website: https://www.gnu.org/software/wget/faq.html#download. To install wget on a Mac computer, first install the Homebrew package manager (https://brew.sh), by running the following command in the terminal:

```
$ ruby -e "$(curl -fsSL https://raw.githubusercontent.com/
    Homebrew/install/master/install)"
```

After Homebrew has been installed, run the following command to install the wget package:

```
$ brew install wget
```

The plot2DO project's repository is freely available on GitHub, at the following address: https://github.com/rchereji/plot2DO. The most recent version of plot2DO can be downloaded by clicking the "Clone or download" button on the GitHub page indicated above. Alternatively, you can download the package directly from the terminal. In order to do this, install a git client from https://git-scm.com/downloads. After the git client has been installed, clone the plot2DO repository to the local machine by running the following command in the terminal:

```
$ git clone https://github.com/rchereji/plot2DO.git
```

After downloading the plot2DO package, install all the required R packages by starting R (or RStudio) and executing the `plot2DO_setup.R` script. Alternatively, this script can also be run directly from the terminal using the following commands:

```
$ Rscript plot2DO_setup.R
```

2.4 Input File Formats

In general, the starting point for all analyses involving sequencing data are the FASTQ files—the typical format for storing information about the reads obtained from Illumina sequencers. In order to get information about the lengths of DNA fragments obtained in an experiment, paired-end sequencing is required. We assume that users have already aligned the raw sequencing reads to a reference genome, and the aligned reads are available in BAM- or BED-formatted files. If the paired-end sequencing data is available as a BED file, then the first three BED fields (columns) are required: (1) the name of the chromosome; (2) the starting position of the paired-end read; and (3) the ending position of the paired-end read. Note that each genomic interval stored in the BED file needs to include the full length of the original DNA fragment, that is, the lengths of the two sequenced ends plus the length of the insert, which is not sequenced if the original DNA fragment is longer than the combined length of the two sequenced ends.

2.5 Example Data

To demonstrate the available features of plot2DO, we have used sequencing datasets that are freely available on Gene Expression Omnibus (GEO; https://www.ncbi.nlm.nih.gov/geo/). The GEO accession numbers corresponding to these datasets are

Table 1
GEO accession numbers for the datasets that were used in this chapter to illustrate the features of plot2DO

Filename	GEO accession number	Source
yeast_SRR2045610.bam	GSM1700669	[14]
yeast_50U_MNase_SRR3649301.bam	GSM2193286	[7]
yeast_100U_MNase_SRR3649296.bam	GSM2193281	[7]
yeast_200U_MNase_SRR3649297.bam	GSM2193282	[7]
yeast_300U_MNase_SRR3649298.bam	GSM2193283	[7]
yeast_400U_MNase_SRR3649299.bam	GSM2193284	[7]
fly_SRR2038265.bam	GSM1694824	[6]
mouse_SRR5116996.bam	GSM2430825	[15]
human_SRR1781839.bam	GSM1425454	[16]

provided in Table 1. For testing purposes, we also provide a collection of BAM files containing subsets of five million random paired-end reads originating from the datasets used in this chapter, at the following website: bit.ly/plot2DO_BAM_files.

3 Methods

In this section we demonstrate the use of plot2DO and explain the outputs generated by the software. We will go through a series of examples in which we use plot2DO to analyze sequencing data obtained from different experiments.

After the required R packages have been installed as described above, plot2DO can be launched from the terminal. To check that plot2DO was correctly downloaded and is available in the current directory, start a terminal and run the following command:

```
$ Rscript plot2DO.R --help
```

If plot2DO was correctly downloaded from GitHub and unpacked in the current folder, the above command will display a short help message containing all the available options and arguments (or flags) of plot2DO (Fig. 2). As shown in Fig. 2, most of the plot2DO arguments can be specified either by providing a "short" or a "long" flag, followed by the corresponding value. The "short" flags (preceded by a single dash) are generally one character long and are separated from the corresponding value by a space (e.g., `-f <filename.bam>`). The "long" flags (preceded by

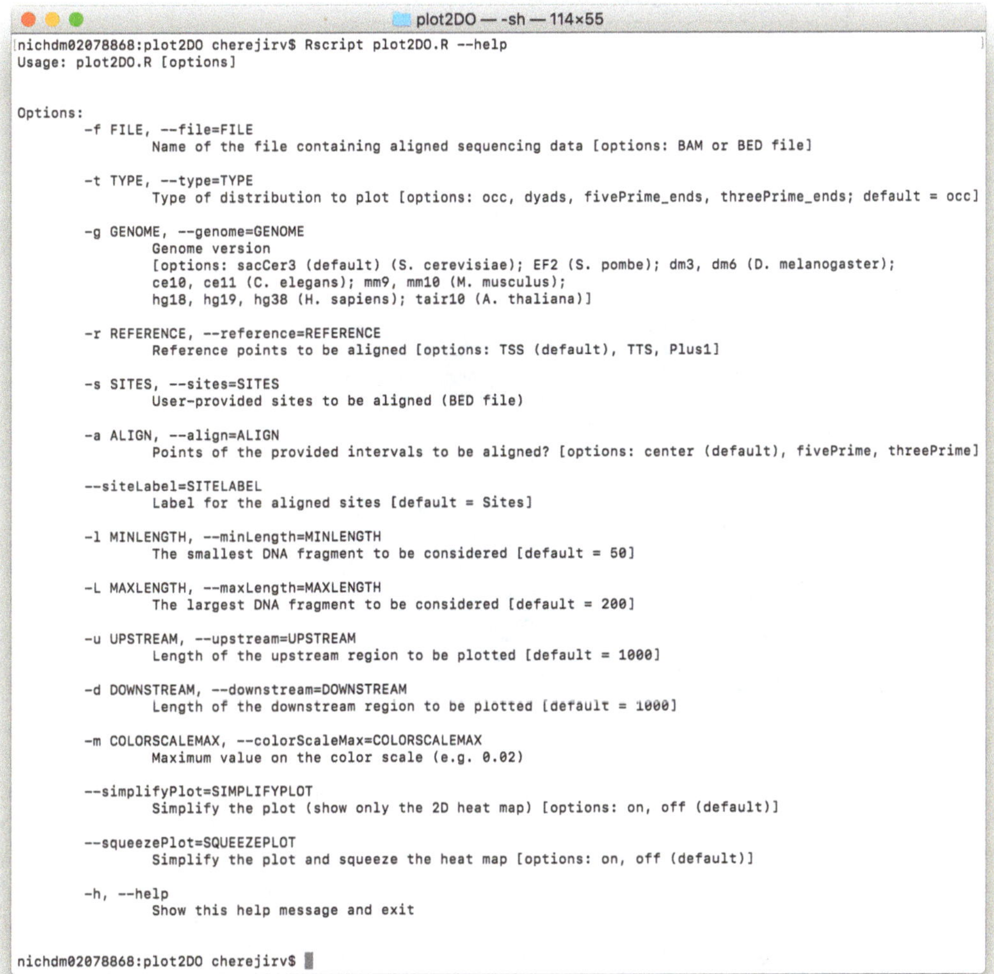

Fig. 2 Help screen showing all the options and arguments that are available in plot2DO

two dashes) are represented by multicharacter strings and are separated from their corresponding values by an equals (=) sign (e.g., `--file=<filename.bed>`). Here, `<filename.bam>` and `<filename.bed>` represent placeholders for the names of the input files containing the aligned sequencing data.

The figures generated by plot2DO contain three panels (Fig. 3). The most important panel—the 2DO plot—is represented by a heat map that indicates the relative coverage of the DNA fragments of specified lengths, at different locations relative to the selected references (TSSs by default). Each row of the 2DO plot indicates the average occupancy generated by DNA fragments of specified lengths (shown on the right side of the heat map) as a

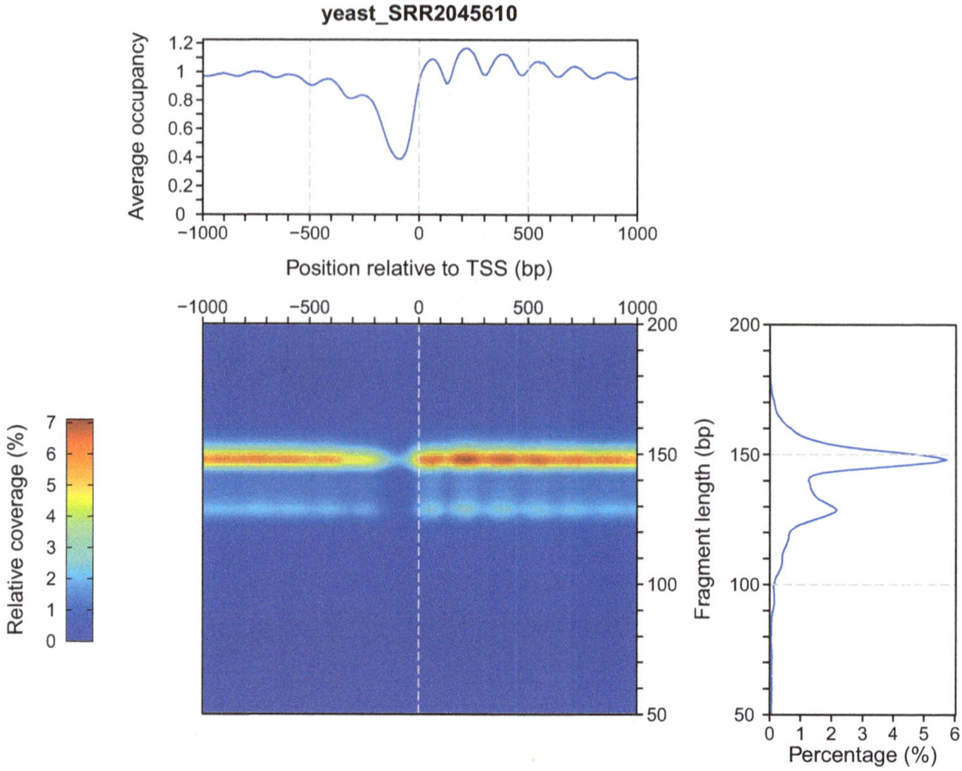

Fig. 3 The three panels that are generated by plot2DO: (1) 2D occupancy (2DO) plot (central heat map) indicating the relative coverage of the DNA fragments of specified lengths, at different locations relative to the reference points (transcription start sites (TSSs) in this case). The red color indicates a high coverage, while dark blue indicates zero coverage. (2) One-dimensional occupancy (top panel), generated by stacking DNA fragments of all lengths shown in the heat map. (3) Fragment length histogram (right panel) indicating the percentage corresponding to each DNA fragment size, from the whole sample of sequenced reads

function of the position relative to the reference points (shown on top of the heat map). Obviously, the heat map is a color-coded representation of a matrix, which we call the 2DO matrix. Each row of the 2DO matrix represents the relative occupancy produced by the DNA fragments of a given size. Each column of the 2DO matrix indicates the relative occupancy generated by DNA fragments of different sizes at a given position relative to the reference points.

The second panel—the overall average occupancy (one-dimensional plot shown above the 2DO heat map)—is generated by stacking DNA fragments of all lengths in the specified size range (`minLength` $\leq$ L $\leq$ `maxLength`). This plot represents the sum of the elements from each column of the 2DO matrix, shown in the heat map. To compute these relative occupancy profiles, the raw sequencing data is normalized such that the average occupancy is 1 for each chromosome.

The third panel that is generated by plot2DO is the fragment length histogram, which is shown to the right of the 2DO plot. This plot represents the percentages of paired-end sequencing reads that have a given length (which is the sum of the lengths of the two sequenced ends plus the insert size that was not sequenced). Note that this histogram considers all the sequencing reads, not only the ones that are mapped to the regions shown in the heat map (i.e., the reads far from the reference points are also considered).

3.1 Plot Using the Default Parameters

Almost all arguments of plot2DO are optional. If an argument is not specified, it will take its default value, as indicated in the help screen (Fig. 2). The only required argument for plot2DO is the name of the file that contains the aligned genomic data (in BAM or BED format), which is provided using the `-f/--file` argument. A minimal example is given below,

```
$ Rscript plot2DO.R -f yeast_SRR2045610.bam
```

or equivalently,

```
$ Rscript plot2DO.R --file=yeast_SRR2045610.bam
```

where `yeast_SRR2045610.bam` is a BAM file, containing paired-end reads that were aligned to the *sacCer3* reference genome. The plot generated by any of the previous two commands is shown in Fig. 3. Without specifying other arguments, it is assumed that the sequencing data corresponds to DNA fragments from the *S. cerevisiae* genome, which were aligned to the *sacCer3* reference genome. For clarity we recommend to always specify the reference genome, using the `-g/--genome` argument, as follows:

```
$ Rscript plot2DO.R -f yeast_SRR2045610.bam -g
  sacCer3
```

The default 2DO plot, obtained by running one of the previous three commands, will show the distribution of all sequenced DNA fragments with the lengths between 50 bp and 200 bp, which are found in 2 kb windows centered on the transcription start sites (TSSs).

From the panels produced by plot2DO (Fig. 3), we can obtain the following information. The top panel shows the traditional average occupancy plot, obtained by stacking all the sequencing reads and averaging the obtained coverage for all genes, aligned at their TSSs. In this panel we can observe the nucleosome depleted region (NDR) that forms at the gene promoters, and we can see the degree of nucleosome phasing [9]—how consistent this regular nucleosome organization is near TSSs, among all genes. In the right panel, we can observe the degree of chromatin digestion. As expected, most DNA fragments have a length of about 147 bp—

the size of a nucleosome core particle, but we can also observe that some of the fragments correspond to overdigested nucleosomes, which have lost about 10 bp of DNA from each end, and the reduced footprint that was obtained is about 130 bp long. The most important part of Fig. 3 is represented by the 2DO plot, which contains all the information of the previous 1D plots, in a disentangled way. Therefore, one can now distinguish the contribution to occupancy of DNA fragments of different sizes occupying the same positions in different genes or cells, which is not possible from the typical 1D occupancy plot. Furthermore, it is also possible to observe the distribution near promoters of all fragments corresponding to a given size, which is not possible from the usual 1D fragment length distribution.

3.2 Plot the Distribution of DNA Fragment Centers Near +1 Nucleosomes

By default, plot2DO aligns the TSSs of all genes and plots the relative coverage obtained by stacking the full footprints of the sequenced DNA fragments. One may want to plot the distribution of DNA fragments at other locations from the genome (e.g., transcription termination sites (TTSs)) or to align the typical positions of the +1 nucleosomes—the first nucleosomes downstream from the gene promoters. This can be done using the `-r/--reference` argument. Moreover, it is also possible to plot the distribution of the fragment centers (corresponding to nucleosome dyads in MNase-seq experiments) instead of the coverage produced by the whole DNA fragments, similar to the V-plots introduced by Henikoff et al. [10]. To change from the default 2D occupancy plot to a 2D dyad plot, we use the `-t/--type` argument, as in the following example:

```
$ Rscript plot2DO.R -f yeast_SRR2045610.bam -g sac-
   Cer3 --reference=Plus1 --type=dyads
```

Because in higher organisms the coordinates of the +1 nucleosomes vary among different cell types, the `--reference=Plus1` option is only available for *S. cerevisiae*, for which the +1 nucleosome positions have been previously identified with high resolution using the chemical cleavage mapping method [11].

3.3 Compare DNA Distributions from Multiple Samples

To compare DNA distributions obtained in different experiments, it is very important to set the color scale in a consistent way in all plots. The maximum value of the color scale can be set using the `-m/--colorScaleMax` argument,

```
$ Rscript plot2DO.R -f yeast_SRR2045610.bam -g sac-
   Cer3 -r Plus1 -t dyads --colorScaleMax=0.15
```

The plots generated by the above command are shown in Fig. 4.

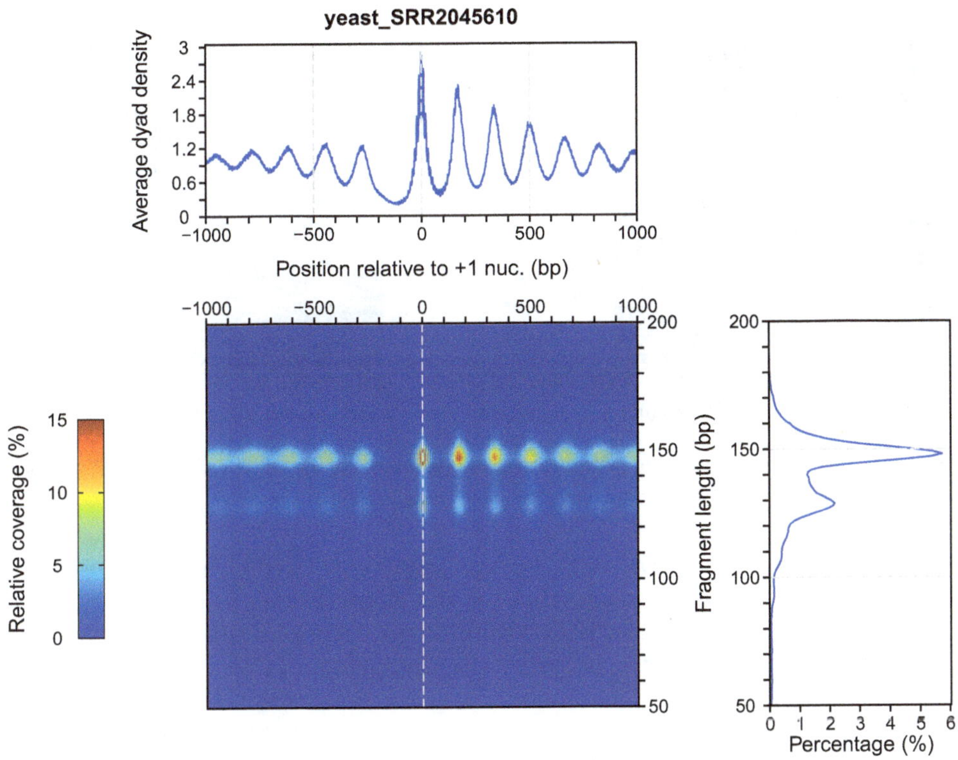

Fig. 4 Distribution of nucleosome dyads near the typical positions of +1 nucleosomes. This special case of 2DO plot is also known as a V-plot [10].

3.4 Plot the Distribution of 5′ or 3′ Ends of DNA Fragments

Similarly, one can plot the distribution of the 5′ and 3′ ends of DNA fragments, using the other available options of the `-t/--type` argument:

```
$ Rscript plot2DO.R -f yeast_SRR2045610.bam -g sacCer3 -r
Plus1 --type=fivePrime_ends
$ Rscript plot2DO.R -f yeast_SRR2045610.bam -g sacCer3 -r
Plus1 --type=threePrime_ends
```

3.5 Align Custom Lists of Genomic Loci Using User-Provided BED Files

Apart from the usual alignments of transcription start sites (`--reference=TSS`), transcription termination sites (`--reference=TTS`), and +1 nucleosomes (`--reference=Plus1`), we can also align lists of different reference points, provided in BED-formatted files. For convenience, we provide a few lists of yeast features, such as tRNA genes, origins of replication, and centromeres. To align user-provided regions, we use the argument `-s/--sites` to provide the coordinates of these genomic loci in a BED-formatted file. Because BED files can store regions of different sizes and the strands corresponding to these genomic features, we allow the user to also specify the exact points of these regions that need to be aligned (centers, 5′ ends or 3′ ends). This is done

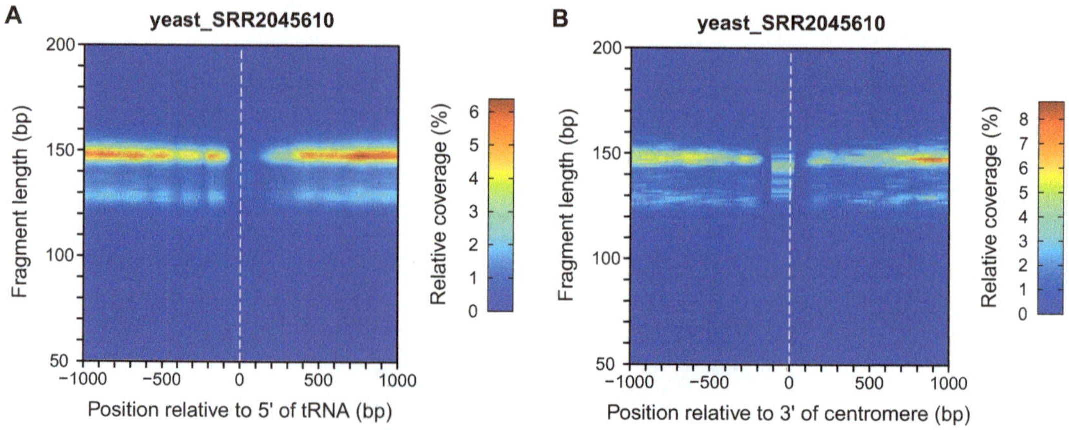

Fig. 5 2D occupancy plots showing the distribution of sequencing data near the yeast (**a**) tRNA genes, aligned at their 5′ ends, and (**b**) centromeres, aligned at their 3′ ends

using the `-a/--align` argument. Furthermore, the user can specify a custom label for the aligned sites, using the `--siteLabel` argument. Below we give two examples that use these three arguments to customize the alignments:

```
$ Rscript plot2DO.R -f yeast_SRR2045610.bam -g sacCer3 --sites=annotations/Yeast_tRNA_genes.bed --align=fivePrime --siteLabel=tRNA --simplifyPlot=on
$ Rscript plot2DO.R -f yeast_SRR2045610.bam -g sacCer3 --sites=annotations/Yeast_centromeres.bed --align=threePrime --siteLabel=centromere --simplifyPlot=on
```

3.6 Simplify Figures by Plotting Only 2DO Heat Maps

Notice that in the above examples we have also used the `--simplifyPlot=on` flag, which simplifies the generated figure, by printing only the 2DO heat map, without the extra panels that show the one-dimensional occupancy and the histogram of DNA fragment lengths. This option allows the user to easily combine multiple 2DO plots in a single figure (Fig. 5).

3.7 Multiple Organisms Are Supported

Plot2DO can process paired-end sequencing data originating from a variety of organisms (animals and plants). Currently the following organisms are supported: *S. cerevisiae*, *S. pombe*, *D. melanogaster*, *C. elegans*, *M. musculus*, *H. sapiens*, and *A. thaliana*. The corresponding reference genomes that are currently supported are: *sacCer3*; *dm3*, *dm6*; *ce10*, *ce11*; *mm9*, *mm10*; *hg18*, *hg19*, *hg38*; *tair10*. As mentioned above, for multicellular organisms the alignment of +1 nucleosomes is not possible by using the `--reference=Plus1` argument, as the locations of these nucleosomes could vary from cell type to cell type. The alignment of the +1 nucleosomes for these organisms is still possible, if these coordinates are identified and provided as a separate BED file, using the `-s/--sites` argument.

Next we provide a few command examples used to investigate the distribution of reads at the TSSs of some of the aforementioned organisms, and the quality (size distribution) of DNA fragments resulted from chromatin digestion using MNase:

```
$ Rscript plot2DO.R -f fly_SRR2038265.bam -g dm3 -m 0.015 --
simplifyPlot=on --minLength=100 --maxLength=200
$ Rscript plot2DO.R -f mouse_SRR572708.bam -g mm10 -m 0.015 --
simplifyPlot=on --minLength=100 --maxLength=200
$ Rscript plot2DO.R -f human_SRR1781839.bam -g hg19 -m 0.015 -
-simplifyPlot=on --minLength=100 --maxLength=200
```

3.8 Perform In-Silico Size Selection of DNA Fragments

In the previous three examples we have also used the following optional arguments: `-l/--minLength` and `-L/--maxLength`, which allow the user to select the DNA fragments by size within a specified interval, and to plot the distributions of these fragments alone. By default, the range of 50–200 bp is used in the 2DO plots. To illustrate the size selection option, in the above examples we have used a reduced range of 100–200 bp, which is appropriate for investigating the nucleosome footprints to deduce the degree of chromatin digestion. The shorter fragments (<147 bp) indicate nucleosomes that are overdigested, while the longer fragments (>147) represent nucleosomes that remained with linker DNA attached to their cores, due to under-digestion. The plots produced by the previous commands are shown as individual panels in Fig. 6.

3.9 Customize the Displayed Windows Near the Reference Points

If we want to investigate small genomic regions such as the tRNA genes, origins of replication, or the locations of +1 nucleosomes, the arguments `-u/--upstream` and `-d/--downstream` can be used to modify the size of the window that is displayed near the reference points. If the displayed windows are very small, then users have the option to reduce the heat map to a vertical strip, using the `--squeezePlot=on` flag. Below we provide a few examples in which we have used these three arguments:

```
$ Rscript plot2DO.R -f yeast_50U_MNase_SRR3649301.bam -g
sacCer3 -r Plus1 -m 0.03 -u 100 -d 100 --squeezePlot=on
$ Rscript plot2DO.R -f yeast_100U_MNase_SRR3649296.bam -g
sacCer3 -r Plus1 -m 0.03 -u 100 -d 100 --squeezePlot=on
$ Rscript plot2DO.R -f yeast_200U_MNase_SRR3649297.bam -g
sacCer3 -r Plus1 -m 0.03 -u 100 -d 100 --squeezePlot=on
$ Rscript plot2DO.R -f yeast_300U_MNase_SRR3649298.bam -g
sacCer3 -r Plus1 -m 0.03 -u 100 -d 100 --squeezePlot=on
$ Rscript plot2DO.R -f yeast_400U_MNase_SRR3649299.bam -g
sacCer3 -r Plus1 -m 0.03 -u 100 -d 100 --squeezePlot=on
```

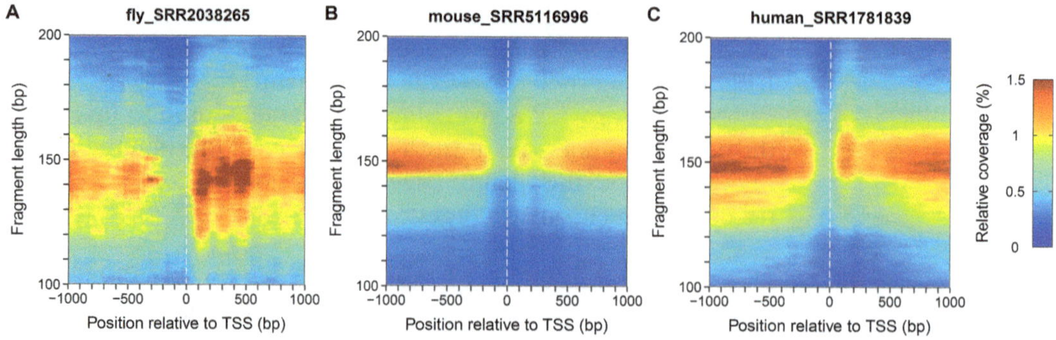

Fig. 6 2D occupancy plots showing the degree of chromatin digestion and the distribution of sequencing data near the TSS in (**a**) fly, (**b**) mouse, and (**c**) human

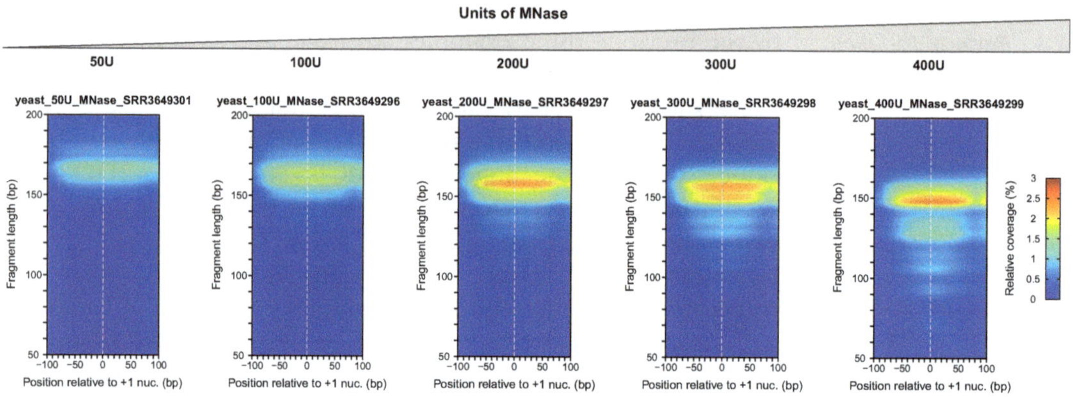

Fig. 7 A titration of MNase produces different levels of chromatin digestion, and different sizes for the undigested fragments of DNA. Plus 1 nucleosomes are particularly sensitive to MNase, as they are located adjacent to the nucleosome free regions (the most accessible regions of the genome)

The narrow strips resulted from these commands, can be easily combined into a multipanel figure as shown in the example from Fig. 7.

4 Conclusions

The plots generated by plot2DO are particularly useful for investigating the level of digestion and the effect of MNase on the subset of nucleosomes that are obtained in MNase-seq experiments. Figure 7 shows the effect of MNase on the subset of +1 nucleosomes that were obtained in an MNase titration experiment [7] at different levels of digestion. It is obvious that extensively digested samples contain shorter protected DNA fragments, which are eventually destroyed by MNase, if the chromatin digestion is not stopped early enough. Similarly, one can analyze the amount of sequencing data that is obtained from other regions of the genome in a series of digestions and study the kinetics of chromatin digestion for different regions of the genome.

Plot2DO is a very versatile tool. Apart from MNase-seq data analysis, plot2DO was also used to investigate DNA fragments obtained in other types of experiments, such as ChIP-seq [7, 12, 13], ATAC-seq [7], and chemical cleavage mapping [11]. In principle, plot2DO can be used to analyze the amount of DNA that is obtained from different regions of the genome, and the corresponding distribution of fragment sizes for each region, independent of the protocol that is used to generate the short DNA fragments.

Nucleosome mapping experiments are relatively expensive, especially when long genomes are considered. Therefore, it is very important to assess the level of MNase digestion and the quality of the resulting sequencing library, before a large amount of money is spent for sequencing. Before committing to a large sequencing experiment that results in hundreds of millions or billions of reads, a trial sequencing round could be performed (e.g., getting only a few million reads) and the level of digestion could be assessed using plot2DO. We recommend that in all MNase-seq, MNase-ChIP-seq, and perhaps in other genome-wide techniques that break the chromatin to produce short DNA fragments, plot2DO should be the first *plot to do*, in order to assess the level of chromatin fragmentation and the quality of the genomic data.

Acknowledgement

We thank Natalia Petrenko, Burke Squires, Ming-an Sun, and Yashpal Rawal for insightful discussions and for testing the software. R.V.C. was supported by the Intramural Research Program of the National Institute of Child Health and Human Development, National Institutes of Health. This work utilized the computational resources of the NIH HPC Biowulf cluster (http://hpc.nih.gov).

References

1. Luger K, Mader AW, Richmond RK, Sargent DF, Richmond TJ (1997) Crystal structure of the nucleosome core particle at 2.8 A resolution. Nature 389(6648):251–260. https://doi.org/10.1038/38444
2. Liu X, Lee CK, Granek JA, Clarke ND, Lieb JD (2006) Whole-genome comparison of Leu3 binding in vitro and in vivo reveals the importance of nucleosome occupancy in target site selection. Genome Res 16(12):1517–1528. https://doi.org/10.1101/gr.5655606
3. Dixon JR, Jung I, Selvaraj S, Shen Y, Antosiewicz-Bourget JE, Lee AY, Ye Z, Kim A, Rajagopal N, Xie W, Diao Y, Liang J, Zhao H, Lobanenkov VV, Ecker JR, Thomson JA, Ren B (2015) Chromatin architecture reorganization during stem cell differentiation. Nature 518(7539):331–336. https://doi.org/10.1038/nature14222
4. Dingwall C, Lomonossoff GP, Laskey RA (1981) High sequence specificity of micrococcal nuclease. Nucleic Acids Res 9 (12):2659–2673
5. Horz W, Altenburger W (1981) Sequence specific cleavage of DNA by micrococcal nuclease. Nucleic Acids Res 9(12):2643–2658

6. Chereji RV, Kan TW, Grudniewska MK, Romashchenko AV, Berezikov E, Zhimulev IF, Guryev V, Morozov AV, Moshkin YM (2016) Genome-wide profiling of nucleosome sensitivity and chromatin accessibility in Drosophila melanogaster. Nucleic Acids Res 44 (3):1036–1051. https://doi.org/10.1093/nar/gkv978
7. Chereji RV, Ocampo J, Clark DJ (2017) MNase-sensitive complexes in yeast: nucleosomes and non-histone barriers. Molecular cell 65(3):565–577. e563. https://doi.org/10.1016/j.molcel.2016.12.009
8. Gentleman RC, Carey VJ, Bates DM, Bolstad B, Dettling M, Dudoit S, Ellis B, Gautier L, Ge Y, Gentry J, Hornik K, Hothorn T, Huber W, Iacus S, Irizarry R, Leisch F, Li C, Maechler M, Rossini AJ, Sawitzki G, Smith C, Smyth G, Tierney L, Yang JY, Zhang J (2004) Bioconductor: open software development for computational biology and bioinformatics. Genome Biol 5(10): R80. https://doi.org/10.1186/gb-2004-5-10-r80
9. Chereji RV, Clark DJ (2018) Major determinants of nucleosome positioning. Biophys J 114(10):2279–2289. https://doi.org/10.1016/j.bpj.2018.03.015
10. Henikoff JG, Belsky JA, Krassovsky K, Macalpine DM, Henikoff S (2011) Epigenome characterization at single base-pair resolution. Proc Natl Acad Sci U S A 108(45):18318–18323. https://doi.org/10.1073/pnas.1110731108
11. Chereji RV, Ramachandran S, Bryson TD, Henikoff S (2018) Precise genome-wide mapping of single nucleosomes and linkers in vivo. Genome Biol 19(1):19. https://doi.org/10.1186/s13059-018-1398-0
12. Lawrence MBD, Coutin N, Choi JK, Martin BJE, Irwin NAT, Young B, Loewen C, Howe LJ (2017) Histone acetylation, not stoichiometry, regulates linker histone binding in Saccharomyces cerevisiae. Genetics 207(1):347–355. https://doi.org/10.1534/genetics.117.1132
13. Rawal Y, Chereji RV, Qiu H, Ananthakrishnan S, Govind CK, Clark DJ, Hinnebusch AG (2018) SWI/SNF and RSC cooperate to reposition and evict promoter nucleosomes at highly expressed genes in yeast. Genes Dev 32(9-10):695–710. https://doi.org/10.1101/gad.312850.118
14. Ocampo J, Chereji RV, Eriksson PR, Clark DJ (2016) The ISW1 and CHD1 ATP-dependent chromatin remodelers compete to set nucleosome spacing in vivo. Nucleic Acids Res 44 (10):4625–4635. https://doi.org/10.1093/nar/gkw068
15. Johnson TA, Chereji RV, Stavreva DA, Morris SA, Hager GL, Clark DJ (2018) Conventional and pioneer modes of glucocorticoid receptor interaction with enhancer chromatin in vivo. Nucleic Acids Res 46(1):203–214. https://doi.org/10.1093/nar/gkx1044
16. West JA, Cook A, Alver BH, Stadtfeld M, Deaton AM, Hochedlinger K, Park PJ, Tolstorukov MY, Kingston RE (2014) Nucleosomal occupancy changes locally over key regulatory regions during cell differentiation and reprogramming. Nat Commun 5:4719. https://doi.org/10.1038/ncomms5719

Chapter 6

Detection of Epigenetic Field Defects Using a Weighted Epigenetic Distance-Based Method

Ya Wang, Min Qian, Peifeng Ruan, Andrew E. Teschendorff, and Shuang Wang

Abstract

Identifying epigenetic field defects, notably early DNA methylation alterations, is important for early cancer detection. Research has suggested these early methylation alterations are infrequent across samples and identifiable as outlier samples. Here we developed a weighted epigenetic distance-based method characterizing (dis)similarity in methylation measures at multiple CpGs in a gene or a genetic region between pairwise samples, with weights to up-weight signal CpGs and down-weight noise CpGs. Using distance-based approaches, weak signals that might be filtered out in a CpG site-level analysis could be accumulated and therefore boost the overall study power. In constructing epigenetic distances, we considered both differential methylation (DM) and differential variability (DV) signals. We demonstrated the superior performance of the proposed weighted epigenetic distance-based method over nonweighted versions and site-level EWAS (epigenome-wide association studies) methods in simulation studies. Application to breast cancer methylation data from Gene Expression Omnibus (GEO) comparing normal-adjacent tissue to tumor of breast cancer patients and normal tissue of independent age-matched cancer-free women identified novel epigenetic field defects that were missed by EWAS methods, when majority were previously reported to be associated with breast cancer and were confirmed the progression to breast cancer. We further replicated some of the identified epigenetic field defects.

Key words DNA methylation, Epigenetic field defects, Weighted distance-based method

1 Introduction

Identifying molecular alterations that happen early in carcinogenesis, known as field defects, is important for early cancer detection. One common approach is to compare normal tissue of healthy individuals to normal tissue adjacent to tumor (normal-adjacent tissue) of cancer patients as a surrogate of precancer tissue that are difficult to collect. There have been studies in identifying epigenetic field defects [1–4], notably early DNA methylation alterations. DNA methylation is an epigenetic modification that has been shown to be crucial in gene expression [5–8] and cancers [9–

Benjamin L. Kidder (ed.), *Stem Cell Transcriptional Networks: Methods and Protocols*, Methods in Molecular Biology, vol. 2117, https://doi.org/10.1007/978-1-0716-0301-7_6, © Springer Science+Business Media, LLC, part of Springer Nature 2020

12]. There are mainly two types of aberrant DNA methylation in cancers, local hyper methylation in promoter-related CpGs that leads to the silencing of downstream tumor suppressor genes [13–17], and global hypomethylation that leads to chromosome instability [17–20]. Studies have successfully identified epigenetic field defects in breast cancer by comparing normal-adjacent tissue of breast cancer patients to normal tissue from healthy individuals. Teschendorff et al. identified epigenetic field defects in breast cancer based on differential variability (DV), that is, variance signals in DNA methylation [3], using methylation site-level analyses. Our previous work [21] identified epigenetic field defects in breast cancer based on both differential methylation (DM) (i.e., mean signals) and DV, using methylation region-level analyses. In both studies, epigenetic field defects were found to be mainly driven by increased variation in methylation due to several outlier normal-adjacent tissue samples.

Due to the fact that CpG site-level signals for epigenetic field defects may be very small, existing methods based on differences (DM or DV or both) on CpG site-level may not have good power. Standard epigenome-wide association studies (EWAS) that focus on mean signals (EWAS-DM) perform CpG site-level tests to identify differentially methylated CpGs between two experimental groups using standard tests such as a *t*-test, a regression-based test or its regularized versions [22–24], or a nonparametric Wilcoxon rank sum test [25]. EWAS that focus on variance signals (EWAS-DV) perform CpG site-level tests to identify differential variation CpGs between two experimental groups using standard tests such as the *F*-test [26, 27], the Bartlett's test or its regularized version [3, 4], or an empirical Bayes extension of the Levene's test [28]. The *F*-test and Bartlett's test are sensitive to departures from normality which is usually the case for methylation data, while the Levene's test is more robust to nonnormality. On the other hand, distance-based methods that characterize (dis)similarity between pairwise samples across a gene, a genetic region, a pathway or an entire genome have been proven to be powerful in genetic and gene expression studies [29–33]. While standard EWAS perform CpG site-level tests with stringent multiple comparisons adjustment, in a gene or a genetic region level, the common practice using non-distance-based methods is to select the minimum *p*-value out of all CpGs in that region. These methods will not be powerful when site-level effects are very small. Alternatively, the distance-based methods accumulate any CpG site-level signals from a gene or a genetic region via the (dis)similarity matrix thus boost the overall association power, making them the ideal methods for detection of epigenetic field defects.

Here we developed a weighted epigenetic distance-based method to identify epigenetic field defects at gene or genetic-region levels using both DM and DV signals. CpG site-level weights were incorporated in the calculation of (dis)similarity matrix to further boost signals and reduce noises. Specifically, we used original DNA methylation measures to examine DM and centered quadratic methylation measures to examine DV and considered site-level weights based on strengths of site-level DM and DV signals. Simulation studies showed much improved performance of the proposed weighted epigenetic distance-based method over several comparison methods including nonweighted versions and methods that use either DM or DV signals as well as standard EWAS methods. We further demonstrated the performance of the proposed method through an application to the 450K DNA methylation data of normal-adjacent tissue of breast invasive carcinoma (BRCA) patients and normal tissue from independent age-matched cancer-free women from Gene Expression Omnibus (GEO). The proposed method that accumulates weighted DM and DV signals identified genes with epigenetic field defects that were missed by standard EWAS methods and nonweighted distance-based methods. Many of these epigenetic field defects were previously reported to be associated with breast cancer. Further examination confirmed their enrichment in the progression to breast cancer and replicated some of these identified epigenetic field defects.

2 Methods

Case-control designs using normal tissue from healthy individuals ($\Upsilon = 0$) and normal tissue adjacent to tumor from cancer patients ($\Upsilon = 1$) as a surrogate of precancer tissue are widely used to identify epigenetic field defects in cancers. We therefore focused on case-control designs and illustrated and applied the proposed weighted epigenetic distance-based method on gene level. However, the proposed method can be easily adapted to other types of design and on genetic region or genome levels.

2.1 The Proposed Distance-Based Method

There are three steps in the proposed distance-based method: (1) to define gene-level weighted epigenetic distance matrix; (2) to calculate pseudo-F statistic; and (3) to assess statistical significance using permutations.

2.1.1 Step 1: Define Gene-Level Weighted Epigenetic Distance Matrix

(a) *Define epigenetic distance matrix.* For each gene, let $\mathbf{X}^{\mathrm{m}}$ be a $2N \times n$ matrix with original DNA methylation measures for N cases and N controls of n CpG sites in a gene, where element $x_{\mathrm{ij}}^{\mathrm{m}}$ harbors DNA methylation measure of the jth CpG site, $j = 1, \ldots, n$ in the gene, for the ith subject, $i = 1, \ldots, N$. This $\mathbf{X}^{\mathrm{m}}$ matrix will be used to examine differential methylation (DM) capturing methylation mean signals. Let $\mathbf{X}^{\mathrm{v}}$ be a $2N \times n$ pseudo data matrix of variability score capturing methylation variance signals, which will be used to examine differential variation (DV). The element $x_{ij}^{v} = (x_{ij}^{m} - \overline{x}_{j}^{m})^2$ harbors centered quadratic methylation measure of the same *j*th CpG site for the *i*th subject. Here $\overline{x}_{j}^{m} = \frac{1}{N}\sum_{i=1}^{N} x_{ij}^{m}$ is the mean methylation measure of the *j*th CpG site across N cases and N controls separately. The quadratic terms are centered to better capture variance signals. By using $\mathbf{X}^{mv} = [\mathbf{X}^{m}, \mathbf{X}^{v}]$, a $2N \times 2n$ matrix, we will be able to capture both methylation mean and methylation variance signals of the n CpG sites. Before constructing the epigenetic distance between any pair of subjects, we performed normalization on each column of $\mathbf{X}^{mv}$ such that each column has mean zero and unit standard deviation.

We define the $2N \times 2N$ epigenetic distance matrix $\mathbf{D}^{DM-DV}$ with element d_{st}^{DM-DV} that captures dissimilarities between any given pair of individuals s and t, $s, t = 1, \ldots, 2N$ as

$$d_{st}^{DM-DV} = \sqrt{\sum_{j=1}^{n} \left\{ \frac{1}{2n} (x_{sj}^{m} - x_{tj}^{m})^2 + \frac{1}{2n} (x_{sj}^{v} - x_{tj}^{v})^2 \right\}} \quad (1)$$

(b) *Incorporate CpG site-level weights into epigenetic distance matrix.* We construct CpG site-level weights aiming to up-weight signal CpGs (mean or variance) and to down-weight noise CpGs in calculating distances between pairs of subjects. Therefore, we define weights for mean and variance signals at CpG site j as follows:

$$w_{j}^{m} = \frac{-\log_{10}(p_{j}^{m})}{\sum_{j=1}^{n} -\log_{10}(p_{j}^{m})}, \quad w_{j}^{v} = \frac{-\log_{10}(p_{j}^{v})}{\sum_{j=1}^{n} -\log_{10}(p_{j}^{v})} \quad (2)$$

where p_{j}^{m} and p_{j}^{v} are the p-values from the two-sided two-sample t-test testing if the mean methylation measures are the same between cases and controls and from the one-sided Levene's test testing if the variance of the methylation measures in cases is greater than that in controls at CpG site j, $j = 1, \ldots, n$ in a gene. Note that $\sum_{j=1}^{n} w_{j}^{m} = \sum_{j=1}^{n} w_{j}^{v} = 1$.

The corresponding $2N \times 2N$ weighted epigenetic distance matrix $\mathbf{D}^{w-DM-DV}$ with element $d_{st}^{w-DM-DV}$ that captures weighted dissimilarities between individuals s and t, $s, t = 1, \ldots, 2N$ can be defined as

$$d_{\mathrm{st}}^{\mathrm{w-DM-DV}} = \sqrt{\sum_{j=1}^{n} \{\frac{w_{\mathrm{j}}^{\mathrm{m}}}{2}(x_{\mathrm{sj}}^{\mathrm{m}} - x_{\mathrm{tj}}^{\mathrm{m}})^2 + \frac{w_{\mathrm{j}}^{\mathrm{v}}}{2}(x_{\mathrm{sj}}^{\mathrm{v}} - x_{\mathrm{tj}}^{\mathrm{v}})^2\}} \quad (3)$$

2.1.2 Step 2: Calculate Pseudo-F Statistic

We apply distance-based regression originally developed in the field of ecology [31, 32] to test if DNA methylation measures in a gene is associated with the case-control status. Specifically, for each gene, we calculate a pseudo-F statistic based on the weighted epigenetic distance matrix $\mathbf{D}^{\mathrm{w-DM-DV}}$ introduced above

$$F = \frac{\mathrm{tr}(\mathbf{HGH})}{\mathrm{tr}[(\mathbf{I} - \mathbf{H})\mathbf{G}(\mathbf{I} - \mathbf{H})]} \quad (4)$$

where $\mathbf{H} = \mathbf{Y}(\mathbf{Y}^T\mathbf{Y})^{-1}\mathbf{Y}^T$ is a $2N \times 2N$ projection matrix, Υ is a $2N \times 1$ vector with case ($\Upsilon = 1$) and control ($\Upsilon = 0$) status, $\mathbf{G} = (\mathbf{I} - \frac{1}{2N}\mathbf{1}\mathbf{1}^T)\mathbf{A}(\mathbf{I} - \frac{1}{2N}\mathbf{1}\mathbf{1}^T)$ is the Gower's centered matrix, $\mathbf{A} = (a_{\mathrm{st}}) = (-\frac{1}{2}(d_{\mathrm{st}}^{\mathrm{w-DM-DV}})^2)$, $\mathbf{1}$ is a $2N$-dimensional column vector with elements 1, and $\mathbf{I}$ is a $2N \times 2N$ identity matrix. The pseudo-F statistic is used to evaluate the association between epigenetic similarity of a gene with n CpG sites and the case/control status.

2.1.3 Step 3: Assess Statistical Significance Using Permutations

To access significance of all G genes tested, we use permutation procedures, where we randomly shuffle cases ($\Upsilon = 1$) and controls ($\Upsilon = 0$) and repeat Steps 1–2 on the permuted data. In order to have more granular p-values, we pool pseudo-F statistics of all G genes from all permutations, as well as those from the observed data, to compute the empirical p-value [34]. We repeat the permutation procedure 999 times, and calculate the empirical p-value for gene g, $g = 1, \ldots, G$, as follows

$$P_g = \frac{\sum_{g'=1}^{G}\left\{I\left(F_{g',\mathrm{obs}} \geq F_{g,\mathrm{obs}}\right) + \sum_{\mathrm{perm}=1}^{999} I\left(F_{g',\mathrm{perm}} \geq F_{g,\mathrm{obs}}\right)\right\}}{G \times (1 + 999)}. \quad (5)$$

In the real data application, we have $G = 19{,}271$ genes, which helps to have high resolution gene-level empirical p-values.

2.2 Comparison Methods

We compare the performance of the proposed method $\mathbf{D}^{\mathrm{w-DM-DV}}$ that considers site-level weights for mean and variance signals to that of several comparison methods, including the weighted distance-based methods that consider mean signals only $\mathbf{D}^{\mathrm{w-DM}}$ or variance signals only $\mathbf{D}^{\mathrm{w-DV}}$, and distance-based methods without weights that consider both mean and variance signals $\mathbf{D}^{\mathrm{DM-DV}}$, mean signals only $\mathbf{D}^{\mathrm{DM}}$, variance signals only $\mathbf{D}^{\mathrm{DV}}$, and standard

EWAS methods on each CpG site with multiple comparisons adjustment of number of CpGs in a gene based on mean signals $EWAS^{DM}$ or variance signals $EWAS^{DV}$.

2.3 Simulation Study

We conducted simulation studies to evaluate type I error rate and power of the proposed method $\mathbf{D}^{W-DM-DV}$ and those of the comparison methods described above. Type I error rate is defined as the proportion of simulations with any significant genes when the data are generated under the null hypothesis of no genes are associated with case-control status. Power is defined as the proportion of simulations with any significant genes when the data are generated with one/multiple genes with one gene having signals.

2.3.1 Simulation Setup

We simulated methylation measures X of cases ($\Upsilon = 1$) and controls ($\Upsilon = 0$) at every CpG site in a gene from beta distributions:

$$X \mid \Upsilon = 0 \sim \text{Beta}(a_0, b_0)$$

$$X \mid \Upsilon = 1 \sim \text{Beta}(a_1, b_1)$$

where shape parameters a_0 and b_0 for samples in the control group were chosen based on estimates from the 50 normal tissue samples from cancer-free women in the GEO BRCA data (GSE69914), and shape parameters a_1 and b_1 for samples in the case group were chosen based on estimates from the 42 normal-adjacent tissues in the GEO BRCA data. More specifically, the average of the methylation means and standard deviations (SDs) of all CpG sites with gene information for the 50 normal tissue samples is 0.47 and 0.05, respectively. Therefore, we set $a_0 = 46.36$ and $b_0 = 52.28$ for noise CpGs such that the corresponding mean and SD of the beta distribution are 0.47 and 0.05, respectively. We generated methylation measures for 40 cases and 40 controls to mimic the size of the GEO BRCA study. We set $a_1 = a_0$ and $b_1 = b_0$ for all CpG sites in case and control groups to evaluate type I error rates. For power scenarios, we considered scenarios when signal CpGs have different mean or variance signals through varying shape parameters a_1 and b_1. We conducted 1000 simulations in each simulation setting.

Simulation settings with one gene. We first considered one gene with different number of CpGs with different signal-to-noise ratios of the CpGs. That is, the ratio between number of signal CpGs and number of noise CpGs in this gene ranges from 1:0, 1:24, 1:49, 3:47, to 5:45. We considered scenarios when signal CpGs have different mean or variance signals by varying shape parameters a_1 and b_1 such that the mean differences in methylation measures between cases and controls are 0.02, 0.04 0.06, 0.08, and 0.1, and the ratios of SDs for cases and controls are 1.25, 1.50, 1.75, 2, 2.25, and 2.50, respectively. We also investigated the impact of correlations among neighboring CpGs on the performance of the proposed method (*see* **Note 1**).

Simulation settings with ten genes. We then considered ten genes with one gene having signals when there are 25 CpGs in each of the ten genes. In the signal gene, we set one CpG to have mean or/and variance signals with different effect sizes. Here we test for the global null and consider a simulation study to be significant if any gene is significant after Bonferroni adjustment for testing ten genes. The empirical p-value for each gene is calculated using formula 5, where $G = 10$.

Simulation settings with outliers. Since epigenetic field defects are often characterized by increased variation in DNA methylation due to a few outlier normal-adjacent tissue samples [3, 21], we considered simulation scenarios with outlier samples. Here we only considered one gene with 50 CpGs for illustration purposes. We considered two signal-to-noise ratios in this gene to be either 5:45 or 10:40. We set 10%, 15%, or 20% of cases to be outlier samples with DNA methylation alterations at some signal CpGs, while the rest cases have the same methylation measures as controls at those signal CpGs when different outlier samples could have DNA methylation alterations at different signal CpGs, For each signal CpG, we generated methylation measures X for cases from a mixture distribution $X = (1 - Z)X_1 + ZX_2$, and methylation measures for controls from $X_1 \sim \text{Beta}(a_0, b_0)$. Specifically, at each signal CpG, we randomly assigned 40 cases to be either outlier samples ($Z = 1$) or nonoutlier samples ($Z = 0$) by $Z \sim \text{Bernoulli}(p)$, where p is the probability of any case being an outlier sample. We then generated methylation measures of outlier samples from $X_2 \sim \text{Beta}(a_2, b_2)$ and nonoutlier samples from $X_1 \sim \text{Beta}(a_0, b_0)$.

2.3.2 Simulation Results

Type I error rate. Type I error rates are well controlled at the 0.05 significance level in settings with one gene and ten genes after Bonferroni adjustment for multiple comparisons (Table 1), respectively.

Power for simulation settings with one gene. Power results for simulation settings with one gene are summarized in Fig. 1. When there are only mean signals at signal CpGs, $\mathbf{D}^{\text{w-DV}}$, $\mathbf{D}^{\text{DV}}$ and EWAS^{DV} that consider variance signals only do not have any power as expected. When there is only one CpG in the gene, the nonweighted distance-based methods are the same as the weighted versions, as well as the EWAS method as expected. When there is one signal CpG and increasing number of noise CpGs in the gene, power of $\mathbf{D}^{\text{DM}}$ decreases drastically while power of the weighted version $\mathbf{D}^{\text{w-DM}}$ are well maintained. This suggests that incorporating weights to CpGs indeed helps to up-weight signal CpGs and down-weight noise CpGs in constructing the distance matrix, thus improves the performance. When the size of a gene, that is, number of CpGs in a gene, is fixed, among which when the number of signal CpGs increases, power of $\mathbf{D}^{\text{w-DM}}$ increases much slower than that

Table 1
Type I error rates

Methods	One gene 1 CpG[b]	25 CpGs	50 CpGs	10 genes[a] 25 CpGs
$\mathbf{D}^{w-DM-DV}$	0.044	0.044	0.037	0.050
$\mathbf{D}^{w-DM}$	0.046	0.032	0.048	0.053
$\mathbf{D}^{w-DV}$	0.048	0.056	0.048	0.049
$\mathbf{D}^{DM-DV}$	0.044	0.052	0.045	0.054
$\mathbf{D}^{DM}$	0.046	0.043	0.041	0.057
$\mathbf{D}^{DV}$	0.044	0.052	0.054	0.045
EWAS^{DM}	0.046	0.030	0.039	0.050
EWAS^{DV}	0.044	0.047	0.040	0.037

[a]Type I error rates after Bonferroni adjustment for ten genes
[b]Number of CpG sites in a gene.

of $\mathbf{D}^{DM}$ while $\mathbf{D}^{w-DM}$ always has greater power than that of $\mathbf{D}^{DM}$. This implies that adding weights is most effective when a small percent of CpGs in a gene are signals. Similar power patterns are observed between weighted and nonweighted versions of the distance-based methods that consider both mean and variance signals, $\mathbf{D}^{w-DM-DV}$ and $\mathbf{D}^{DM-DV}$. We also notice that $\mathbf{D}^{w-DM-DV}$ is slightly less powerful than $\mathbf{D}^{w-DM}$ because the overall mean signals are diluted by the inclusion of pseudosites for variance when there are only mean signals in the data. Moreover, $\mathbf{D}^{w-DM}$ slightly outperform EWAS^{DM} when there are several signal CpGs. This is because the distance-based method has the advantage to accumulate weak signals and thus boost the overall power.

Similar power patterns are observed when signal CpGs are set to have variance signals only. $\mathbf{D}^{w-DM}$, $\mathbf{D}^{DM}$ and EWAS^{DM} that consider mean signals only do not have any power, and the weighted distance-based methods outperform the nonweighted versions in the presence of noise CpGs, and $\mathbf{D}^{w-DV}$ performs better than $\mathbf{D}^{w-DM-DV}$, and $\mathbf{D}^{w-DV}$ outperforms EWAS^{DV} when there are several signal CpGs.

Power for simulation settings with ten genes. Power results for simulation settings with ten genes are summarized in Fig. 2. When signal CpGs have either mean or variance signals, we observed similar patterns as in the simulation settings with one gene. When signal CpGs have nonnegligible mean signals and variance signals ranging from weak to strong, $\mathbf{D}^{w-DM-DV}$ performs the best when variance signals are also weak to moderate as expected. This confirms that the potential area of usage for distance-based methods to

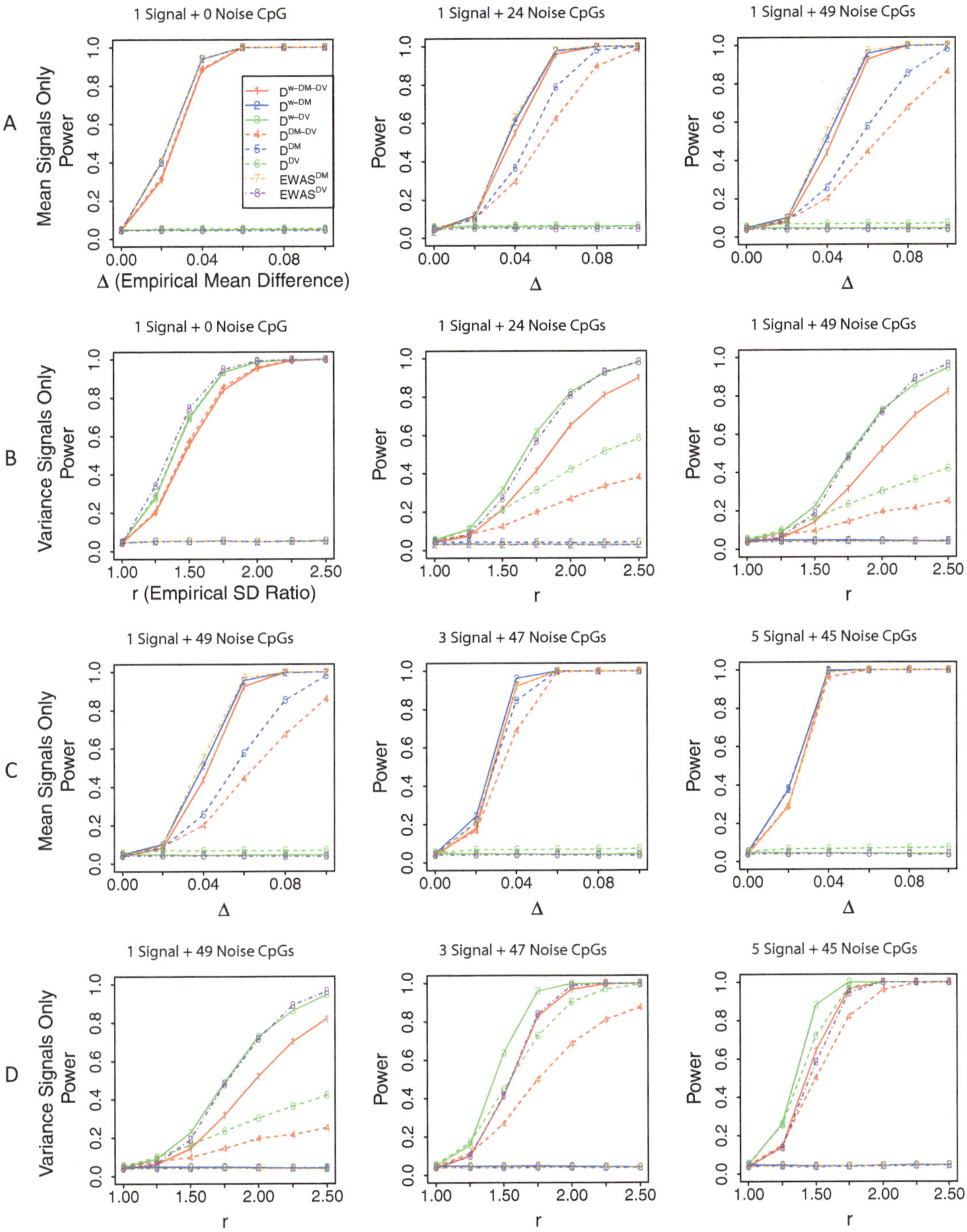

Fig. 1 Power results for simulation settings with one gene. The signal gene has one signal CpG and increasing number of total CpGs, that is, decreasing signal-to-noise ratios from 1:0, 1:24–1:49 (panel **A** for mean signals only, panel **B** for variance signals only), or with a fixed total number of CpGs 50 and increasing signal-to-noise ratios from 1:49, 3:47, to 5:45 (panel **C** for mean signals only, panel **D** for variance signals only)

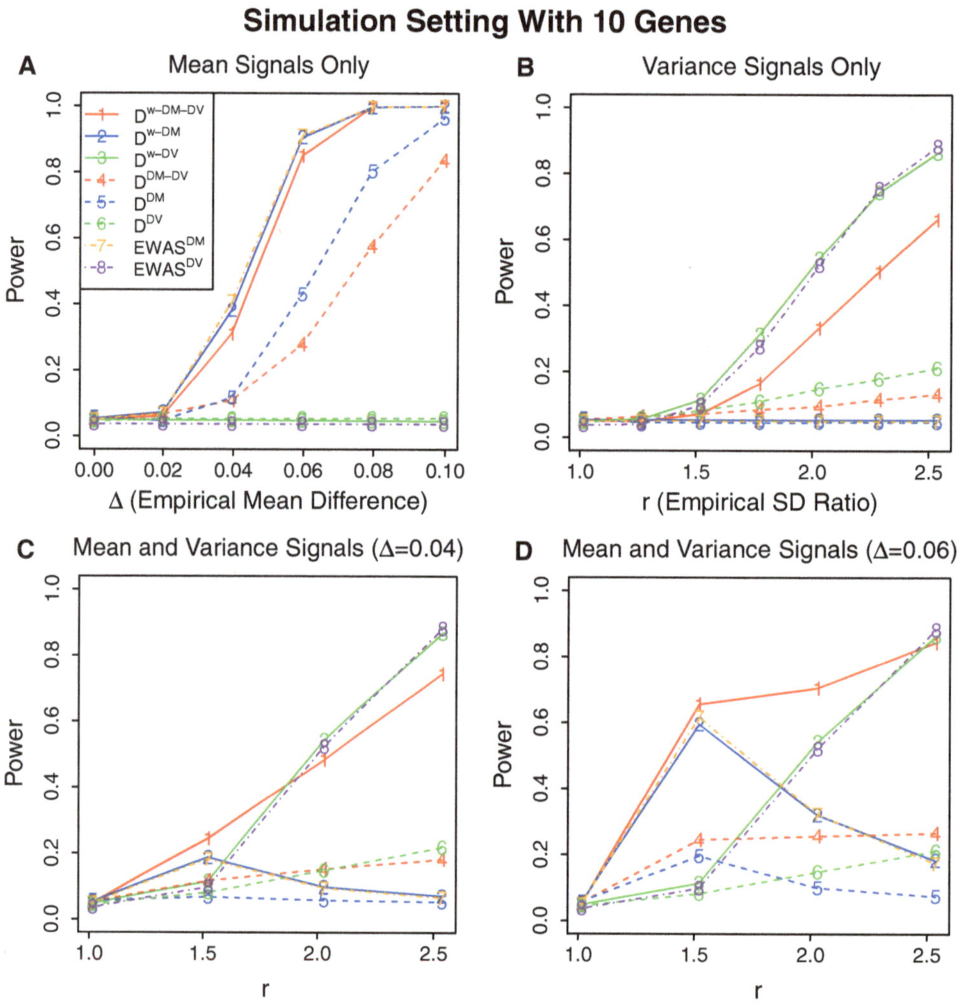

Fig. 2 Power results for simulation settings with ten genes. We set each gene to have 25 CpGs and only one gene to have signals. The signal gene has 1 signal CpG and 24 noise CpGs, with signal CpG having mean signal only (panel **A**), variance signal only (panel **B**), and mean and variance signals with different sizes of mean signals (panels **C** and **D**)

be most effective is when there are weak signals that could be accumulated to boost the study power. When there are very strong signals at some sites, any methods will perform well. One observation that we need to point out is, powers of $\mathbf{D}^{w-DM}$, $\mathbf{D}^{DM}$ and EWAS^{DM} that only consider mean signals actually decrease as variance signals increase when mean signals exist. This is due to the fact that we worked on the standardized data in $\mathbf{X}^{mv} = [\mathbf{X}^m, \mathbf{X}^v]$, and the effect sizes of mean signals (standardized mean difference) decrease as the effect sizes of variance signals (ratio of standard deviation for cases and controls) increase after standardization.

Power for simulation settings with outliers. Power results for simulation settings with outlier samples are summarized in Fig. 3. We observe that power of all methods increases as the signal-to-

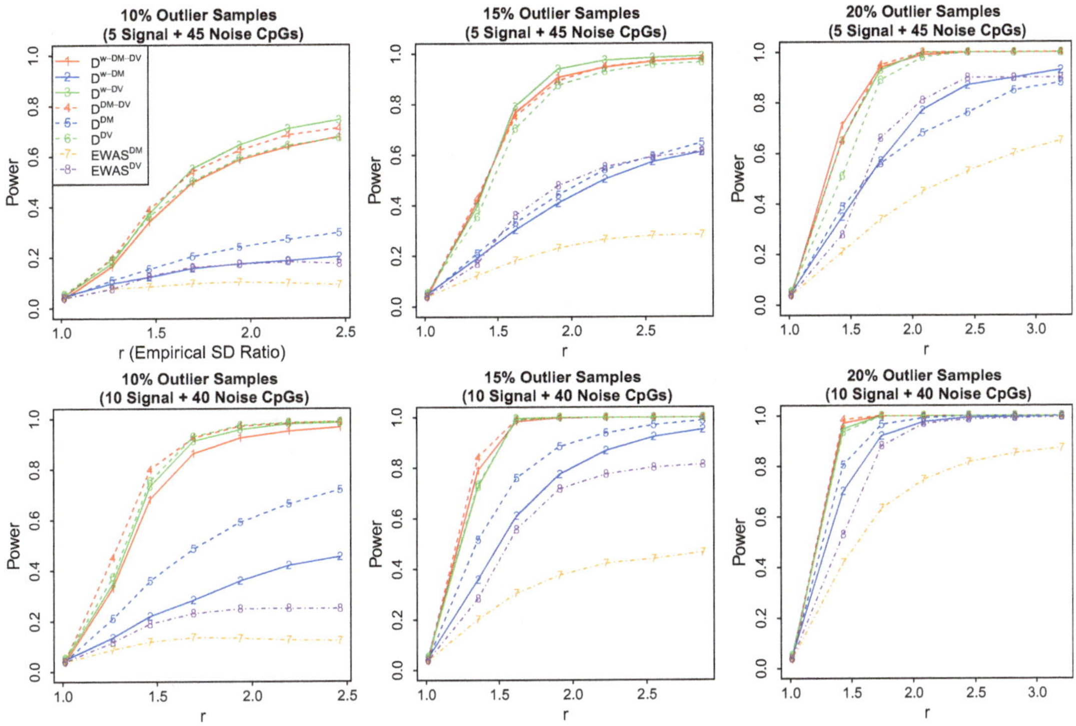

Fig. 3 Power results for simulation settings with outlier samples. We set to have 10%, 15% and 20% outlier samples and two different signal-to-noise ratios 5:45 and 10:40

noise ratio increases and as the proportion of outlier samples increases as expected, and distance-based methods outperform non–distance-based methods while EWAS^{DM} and EWAS^{DV} have very little power when there are only 10% outlier samples. Among distance-based methods, $\mathbf{D}^{\text{w}-\text{DM}}$ and $\mathbf{D}^{\text{DM}}$ that consider mean signals only have lower power compare to other methods as the mean signals introduced by a few outlier samples are usually too weak to be detected by methods that consider mean signals only. On the other hand, $\mathbf{D}^{\text{DM}-\text{DV}}$ that considers both mean and variance signals outperforms methods that consider variance signals only, $\mathbf{D}^{\text{DV}}$.The two weighted distance-based methods $\mathbf{D}^{\text{w}-\text{DM}-\text{DV}}$ and $\mathbf{D}^{\text{w}-\text{DV}}$ are among the best performed methods consistently. This implies the superiority of $\mathbf{D}^{\text{w}-\text{DM}-\text{DV}}$ in the presence of weak signals in both DM and DV.

2.4 Real Data Application

We applied the proposed method $\mathbf{D}^{\text{w}-\text{DM}-\text{DV}}$ and all the comparison methods to two GEO 450K DNA methylation data of breast invasive carcinoma (BRCA) (GSE69914 and GSE67919). As we have demonstrated the superior power of $\mathbf{D}^{\text{w}-\text{DM}-\text{DV}}$ over other distance-based methods in the simulation studies, we focused on $\mathbf{D}^{\text{w}-\text{DM}-\text{DV}}$ in the real data application and compared its performance to that of the EWAS method.

In order for the two EWAS methods, $EWAS^{DM}$ and $EWAS^{DV}$, to have a fair comparison with $\mathbf{D}^{w-DM-DV}$, we first adjusted multiple comparisons for the number of CpGs in a gene by multiplying the site-level *p*-values based on DM and DV with the number of CpGs in the gene, and then selected the minimum adjusted DM and DV *p*-value across all *p*-values in the gene as the gene-level *p*-value. We refer to this method as $EWAS^{\min-P}$.

2.4.1 Discovery Analysis Using the GEO BRCA Data

We applied the proposed method $\mathbf{D}^{w-DM-DV}$ and the comparison methods to the GEO 450K DNA methylation data of normal-adjacent tissue of breast invasive carcinoma (BRCA) patients and normal tissue from independent age-matched cancer-free women (GSE69914). In the original GEO BRCA data, there are DNA methylation measures on 485,512 CpGs for 42 tumor and normal-adjacent pairs from breast cancer patients, 50 normal tissue of independent age-matched cancer-free women and 263 additional tumor tissue of independent breast cancer patients. We conducted standard quality control steps where we removed CpGs on sex chromosomes and those contain either a SNP at the CpG interrogation or at the single nucleotide extension (SBE) based on UCSC dbSNP table version 147 using the R package "IlluminaHumanMethylation450kanno.ilmn12.hg19" [35]. We also required at least 95% CpG coverage per sample and 70% sample coverage per CpG, and only kept CpGs with gene annotations. We ended up with 344,947 CpGs, covering 19,271 genes, from 42 normal-adjacent tissues, 50 normal tissues and 263 independent tumor tissues.

Since Bonferroni adjustment for multiple comparisons of the 19,271 genes is too conservative, especially with the small sample size in the GEO BRCA dataset, we used a less stringent threshold 0.0005 on empirical gene-level *p*-values obtained from the permutation procedure (Fig. 4). Our main purpose is to demonstrate the superior performance of the proposed method $\mathbf{D}^{w-DM-DV}$ over several comparison methods, especially the EWAS methods. Results using $\mathbf{D}^{w-DM-DV}$ and $EWAS^{\min-P}$ comparing 42 normal-adjacent tissues to 50 normal tissues are shown in the Manhattan plots in Fig. 4. At the 0.0005 threshold for gene-level *p*-values, $\mathbf{D}^{w-DM-DV}$ identified 21 genes (Table 2), of which 18 were previously reported to be associated with breast cancer; $EWAS^{\min-P}$ identified 14 genes (Table 3), of which 9 were previously reported to be associated with breast cancer. There are 7 overlapping genes, *TMC4*, *NAA35*, *THY1*, *CXCL6*, *KDM5A*, *FKBP4*, and *TMEM200B* that were identified by both methods. Except for the *PLS1* gene, the 7 genes uniquely identified by $EWAS^{\min-P}$ all rank very high in $\mathbf{D}^{w-DM-DV}$ results out of the 19,271 genes (Table 3). Except for the *CFTR* gene, the 14 genes uniquely identified by $\mathbf{D}^{w-DM-DV}$ also all rank very high in $EWAS^{\min-P}$ results. This

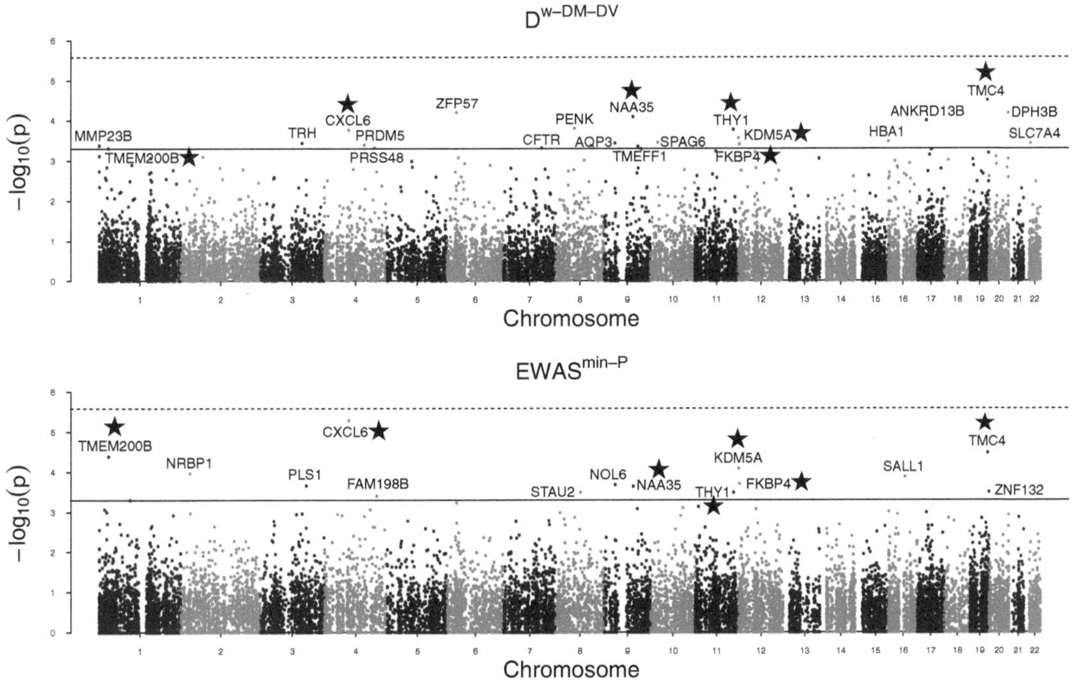

Fig. 4 Manhattan plots with results from $\mathbf{D}^{w-DM-DV}$ and EWAS^{min-P}. The solid horizontal line is the 0.0005 gene-level *p*-value threshold. The dashed horizontal line is the Bonferroni adjusted 0.05 significance level (0.05/19,271 genes = 0.0000026 adjusted gene-level *p*-value threshold). Genes annotated with stars are genes identified by both methods at the 0.0005 gene-level *p*-value threshold

suggests an overall good consistency between results of $\mathbf{D}^{w-DM-DV}$ and EWAS^{min-P}. At the same 0.0005 gene-level *p*-value threshold, other comparison methods $\mathbf{D}^{w-DM}$, $\mathbf{D}^{w-DV}$, $\mathbf{D}^{DM-DV}$, $\mathbf{D}^{DM}$ and $\mathbf{D}^{DV}$ identified 11, 9, 2, 6, and 4 genes, of which 6, 7, 1, 3, and 1 genes were also identified by the proposed $\mathbf{D}^{w-DM-DV}$, respectively.

We further examined the 14 and 7 genes uniquely identified by $\mathbf{D}^{w-DM-DV}$ and EWAS^{min-P}, respectively. We plotted heatmaps of the original DNA methylation measures of CpG sites on these genes for the 50 normal tissues, 42 normal-adjacent tissues together with the 42 matched tumor tissues. In general, the 14 genes uniquely identified by $\mathbf{D}^{w-DM-DV}$ are genes with multiple CpGs of weak signals, that is, weak dense signals. Moreover, some of these weak dense signals were mainly due to a few outlier normal-adjacent tissue samples, thus were missed by EWAS^{min-P}. The 7 genes uniquely identified by EWAS^{min-P} are those with just one or two CpGs with very strong signals, that is, strong sparse signals.

We then investigated the two genes, *CFTR* and *PLS1*, that were uniquely identified by $\mathbf{D}^{w-DM-DV}$ and EWAS^{min-P}, respectively, but ranked the last using the other method among all uniquely

Table 2
Twenty-one genes identified by $D^{w-DM-DV}$ at the 0.0005 gene-level *p*-value threshold using the GEO BRCA Data

Rank	Gene	# CpG	Cancer	Rank in $EWAS^{min-P}$
1	*TMC4*[a]	13	Breast cancer [37]	2
2	*ZFP57*	5	Breast cancer [38]	16
3	*DPH3B*	5	–	61
4	*NAA35*[a]	7	Breast cancer [39]	10
5	*ANKRD13B*	22	Breast cancer [40]	25
6	*PENK*	23	Breast cancer [41]	37
7	*THY1*[a]	19	Breast cancer [42]	13
8	*CXCL6*[a]	7	Breast cancer [43]	1
9	*KDM5A*[a]	2	Breast cancer [44]	4
10	*HBA1*	7	Breast cancer [45]	23
11	*SPAG6*	16	Acute myeloid leukemia [46]	170
12	*AQP3*	7	Breast cancer [47]	140
13	*TRH*	16	Breast cancer [48]	28
14	*SLC7A4*	12	Breast cancer [49]	175
15	*FKBP4*[a]	18	Breast cancer [50]	7
16	*PRDM5*	18	Breast cancer [51]	36
17	*MMP23B*	2	Breast cancer [52]	80
18	*TMEFF1*	5	Breast Cancer [53]	156
19	*PRSS48*	7	–	64
20	*CFTR*	16	Breast cancer [54]	1055
21	*TMEM200B*[a]	20	Breast cancer [55]	3

[a]Genes identified by both $\mathbf{D}^{w-DM-DV}$ and $EWAS^{min-P}$

identified genes. We similarly plotted the heatmap of the original DNA methylation measures of CpG sites in these two genes (Fig. 5a). For the *CFTR* gene that has 16 CpGs, it is clear that variation in methylation measures increases in the progression from normal tissues to normal-adjacent tissues and to tumor tissues in multiple CpGs when there are several samples among the 42 normal-adjacent tissue samples that are very different from the normal samples. On the other hand, for the *PLS1* gene that also has 16 CpGs, it was identified uniquely by $EWAS^{min-P}$ because of one signal CpG site cg00137209 (Fig. 5a), mainly due to the very small variation in the methylation measures of the normal

Table 3
Fourteen genes identified by $EWAS^{min-P}$ at the 0.0005 gene-level *p*-value threshold using the GEO BRCA Data

Rank	Gene	# CpG	Top CpG Signal[a]	Cancer	Rank in $D^{w-DM-DV}$
1	*CXCL6*[b]	7	Variance	Breast cancer [43]	11
2	*TMC4*[b]	13	Variance	Breast cancer [37]	1
3	*TMEM200B*[b]	20	Variance	Acute myeloid leukemia [56]	41
4	*KDM5A*[b]	2	Variance	Breast cancer [44]	4
5	*NRBP1*	12	Variance	Breast cancer [57]	110
6	*SALL1*	44	Variance	Breast cancer [58]	887
7	*FKBP4*[b]	18	Variance	Breast cancer [50]	32
8	*NOL6*	5	Variance	–	160
9	*PLS1*	16	Variance	Bladder cancer [59]	1069
10	*NAA35*[b]	7	Variance	Breast cancer [39]	6
11	*ZNF132*	12	Mean	Prostate cancer [60]	118
12	*STAU2*	39	Variance	Hepatocellular carcinoma [61]	666
13	*THY1*[b]	19	Variance	Breast cancer [42]	14
14	*FAM198B*	14	Variance	Breast cancer [62]	84

[a]Mean or variance tests with smaller *p*-value at the most significant CpG in a gene
[b]Genes identified by both $\mathbf{D}^{w-DM-DV}$ and $EWAS^{min-P}$

tissues. We then plotted DNA methylation measures of the top four *p*-value ranked CpGs, ranked by CpG site-level *p*-values from both mean and variance tests each after adjusting for multiple comparisons for the number of CpGs in the *CFTR* gene (Fig. 5b), which clearly shows elevated methylation levels in the progression to tumor. For the *PLS1* gene, we similarly plotted the DNA methylation measures of the top two *p*-value ranked CpGs (Fig. 5b), where the #1 ranked CpG cg00137209 is the one that shows strong variance signal due to very small variation in the methylation measures of the normal tissues, when neither CpGs showed any enrichment in methylation measures in the progression to tumor. This suggests that genes uniquely identified by $EWAS^{min-P}$ due to extreme *p*-values at one or two CpGs may not be reliable, while genes identified uniquely by $\mathbf{D}^{w-DM-DV}$ are generally characterized by multiple signal CpGs, thus are more reliable.

We also plotted the DNA methylation measures of all CpGs in these two genes *CFTR* and *PLS1*. It is again clear that almost half of the CpGs in the *CFTR* gene have weak mean signals and weak variance signals, thus missed by $EWAS^{min-P}$ due to stringent multiple comparisons adjustment. In addition, we plotted the

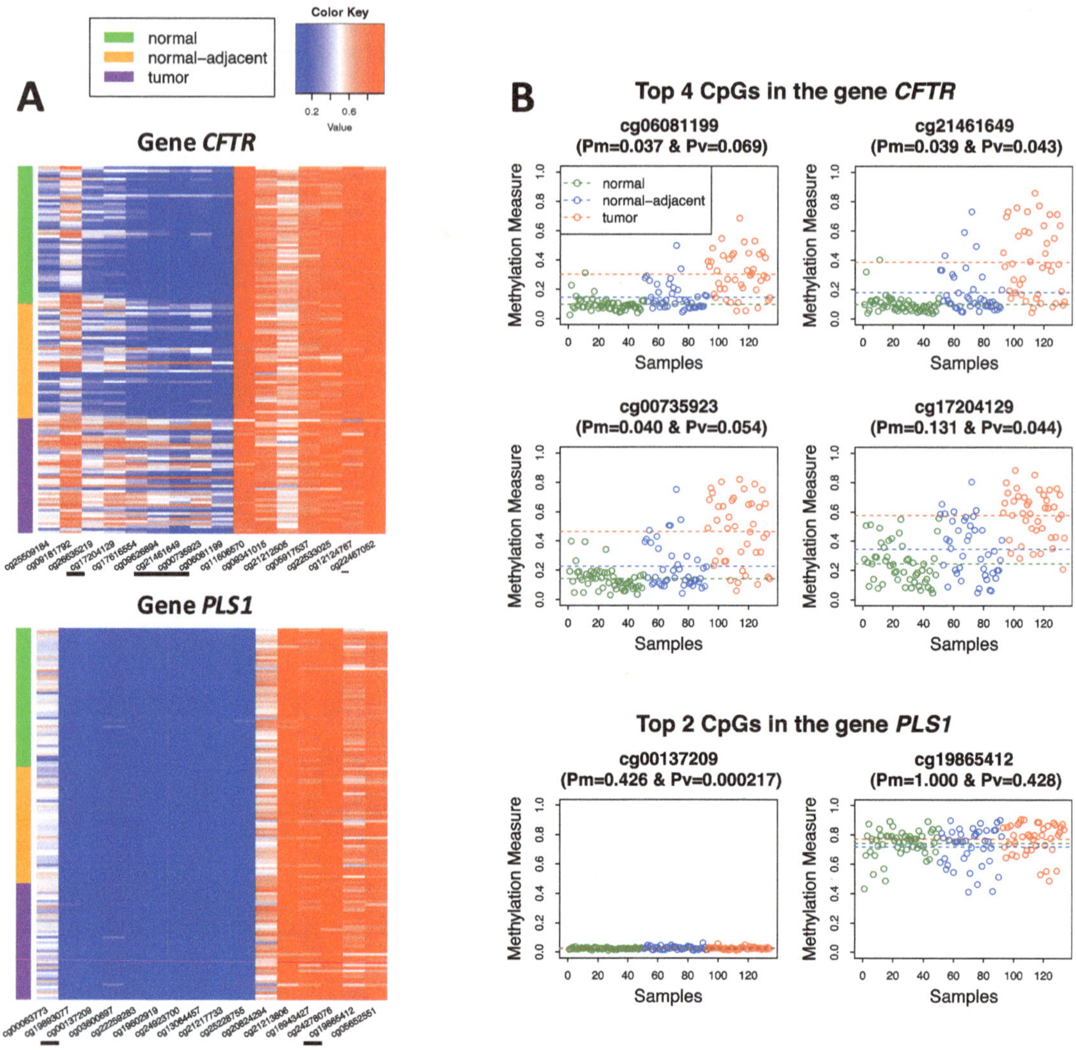

Fig. 5 (**A**) Heatmaps of DNA methylation measures of CpGs in the *CFTR* and *PLS1* genes. The CpGs underlined are the top four *p*-value ranked CpGs in the *CFTR* gene and the top two *p*-value ranked CpGs in the *PLS1* gene. (**B**) DNA methylation measures of 50 normal tissues, 42 normal-adjacent tissues and 42 matched tumors of the top four *p*-value ranked CpGs in the *CFTR* gene and the top two *p*-value ranked CpGs in the *PLS1* gene. Pm and Pv are *p*-values from CpG site-level mean and variance tests adjusted for multiple comparisons for the number of CpGs in the gene. The three horizontal lines represent mean methylation levels of the three groups of normal tissues, normal-adjacent tissues, and matched tumors

weighted distance matrices of the 50 normal tissues and the 42 normal-adjacent tissues for the *CFTR* gene and the *PLS1* gene. For the *CFTR* gene, we observe little variation in distances among normal samples and increased variation in distances between several pairs of normal and normal-adjacent samples, while for the *PLS1* gene, we observe no clear pattern. We also plotted the DNA methylation measures of CpGs in the *TMC4* gene that was identified by both

$\mathbf{D}^{w-DM-DV}$ and EWAS$^{\min-P}$ and ranked #1 and #2 in the two methods, respectively. There are 13 CpGs in the *TMC4* gene, three CpGs have strong variance signals when two of the three CpGs also have mean signals. Thus, the *TMC4* gene was identified by both $\mathbf{D}^{w-DM-DV}$ and EWAS$^{\min-P}$.

In our previous work on differentially methylated regions (DMRs) using the same GEO BRCA data, we identified two DMRs of epigenetic field defects using both mean and variance signals [21]. The two DMRs cover two genes, *NKX6–2* and *CCND2*, which rank #113 and #359 in the $\mathbf{D}^{w-DM-DV}$ results. Further investigation revealed that the two DMRs only cover part of the two genes. We therefore broke down the two genes into smaller parts so that there is one part that covers exactly the identified DMR. We then treated these smaller parts as individual regions and repeated $\mathbf{D}^{w-DM-DV}$ across the whole genome. The rank of the *NKX6-2* part that matches with the DMR moved up to #90 from #113 while the other two parts rank #107 and #4855, respectively. The rank of the *CCND2* part that matches with the other DMR moved up to #154 from #359 and the other part ranks #1116. Overall, the two DMR-covered genes previously identified as epigenetic field defects also rank on top in the results of $\mathbf{D}^{w-DM-DV}$. This suggests that we may combine DMR detection techniques with distance-based methods to first better define "regions of interest" using DMR ideas and then assess significance more powerfully with distance-based methods.

We also investigated the relation between the number of CpGs in a gene and the probability that the gene is selected, where we binned genes based on their sizes and calculated the selection probability of a gene in a bin as the proportion of genes identified out of all genes in the bin. We plotted the selection probabilities against gene sizes and found that the selection probabilities for different versions of the distance-based methods and EWAS$^{\min-P}$ method are not systematically affected by gene sizes.

2.4.2 Validation of the Identified Epigenetic Field Defects in the GEO BRCA Data

We further validated the 21 genes of epigenetic field defects identified by $\mathbf{D}^{w-DM-DV}$ through comparing methylation measures of the 21-gene-covered CpGs between 263 independent tumor tissues and 42 normal-adjacent tissues to examine if the methylation levels at these CpGs exhibit progression to tumor. Specifically, we performed the two-sample *t*-test at each of these CpGs and plotted the $-\log_{10}(p$ - value) from the two comparisons, 50 normal tissues vs. 42 normal-adjacent tissues and 42 normal-adjacent tissues vs. 263 tumor tissues. In general, the majority of these CpGs show more significant signals in the progression from normal tissues to normal-adjacent tissues to tumor.

2.4.3 Replication Analysis Using an Independent Data of Normal Tissues

As epigenetic field defects identified in one set of normal vs. -normal-adjacent comparison maybe driven by a few "outlier" normal-adjacent samples [3, 4, 21], different epigenetic field defects could be identified in a different set of normal vs. normal-adjacent comparison that are driven by different "outlier" normal-adjacent samples. Therefore, we propose to conduct a replication analysis that uses the same normal-adjacent tissue samples but compare to an independent data of normal samples. We used 450K DNA methylation data of 18 normal tissue of 18 breast reduction mammoplasty subjects (GSE67919) [36]. The original data have methylation measures on 485,577 CpG sites. We followed the same quality control steps as for the discovery GEO BRCA data (GSE69914) and kept the same CpG sites for comparison purposes. We ended up with 344,947 CpGs, covering 19,271 genes, from 18 normal tissues. We then compared these normal samples to the same 42 normal-adjacent tissues from the GEO BRCA data in a replication analysis.

At the same 0.0005 threshold for gene-level p-values, 7 out of the 21 previously identified genes with epigenetic field defects in the discovery analysis using the GEO BRCA data were replicated by $\mathbf{D}^{\mathrm{W-DM-DV}}$. The 7 genes are *DPH3B*, *NAA35*, *ANKRD13B*, *CXCL6*, *FKBP4*, *PRSS48*, and *CFTR*. We similarly validated these 7 genes by comparing p-values from the two-sample t-tests comparing the 18 replication normal samples to the 42 GEO BRCA normal-adjacent samples and p-values from the two-sample t-tests comparing the 42 GEO BRCA normal-adjacent samples to the 263 independent GEO BRCA tumor samples. All 7 genes, except the *NAA35* and *FKBP4*, exhibit progression to tumor. We also summarized the replication analysis results for the comparison methods (*see* **Note 2**).

To investigate our hypothesis that different epigenetic field defects maybe identified when comparing normal samples to a different set of normal-adjacent samples, we obtained a new set of BRCA normal-adjacent samples ($n = 90$) from the Cancer Genome Atlas (TCGA) project together with their matched tumor samples ($n = 90$). We plotted DNA methylation measures of CpGs in the 7 replicated genes of the 18 replication normal samples, the 50 discovery GEO BRCA normal samples, the 42 discovery GEO BRCA normal-adjacent samples, the 42 discovery GEO BRCA matched tumor samples, and the 90 TCGA normal-adjacent samples and the 90 TCGA matched tumor samples. It is clear that methylation patterns of the TCGA normal-adjacent tissues are very different from that of the discovery GEO BRCA normal-adjacent tissues in most of these CpGs. This supports our hypothesis that methylation patterns can be very different in different precancer tissues (using normal-adjacent tissue as a surrogate) thus different epigenetic field defects maybe identified when normal samples are compared to different sets of precancer tissues.

2.5 Discussion

In this study, we developed a weighted epigenetic distance-based method $\mathbf{D}^{\mathrm{w-DM-DV}}$ that accumulates both DM (mean) and DV (variance) signals across CpGs in a gene or a genetic region. One known advantage of distance-based methods is, there is no need to preselect outcome-associated features, avoiding the potential to misscreen features with weak signals. In our proposed weighted epigenetic distance-based method $\mathbf{D}^{\mathrm{w-DM-DV}}$, we used CpG site-level association strengths as weights for individual CpGs aiming to up-weight signal CpGs and down-weight noise CpGs. If the feature preselection step could be conducted perfectly, it is equivalent to the case when weight "0" is correctly assigned to noise CpGs and weight "1" is correctly assigned to signal CpGs. Results from simulation studies suggest that when the signal-to-noise ratio in a gene decreases, power of nonweighted epigenetic distance-based methods decreased drastically, while power of the weighted version was well maintained. This suggests that incorporating CpG-site-level association strengths as weights for individual CpGs indeed help to up-weight signal CpGs and down-weight noise CpGs, thus improve the overall study performance. Simulation results also suggest that the weighted epigenetic distance-based methods will be most effective when applied to genes or genetic regions with a small percentage of CpGs having weak signals. This makes the detection of epigenetic field defects, that is, early epigenetic alterations that are usually infrequent across samples and identifiable as outlier samples, the ideal application of the proposed method $\mathbf{D}^{\mathrm{w-DM-DV}}$. Using the GEO BRCA 450K DNA methylation data, $\mathbf{D}^{\mathrm{w-DM-DV}}$ identified 21 genes with epigenetic field defects, when 7 out of the 21 genes overlap with the genes identified by $\mathrm{EWAS}^{\min-P}$. Majority of the genes uniquely identified by $\mathbf{D}^{\mathrm{w-DM-DV}}$ were previously reported to be associated with breast cancer. Most of the genes uniquely identified by $\mathrm{EWAS}^{\min-P}$ also ranked on top in the $\mathbf{D}^{\mathrm{w-DM-DV}}$ results except for the *PLS1* gene. However, further investigations suggested that the *PLS1* gene may not be a real epigenetic field defect. On the other hand, most of the genes uniquely identified by $\mathbf{D}^{\mathrm{w-DM-DV}}$ also ranked on top in the $\mathrm{EWAS}^{\min-P}$ results except for the *CFTR* gene, in which the enrichment in the progression to breast cancer was confirmed in further analyses. This suggests that genes identified by $\mathbf{D}^{\mathrm{w-DM-DV}}$, which are generally characterized by multiple signal CpGs, are more reliable. It is worth noticing that the two DMR-covered genes identified in our previous work [21] also ranked on top in the $\mathbf{D}^{\mathrm{w-DM-DV}}$ results. We validated the identified epigenetic field defects by showing a progression to tumor in an independent dataset of tumor tissues. We also conducted a replication analysis by comparing the same set of normal-adjacent tissues to an independent set of normal tissues, and found that 7 out of the 21 genes of epigenetic field defects identified by $\mathbf{D}^{\mathrm{w-DM-DV}}$ in the discovery analysis were replicated.

In general, distance-based methods have a better performance than that of site-level EWAS methods when site-level signals are weak. As discussed in our previous work [21] and work of others [3, 4], epigenetic field defects are often characterized by increased variation in DNA methylation measures due to a few outlier normal-adjacent tissue samples. So the site-level EWAS methods are usually underpowered due to small mean differences as well as stringent multiple comparisons adjustment. Distance-based methods accumulate weak signals to improve power. Distance-based methods are flexible and can be applied to a CpG site, a gene, a pathway, or an entire genome. A closer investigation on what we identified in our previous work [21] in DMR detection and the current work suggests that we may take advantages of the techniques in DMR detection and combine that with distance-based methods in future works to more efficiently identify regions of epigenetic field defects.

In summary, we proposed a new weighted distance-based method $\mathbf{D}^{\mathrm{w-DM-DV}}$ that considers both DM and DV in DNA methylation and incorporates site-level association strengths as weights on individual CpGs to up-weight signal CpGs and down-weight noise CpGs to further boost the overall study power. The $\mathbf{D}^{\mathrm{w-DM-DV}}$ method is especially powerful in detecting epigenetic field defects when methylation alterations between normal tissues and normal-adjacent tissues are usually minimum.

3 Notes

1. Since neighboring CpGs are known to be correlated, we considered simulation scenarios that assume an AR(1) correlation among CpGs in a gene with a correlation coefficient 0.5. The power patterns are very similar to those observed in simulations ignoring correlations among CpG sites. This implies that the correlations among neighboring CpGs do not have much impact on the performance of the proposed distance-based methods.
2. At the same 0.0005 gene-level p-value threshold, the comparison methods $\mathrm{EWAS}^{\min-P}$, $\mathbf{D}^{\mathrm{w-DM}}$, $\mathbf{D}^{\mathrm{w-DV}}$, $\mathbf{D}^{\mathrm{DM-DV}}$, $\mathbf{D}^{\mathrm{DM}}$, and $\mathbf{D}^{\mathrm{DV}}$ identified 14, 11, 9, 2, 6, and 4 genes in the discovery analysis, of which 11, 4, 1, 2, 2, and 1 genes were replicated in the replication analysis, respectively.

References

1. Katsurano M, Niwa T, Yasui Y, Shigematsu Y, Yamashita S, Takeshima H, Lee M, Kim Y, Tanaka T, Ushijima T (2012) Early-stage formation of an epigenetic field defect in a mouse colitis model, and non-essential roles of T-and B-cells in DNA methylation induction. Oncogene 31(3):342
2. Bernstein C, Nfonsam V, Prasad AR, Bernstein H (2013) Epigenetic field defects in progression to cancer. World J Gastrointest Oncol 5 (3):43
3. Teschendorff AE, Gao Y, Jones A, Ruebner M, Beckmann MW, Wachter DL, Fasching PA, Widschwendter M (2016) DNA methylation outliers in normal breast tissue identify field defects that are enriched in cancer. Nat Commun 7:Article number: 104
4. Teschendorff AE, Jones A, Widschwendter M (2016) Stochastic epigenetic outliers can define field defects in cancer. BMC Bioinformatics 17 (1):1
5. Baylin SB, Esteller M, Rountree MR, Bachman KE, Schuebel K, Herman JG (2001) Aberrant patterns of DNA methylation, chromatin formation and gene expression in cancer. Hum Mol Genet 10(7):687–692
6. Fahrner JA, Eguchi S, Herman JG, Baylin SB (2002) Dependence of histone modifications and gene expression on DNA hypermethylation in cancer. Cancer Res 62(24):7213–7218
7. Jones PA (2012) Functions of DNA methylation: islands, start sites, gene bodies and beyond. Nat Rev Genet 13(7):484–492
8. Phillips T (2008) The role of methylation in gene expression. Nat Educ 1(1):116
9. Das PM, Singal R (2004) DNA methylation and cancer. J Clin Oncol 22(22):4632–4642
10. Ehrlich M (2002) DNA methylation in cancer: too much, but also too little. Oncogene 21 (35):5400–5413
11. Esteller M, Herman JG (2002) Cancer as an epigenetic disease: DNA methylation and chromatin alterations in human tumours. J Pathol 196(1):1–7
12. Kulis M, Esteller M (2010) DNA methylation and cancer. Adv Genet 70(10):27–56
13. Koukoura O, Spandidos DA, Daponte A, Sifakis S (2014) DNA methylation profiles in ovarian cancer: implication in diagnosis and therapy. Mol Med Rep 10(1):3–9
14. Baylin SB (2005) DNA methylation and gene silencing in cancer. Nat Clin Pract Oncol 2: S4–S11
15. Curradi M, Izzo A, Badaracco G, Landsberger N (2002) Molecular mechanisms of gene silencing mediated by DNA methylation. Mol Cell Biol 22(9):3157–3173
16. Herman JG, Baylin SB (2003) Gene silencing in cancer in association with promoter hypermethylation. N Engl J Med 349 (21):2042–2054
17. Robertson KD (2005) DNA methylation and human disease. Nat Rev Genet 6(8):597–610
18. Eden A, Gaudet F, Waghmare A, Jaenisch R (2003) Chromosomal instability and tumors promoted by DNA hypomethylation. Science 300(5618):455–455
19. Feinberg AP, Tycko B (2004) The history of cancer epigenetics. Nat Rev Cancer 4 (2):143–153
20. Jaenisch R, Bird A (2003) Epigenetic regulation of gene expression: how the genome integrates intrinsic and environmental signals. Nat Genet 33:245–254
21. Wang Y, Teschendorff AE, Widschwendter M, Wang S (2017) Accounting for differential variability in detecting differentially methylated regions. Brief Bioinform 20(1):47–57
22. Tusher VG, Tibshirani R, Chu G (2001) Significance analysis of microarrays applied to the ionizing radiation response. Proc Natl Acad Sci 98(9):5116–5121
23. Smyth GK (2004) Linear models and empirical bayes methods for assessing differential expression in microarray experiments. Stat Appl Genet Mol Biol 3(1):1–25
24. Wettenhall JM, Smyth GK (2004) limmaGUI: a graphical user interface for linear modeling of microarray data. Bioinformatics 20 (18):3705–3706
25. Wilcoxon F (1945) Individual comparisons by ranking methods. Biometrics 1(6):80–83
26. Hansen KD, Timp W, Bravo HC, Sabunciyan S, Langmead B, McDonald OG, Wen B, Wu H, Liu Y, Diep D (2011) Increased methylation variation in epigenetic domains across cancer types. Nat Genet 43(8):768–775
27. Ho JW, Stefani M, dos Remedios CG, Charleston MA (2008) Differential variability analysis of gene expression and its application to human diseases. Bioinformatics 24(13):i390–i398
28. Phipson B, Oshlack A (2014) DiffVar: a new method for detecting differential variability with application to methylation in cancer and aging. Genome Biol 15(9):1
29. Zapala MA, Schork NJ (2006) Multivariate regression analysis of distance matrices for

testing associations between gene expression patterns and related variables. Proc Natl Acad Sci 103(51):19430–19435

30. Wessel J, Schork NJ (2006) Generalized genomic distance-based regression methodology for multilocus association analysis. Am J Hum Genet 79(5):792–806
31. McArdle BH, Anderson MJ (2001) Fitting multivariate models to community data: a comment on distance-based redundancy analysis. Ecology 82(1):290–297
32. Anderson MJ (2001) A new method for non-parametric multivariate analysis of variance. Austral Ecol 26(1):32–46
33. Han F, Pan W (2010) Powerful multi-marker association tests: unifying genomic distance-based regression and logistic regression. Genet Epidemiol 34(7):680–688
34. Friedman J, Hastie T, Tibshirani R (2001) The elements of statistical learning, vol 1. Springer series in statistics. Springer, New York, NY
35. Hansen K (2015) Illumina human methylation 450kanno. ilmn12. hg19: annotation for illumina's 450k methylation arrays. R package, version 02 1
36. Hair BY, Xu Z, Kirk EL, Harlid S, Sandhu R, Robinson WR, Wu MC, Olshan AF, Conway K, Taylor JA (2015) Body mass index associated with genome-wide methylation in breast tissue. Breast Cancer Res Treat 151 (2):453–463
37. Krijgsman O, Roepman P, Zwart W, Carroll JS, Tian S, de Snoo FA, Bender RA, Bernards R, Glas AM (2012) A diagnostic gene profile for molecular subtyping of breast cancer associated with treatment response. Breast Cancer Res Treat 133(1):37–47
38. Tada Y, Yamaguchi Y, Kinjo T, Song X, Akagi T, Takamura H, Ohta T, Yokota T, Koide H (2015) The stem cell transcription factor ZFP57 induces IGF2 expression to promote anchorage-independent growth in cancer cells. Oncogene 34(6):752–760
39. Abu-Asab M, Abu-Asab N, Loffredo C, Clarke R, Amri H (2013) Identifying early events of gene expression in breast cancer with systems biology phylogenetics. Cytogenet Genome Res 139(3):206–214
40. Jönsson G, Staaf J, Vallon-Christersson J, Ringnér M, Holm K, Hegardt C, Gunnarsson H, Fagerholm R, Strand C, Agnarsson BA (2010) Genomic subtypes of breast cancer identified by array-comparative genomic hybridization display distinct molecular and clinical characteristics. Breast Cancer Res 12(3):R42
41. Legendre C, Gooden GC, Johnson K, Martinez RA, Liang WS, Salhia B (2015) Whole-genome bisulfite sequencing of cell-free DNA identifies signature associated with metastatic breast cancer. Clin Epigenetics 7(1):100
42. Lehmann BD, Bauer JA, Chen X, Sanders ME, Chakravarthy AB, Shyr Y, Pietenpol JA (2011) Identification of human triple-negative breast cancer subtypes and preclinical models for selection of targeted therapies. J Clin Invest 121(7):2750
43. Bièche I, Chavey C, Andrieu C, Busson M, Vacher S, Le Corre L, Guinebretière J-M, Burlinchon S, Lidereau R, Lazennec G (2007) CXC chemokines located in the 4q21 region are up-regulated in breast cancer. Endocr Relat Cancer 14(4):1039–1052
44. Hou J, Wu J, Dombkowski A, Zhang K, Holowatyj A, Boerner JL, Yang Z-Q (2012) Genomic amplification and a role in drug-resistance for the KDM5A histone demethylase in breast cancer. Am J Transl Res 4(3):247
45. Wolf I, Bose S, Desmond JC, Lin BT, Williamson EA, Karlan BY, Koeffler HP (2007) Unmasking of epigenetically silenced genes reveals DNA promoter methylation and reduced expression of PTCH in breast cancer. Breast Cancer Res Treat 105(2):139–155
46. Steinbach D, Schramm A, Eggert A, Onda M, Dawczynski K, Rump A, Pastan I, Wittig S, Pfaffendorf N, Voigt A (2006) Identification of a set of seven genes for the monitoring of minimal residual disease in pediatric acute myeloid leukemia. Clin Cancer Res 12 (8):2434–2441
47. Cao X-C, Zhang W-R, Cao W-F, Liu B-W, Zhang F, Zhao H-M, Meng R, Zhang L, Niu R-F, Hao X-S (2013) Aquaporin3 is required for FGF-2-induced migration of human breast cancers. PLoS One 8(2):e56735
48. Nicolau M, Levine AJ, Carlsson G (2011) Topology based data analysis identifies a subgroup of breast cancers with a unique mutational profile and excellent survival. Proc Natl Acad Sci 108(17):7265–7270
49. Xia T-S, Wang G-Z, Ding Q, Liu X-A, Zhou W-B, Zhang Y-F, Zha X-M, Du Q, Ni X-J, Wang J (2012) Bone metastasis in a novel breast cancer mouse model containing human breast and human bone. Breast Cancer Res Treat 132(2):471–486
50. Yang WS, Moon H-G, Kim HS, Choi E-J, Yu M-H, Noh D-Y, Lee C (2011) Proteomic approach reveals FKBP4 and S100A9 as potential prediction markers of therapeutic response to neoadjuvant chemotherapy in patients with breast cancer. J Proteome Res 11 (2):1078–1088

51. Deng Q, Huang S (2004) PRDM5 is silenced in human cancers and has growth suppressive activities. Oncogene 23(28):4903
52. Giussani M, Merlino G, Cappelletti V, Tagliabue E, Daidone MG (2015) Tumor-extracellular matrix interactions: Identification of tools associated with breast cancer progression. Sem Cancer Biol 35:3–10
53. Matise LA, Palmer TD, Ashby WJ, Nashabi A, Chytil A, Aakre M, Pickup MW, Gorska AE, Zijlstra A, Moses HL (2012) Lack of transforming growth factor-β signaling promotes collective cancer cell invasion through tumor-stromal crosstalk. Breast Cancer Res 14(4):R98
54. Zhang JT, Jiang XH, Xie C, Cheng H, Da Dong J, Wang Y, Fok KL, Zhang XH, Sun TT, Tsang LL (2013) Downregulation of CFTR promotes epithelial-to-mesenchymal transition and is associated with poor prognosis of breast cancer. Biochim Biophys Acta 1833(12):2961–2969
55. Stirzaker C, Zotenko E, Song JZ, Qu W, Nair SS, Locke WJ, Stone A, Armstong NJ, Robinson MD, Dobrovic A (2015) Methylome sequencing in triple-negative breast cancer reveals distinct methylation clusters with prognostic value. Nat Commun 6:5899
56. Rudenko V, Kazakova S, Tanas A, Popa A, Nemirovchenko V, Kuznetsova E, Zaletaev D, Strelnikov V (2016) Identification of aberrant DNA methylation in pediatric acute myeloid leukaemia by multiplex methylation sensitive PCR. Ann Oncol 27(Suppl_6):939
57. Wei H, Wang H, Ji Q, Sun J, Tao L, Zhou X (2015) NRBP1 is downregulated in breast cancer and NRBP1 overexpression inhibits cancer cell proliferation through Wnt/β-catenin signaling pathway. OncoTargets Ther 8:3721
58. Wolf J, Müller-Decker K, Flechtenmacher C, Zhang F, Shahmoradgoli M, Mills G, Hoheisel J, Boettcher M (2014) An in vivo RNAi screen identifies SALL1 as a tumor suppressor in human breast cancer with a role in CDH1 regulation. Oncogene 33(33):4273
59. Bi D, Ning H, Liu S, Que X, Ding K (2015) Gene expression patterns combined with network analysis identify hub genes associated with bladder cancer. Comput Biol Chem 56:71–83
60. Abildgaard MO, Borre M, Mortensen MM, Ulhøi BP, Tørring N, Wild P, Kristensen H, Mansilla F, Ottosen PD, Dyrskjøt L (2012) Downregulation of zinc finger protein 132 in prostate cancer is associated with aberrant promoter hypermethylation and poor prognosis. Int J Cancer 130(4):885–895
61. Castaneda F, Rosin-Steiner S, Jung K (2007) Functional genomics analysis of low concentration of ethanol in human hepatocellular carcinoma (HepG2) cells. Role of genes involved in transcriptional and translational processes. Int J Med Sci 4(1):28
62. Fidalgo F, Rodrigues TC, Pinilla M, Silva AG, do Socorro Maciel M, Rosenberg C, de Andrade VP, Carraro DM, Krepischi ACV (2015) Lymphovascular invasion and histologic grade are associated with specific genomic profiles in invasive carcinomas of the breast. Tumor Biol 36(3):1835–1848

Part II

Analysis and Visualization of Single-Cell and Bulk RNA-Seq Transcriptome Data

Chapter 7

Profiling Cell Type Abundance and Expression in Bulk Tissues with CIBERSORTx

Chloé B. Steen, Chih Long Liu, Ash A. Alizadeh, and Aaron M. Newman

Abstract

CIBERSORTx is a suite of machine learning tools for the assessment of cellular abundance and cell type-specific gene expression patterns from bulk tissue transcriptome profiles. With this framework, single-cell or bulk-sorted RNA sequencing data can be used to learn molecular signatures of distinct cell types from a small collection of biospecimens. These signatures can then be repeatedly applied to characterize cellular heterogeneity from bulk tissue transcriptomes without physical cell isolation. In this chapter, we provide a detailed primer on CIBERSORTx and demonstrate its capabilities for high-throughput profiling of cell types and cellular states in normal and neoplastic tissues.

Key words Deconvolution, Digital cytometry, Gene expression, scRNA-seq, Cellular heterogeneity, Tumor microenvironment

1 Introduction

Tissue composition is a major determinant of phenotypic variation and a key factor influencing disease outcomes. For example, in malignant tumors, dynamic interactions between heterogeneous subpopulations of cancer cells, stromal elements, and immune subsets differentially impact patient survival [1, 2] and can be leveraged therapeutically [3]. Common assays for studying cellular heterogeneity, such as flow cytometry and immunohistochemistry, require fluorescently labeled antibodies to mark and distinguish diverse cell types of interest. Despite their utility, these approaches are limited by the availability of highly specific antibodies and by the number of cell types that can be simultaneously assessed. While single-cell RNA sequencing (scRNA-seq) has recently emerged as a powerful strategy for resolving cellular heterogeneity at high-resolution and without prior knowledge [4–7], it remains impractical for large-scale analyses and cannot be applied clinically to formalin-fixed, paraffin embedded (FFPE) tissue biopsies.

Benjamin L. Kidder (ed.), *Stem Cell Transcriptional Networks: Methods and Protocols*, Methods in Molecular Biology, vol. 2117, https://doi.org/10.1007/978-1-0716-0301-7_7, © Springer Science+Business Media, LLC, part of Springer Nature 2020

To complement these approaches, especially in settings where tissue is limited, fixed, or challenging to disaggregate, computational methods for dissecting cellular composition directly from bulk tissue gene expression profiles (GEPs) have been previously described [8–20]. These techniques for "digital cytometry" mathematically deconvolve RNA admixtures into their component cell types, allowing significant new insights into bulk tissue expression profiles [21]. Our group previously developed CIBERSORT, an in silico strategy that employs a machine learning model to enumerate cell composition from bulk tissue GEPs [16]. Like other deconvolution techniques [12, 15, 17, 22], CIBERSORT relies on a specialized expression matrix of cell type-specific "barcode" genes, often called a "signature matrix," which provides a reference atlas of known cellular signatures for the deconvolution procedure.

To improve the performance and versatility of digital cytometry, we recently extended CIBERSORT into a new computational framework called CIBERSORTx [23]. Unlike previous methods (Table 1), CIBERSORTx can (1) leverage scRNA-seq-derived

Table 1
Overview of selected bioinformatics tools for studying cellular composition in tissue samples

Reference	Tool/method	scRNA-seq reference	Batch correction	Cell type-specific gene expression
Shen-Orr et al. [19]	csSAM	No	No	Yes
Zhong et al. [20]	DSA	No	No	Yes
Liebner et al. [15]	MMAD	No	No	Yes
Quon et al. [18]	ISOpure	No	No	Yes, 2 component sample-level cell type-specific gene expression
Li et al. [14]	TIMER	No	Yes	No
Baron et al. [22]	BSEQ-sc	Yes	No	Yes, cell type-specific differential expression
Schelker et al. [51]	CIBERSORT	Yes	No	No
Wang et al. [52]	DeMixT	No	No	Yes, 3 component sample-level cell type-specific gene expression
Wang et al. [53]	MuSiC	Yes	No	No
Frishberg et al. [54]	CPM	Yes	No	No
Newman et al. [23]	CIBERSORTx	Yes	Yes	Yes, *n* component sample-level cell type-specific gene expression

The list is restricted to digital cytometry tools that either leverage scRNA-seq reference profiles for deconvolution, handle technical variation between datasets by batch correction, or perform gene expression imputation. The tool/method column refers to the name of the tool presented in each publication or the underlying methodology when no other name is available

DSA digital sorting algorithm, *MMAD* microarray microdissection with analysis of differences, *TIMER* tumor immune estimation resource, *MuSiC* MUlti-Subject SIngle Cell deconvolution, *CPM* cellular population mapping

reference profiles for bulk tissue dissection, (2) overcome technical variation across different platforms (e.g., scRNA-seq, bulk RNA-seq, microarrays) and tissue preservation techniques (e.g., fresh/frozen vs. FFPE), and (3) digitally "purify" cell-type-specific expression profiles from bulk tissues without physical cell isolation. In this chapter, we describe each of these new features and demonstrate how CIBERSORTx can be broadly applied to dissect cellular heterogeneity from complex tissues without antibodies, tissue dissociation, or viable cells.

2 Materials

CIBERSORTx is available as an online tool with a user-friendly interface that does not require prior bioinformatics training or programming experience (http://cibersortx.stanford.edu). Its key functionalities are divided into three main components (Fig. 1):

1. Creation of a custom signature matrix from scRNA-seq or bulk sorted RNA-seq (or microarray) data.
2. Estimation of cell type composition in bulk tissue GEPs.
3. Imputation of cell type-specific expression profiles from bulk tissue GEPs.

In the following sections, we describe each component in detail and provide guidance on how to design and execute a CIBERSORTx analysis. All datasets used in this chapter are available at http://cibersortx.stanford.edu, under "*Menu*" → "*Download.*" The website also offers a series of dedicated tutorials covering a wide range of potential use cases and applications (under "*Menu*" → "*Tutorial*").

3 Methods

3.1 Construction of a Custom Signature Matrix from scRNA-Seq Data

CIBERSORTx requires the use of marker gene reference profiles to enumerate cell subsets in bulk tissue samples. Commonly referred to as a "signature matrix," marker gene profiles can be derived from a variety of sources, including scRNA-seq data and sorted cell populations profiled by bulk RNA-seq or microarrays [21, 23]. We previously described a microarray-derived signature matrix for profiling 22 functionally defined human immune cell types, called LM22 [16]. With scRNA-seq, it is now possible to create a signature matrix that captures all major cell subsets in a tissue without complex sorting experiments (*see* **Note 1**). This can enable large-scale investigation of novel or poorly understood phenotypic states in bulk tissue GEPs. In malignancies, cellular states of

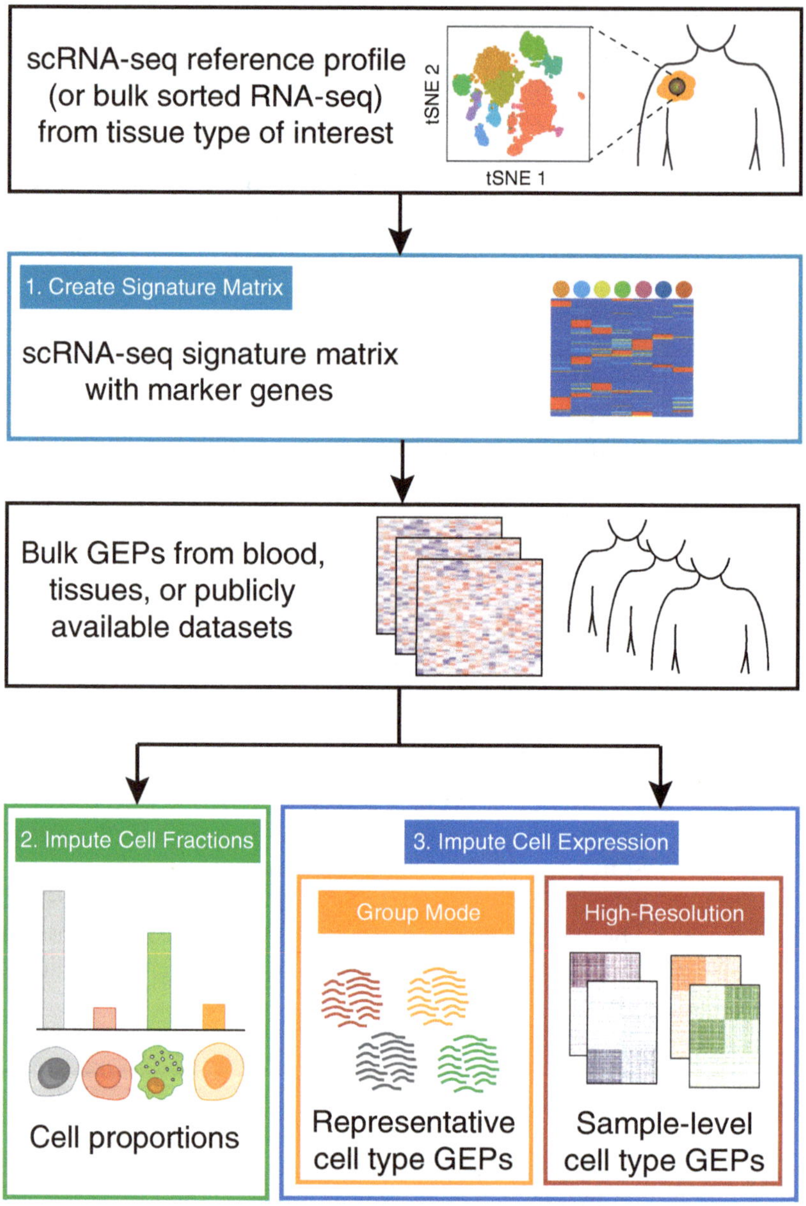

Fig. 1 Overview of CIBERSORTx. Starting from reference profiles generated by scRNA-seq, bulk sorted RNA-seq, or microarrays, CIBERSORTx generates a deconvolution signature matrix, consisting of cell type-specific barcode genes (step 1), which is then repeatedly used to enumerate cell fractions (step 2) or impute cell-type-specific gene expression profiles (step 3) from bulk tissue GEPs. Gene expression imputation can be performed with group-mode, which results in a representative transcriptome profile for each cell type in the signature matrix, or high-resolution mode, which yields sample-level expression estimates for each cell type

interest may include subpopulations of activated, resting, or exhausted T cells [24–26], cancer-associated fibroblasts [27], or cancer cells [28, 29], including tumor initiating cells or cancer stem cells [30]. As further described below, scRNA-seq can also provide a powerful means of validating reference signatures through the use of mixture samples created from single-cell transcriptomes.

In the next section, we describe how to create a signature matrix from scRNA-seq data. Other platforms are described in detail elsewhere (*see* tutorials 6 and 7 at http://cibersortx.stanford.edu/tutorial.php).

3.1.1 Input File

In order to create a custom signature matrix from scRNA-seq data, CIBERSORTx requires a *single cell reference matrix file*. This file consists of a tab-delimited scRNA-seq expression matrix saved as .*txt* or .*tsv*, where each row is a gene and each column is a single-cell transcriptome. The sum of the expression of all genes for a given cell should not be zero. Gene names must populate the first column of the file. CIBERSORTx will append unique numerical identifiers to redundant gene symbols, but we recommend that users remove any redundant gene names prior to uploading their file. In addition, each single cell must be assigned a cell phenotype by the user in row 1 (e.g., "CD8 T cell," "B cell"), and there should be at least three cells per phenotype. The same cell label should be used for each cell belonging to a particular phenotype, and periods should be avoided unless employed to separate the phenotype label from a numerical label. For example, users can label the cells, "CD8 T cell", "CD8 T cell.1", or "CD8 T cell.2", but not "CD8 T.cell" or "CD8.T.cell". Suffixes separated by a period will be removed during the construction of the signature matrix. Any cells without a cell label (i.e., unassigned) should be excluded before uploading the reference matrix file.

Importantly, clustering and de novo identification of cell types from scRNA-seq data are not currently supported by CIBERSORTx. As such, all cell labels must be provided by the user. There are several software tools available to identify cell phenotypes for generating such cell labels when starting from unlabeled scRNA-seq data [31–37] (also, *see* **Note 2**). For more information about the formatting of the scRNA-seq reference matrix file, *see* **Notes 3–5**.

In the following sections, we use a publicly available scRNA-seq dataset consisting of 7688 human peripheral blood mononuclear cells (PBMCs) to illustrate the process of creating a single-cell-derived signature matrix with CIBERSORTx. This dataset is available for download from 10× Genomics (https://www.10xgenomics.com/) and from the CIBERSORTx website ("*Menu*" → "*Download*" → "*10x scRNA-Seq reference matrices (zip)*" → "*Fig2e-5PBMCs_scRNAseq_matrix.txt*"). By applying Seurat [31] along with prior knowledge of key marker genes, we previously identified eight major cell types in this dataset [23]: NK cells, monocytes, B cells, NKT cells, CD4 T cells, CD8 T cells, dendritic cells, and megakaryocytes (Fig. 2a).

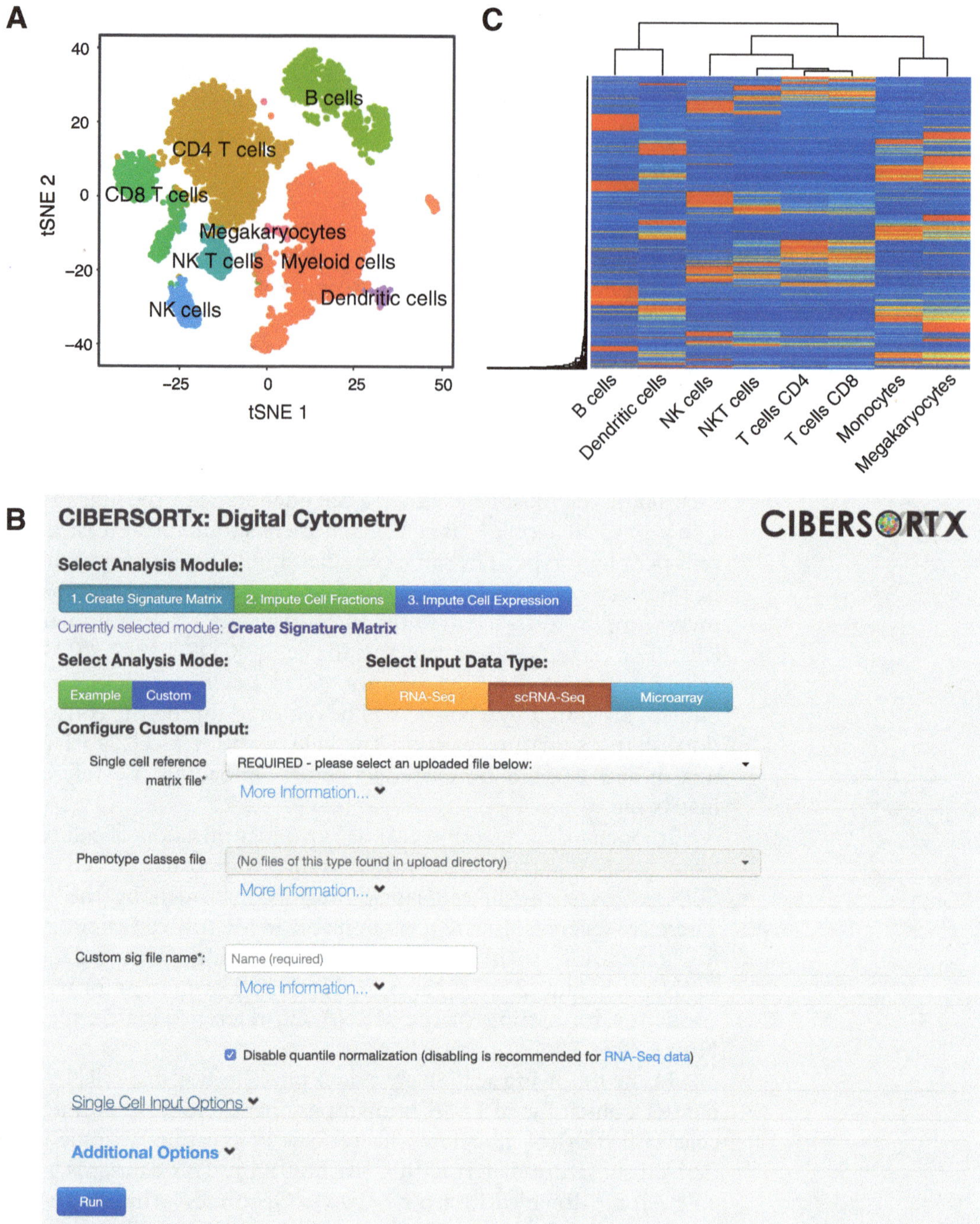

Fig. 2 Building a signature matrix with scRNA-seq data. (**a**) t-SNE projection of PBMCs profiled by scRNA-seq (10× Chromium v2 platform with 5′ chemistry). (**b**) Screen shot of the CIBERSORTx website showing the selection of scRNA-seq data as the input data type. (**c**) Heat map of the signature matrix generated by CIBERSORTx when applied to the dataset shown in panel (**a**)

3.1.2 File Upload

Once the single-cell reference matrix file is formatted according to the instructions above, it may be uploaded by selecting "*Menu*" → "*Upload files*" and "+ *Add file*". The user will then be able to select their input file, provide a title, and select the "*File Type.*" It is important to specify the correct file type to ensure that the input file appears in the appropriate drop-down menu when configuring the CIBERSORTx analysis. For the current application, select the file type, "*Single Cell Reference Matrix,*" under "*Cell Fractions Module.*"

3.1.3 Building the scRNA-Seq Signature Matrix

Having uploaded the single-cell expression matrix, we can now select "*Menu*" → "*Run CIBERSORTx.*" We choose the analysis module, "*1. Create Signature Matrix,*" the "*Custom*" analysis mode, and "*scRNA-seq*" as our input data type (Fig. 2b).

In the "*Single cell reference matrix file*" drop-down menu, our newly uploaded file is available for selection. We provide a name for the signature matrix we are about to create under "*Custom sig file name.*" Providing a file name is required in order to identify the signature matrix in future analyses. *See* **Note 6** for addition details on parameter configuration. Once the single-cell reference file is uploaded and the CIBERSORTx job is configured, click "*Run*" to start creating the signature matrix.

The output of the job will contain (1) the new signature matrix saved as a *.txt* file with the file name provided by the user, (2) the reference sample and phenotypic classes files created by CIBERSORTx as an intermediate step to build the signature matrix, and (3) a heat map of the signature matrix that is organized to show patterns of differentially expressed genes (Fig. 2c). The newly created signature matrix will be automatically available from the "*Signature matrix file*" drop-down menu for future jobs.

3.1.4 Validation of the scRNA-Seq Signature Matrix

When creating a new signature matrix, it is critical to evaluate its reliability for resolving each cell subset in bulk tissue GEPs. Below we provide some suggestions on how users can validate the performance of a single-cell-derived signature matrix before applying it to real tissue specimens.

First, it is important to check whether known marker genes are present and highly expressed in the correct cell subsets. For example, *MS4A1*, a B cell marker gene that encodes CD20, is expected to be selected as a marker gene for the B-cell population in our running example. If the same cell subsets have been profiled elsewhere, either as bulk sorted populations or by scRNA-seq, one can also evaluate the generalizability of genes in the signature matrix for correctly classifying each cellular phenotype (e.g., *see* Newman et al. [16]).

Second, if scRNA-seq data are used to build a signature matrix, it is straightforward to characterize its performance using synthetic tissues created from single-cell transcriptomes. To ensure an unbiased assessment, these source scRNA-seq transcriptomes used for the creation of a synthetic tissue should be held out from the creation of the signature matrix. Moreover, to avoid violating linearity assumptions, each single-cell transcriptome should be represented in nonlog linear space prior to creating synthetic mixtures. By allowing for fine-grained control over the composition of each mixture, this strategy allows one to systematically evaluate both proportion estimation and cellular detection limits without the cost and time associated with profiling new samples with associated ground-truth expectations of compositional representation.

Finally, the gold standard approach for validating a signature matrix is to compare deconvolution performance against orthogonal methods, such as flow cytometry or immunohistochemistry (*see* **Note 7**). This is done by imputing the cell fractions of a bulk GEP dataset with CIBERSORTx and by comparing the imputed cell proportions with known ground truth proportions in each sample (for details of this process, *see* Subheading 3.2).

3.2 Impute Cell Fractions with CIBERSORTx

In this section, we demonstrate how to apply a scRNA-seq-derived signature matrix to resolve cellular composition in 12 whole blood RNA-seq profiles (GEO accession number: GSE127813). This dataset is available from the CIBERSORTx website by selecting "*Menu*" → "*Download*" → "*Expression Datasets (zip)*" ("*Fig2b-WholeBlood_RNAseq.txt*"). Importantly, for each of these 12 samples, ground truth cell proportions determined by flow cytometry and complete blood counts are available ("*Fig2b_ground_truth_-whole_blood.txt*" under "*Expression dataset (zip)*" in "*Download*").

We have preprocessed the mixture dataset into transcript-per-million (TPM) space (also, *see* **Note 8**). Like signature matrices, all bulk mixture GEPs should be in linear space (not $\log_2$) without negative or missing values. In addition, for optimal performance, the signature matrix and mixture files should be normalized identically. The mixture file should be uploaded as described for the single-cell reference matrix file in Subheading 3.1.2 (i.e., by selecting "*Upload Files*" under the "*Menu*" tab). When uploading the file, choose "*Mixture*" under "*All Modules*" as the file type.

Having generated a signature matrix in Subheading 3.1.3 and uploaded our mixture file, we may now select "*Menu*" → "*Run CIBERSORTx*" and choose the "*2. Impute Cell Fractions*" analysis module and "*Custom*" analysis mode (Fig. 3a). In the "*Signature matrix file*" drop-down menu, we select the file that we previously generated (Subheading 3.1.3).

When configuring the analysis, we have the option of selecting "*Batch correction.*" An important caveat with the precursor of CIBERSORTx is that it did not address platform-specific variation (e.g., between scRNA-seq and RNA-seq). In the next section, we describe how CIBERSORTx addresses this important issue.

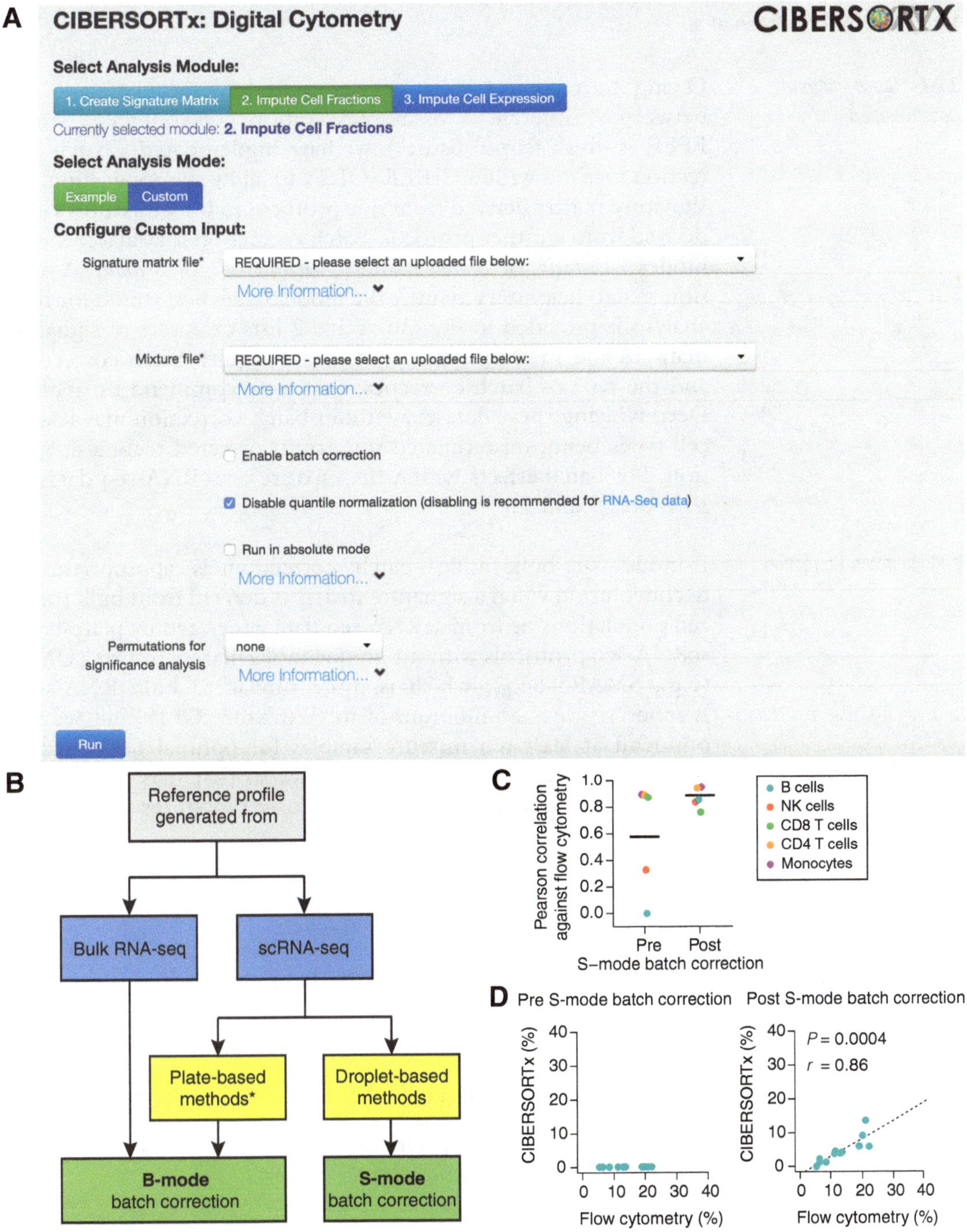

Fig. 3 Imputation of cell fractions and cross-platform deconvolution. (**a**) Screenshot of the CIBERSORTx website showing the selection of "*2. Impute Cell Fractions*" module. (**b**) Decision tree to help CIBERSORTx users identify the batch correction mode that is best suited for their application: (1) single-cell reference mode ("S-mode") or (2) bulk reference mode ("B-mode"). ∗Plate-based methods refer to methods that show little or moderate technical variation compared to bulk RNA-seq, such as SMART-seq2. (**c**) Estimation of cell fractions in RNA-seq GEPs from 12 healthy whole blood samples using the PBMC signature matrix (from Fig. 2) pre- and post-S-mode batch correction. The deconvolution performance is shown as Pearson correlations comparing the CIBERSORTx estimated fractions with the flow cytometry ground truth for five cell types (B cells, NK cells, CD8 T cells, CD4 T cells, and monocytes). Medians are shown as horizontal lines. (**d**) Same as (**c**) but showing imputed proportions of B cells before (left) and after (right) S-mode batch correction. *Right:* Concordance between CIBERSORTx and flow cytometry was determined by Pearson correlation (*r*) and linear regression (dashed line)

3.2.1 Cross-Platform Deconvolution

Owing to technical variation between different platforms and between different tissue-preservation techniques (e.g., FFPE vs. fresh-frozen tissues), we have implemented a batch correction method within CIBERSORTx to allow the application of a signature matrix derived from one protocol to bulk mixtures GEPs derived from another protocol. Batch correction is available in two modes: (1) *bulk*, or *B-mode*, and (2) *single-cell*, or *S-mode*. A decision tree to help users identify the mode that is best suited for their analysis is provided in Fig. 3b. Table 2 lists examples of signature matrices and mixtures pairs that would require batch correction, and the type of batch correction that we recommend be applied. Deconvolving these datasets without batch correction may lead to cell types being misestimated due to uncorrected technical variation. For batch effects within the mixture or scRNA-seq datasets, *see* **Notes 9** and **10**.

B-Mode Batch Correction

B-mode—or bulk-mode—batch correction is appropriate for deconvolution when a signature matrix is derived from bulk sorted cell populations or from scRNA-seq data generated by plate-based scRNA-seq protocols without unique molecular identifiers (UMIs) (e.g., SMART-Seq2, which is more similar to bulk RNA-seq). B-mode requires a minimum of three mixture GEPs, but we recommend at least ten mixture samples for optimal performance. B-mode will adjust the mixture dataset so that it is in the same space as the signature matrix. The adjusted mixtures will then be used for estimating the cell fractions.

Table 2
Pairs of signature matrices and mixture datasets where CIBERSORTx batch correction is strongly recommended

Signature matrix	Mixture dataset	Batch correction method
Microarray of sorted cell populations	Bulk RNA-seq of fresh/frozen tissues NanoString nCounter of FFPE tissues	B-mode
RNA-seq of sorted cell populations	Bulk RNA-seq of FFPE tissues Microarray of fresh/frozen tissues	B-mode
Plate-based scRNA-seq without UMIs (i.e., SMART-Seq2)	Bulk RNA-seq of fresh/frozen tissues Microarray of fresh/frozen tissues Bulk RNA-seq of FFPE tissues NanoString nCounter of FFPE tissues	B-mode
Droplet-based scRNA-seq (i.e., 10× Chromium, Drop-seq)	Bulk RNA-seq of fresh/frozen tissues Microarray of fresh/frozen tissues Bulk RNA-seq of FFPE tissues NanoString nCounter of FFPE tissues	S-mode

S-Mode Batch Correction

S-mode—or single-cell mode—batch correction is tailored for single cell-derived signature matrices generated from droplet-based protocols or protocols that utilize UMIs, such as 10× Chromium [38], InDrop [5], or Drop-Seq [6]. Unlike protocols that profile full transcripts (e.g., bulk RNA-seq and SMART-Seq2), these technologies capture gene expression data from the 3′ or 5′ end of each transcript and employ UMIs to remove PCR duplications and to assign transcripts to individual cells. When running S-mode, an additional input file is required, which is the single-cell reference matrix that was used to build the signature matrix. The drop-down menu for selecting this file automatically appears when the user selects S-mode as an option. S-mode also requires a minimum of three samples in the mixture dataset, but at least ten are recommended.

Unlike B-mode, which adjusts the mixture dataset, S-mode adjusts the signature matrix, which is then used to estimate the cell fractions. The resulting adjusted output will be available for download after running a CIBERSORTx job with batch correction.

3.2.2 Configuring a CIBERSORTx Fractions Job

Since the signature matrix that we created in Subheading 3.1.3 is derived from a UMI-based scRNA-seq dataset generated on the 10× Chromium platform, we select "*S-mode batch correction.*" We then select the corresponding single-cell reference profile that we used for building the signature matrix.

CIBERSORTx provides an empirical *p*-value to evaluate deconvolution performance. The *p*-value is calculated by comparing imputed fractions in a given mixture dataset with fractions that would have been obtained by random chance. The higher the number of permutations, the more reliable the *p*-value estimate will be. For our analysis, we set the number of permutations to 100.

Having configured the CIBERSORTx Fractions jobs, press "*Run.*"

Once the job is complete, results will appear in tabular format as a heatmap. A stacked bar plot representation showing the distribution of cell fractions for each mixture sample will also be produced.

3.2.3 Evaluation of Estimated Cell Fractions

Since ground truth cell proportions are available for the 12 blood samples analyzed in Subheading 3.2.2 through the use of automated blood counting and flow cytometric immunophenotyping, we can directly assess the accuracy of our deconvolution results. Doing so can provide insights into the robustness of the signature matrix and the impact of batch correction. Figure 3c shows the distribution of Pearson correlations between the fractions inferred by CIBERSORTx and ground truth proportions for each cell type, both before and after S-mode batch correction (also, *see* **Note 11**). Clearly, batch correction provides a substantial improvement in the average Pearson correlation across all five cell types. For example, while B cells failed to be imputed without batch correction (Fig. 3d, left), after S-mode batch correction, B cells are successfully imputed (Fig. 3d, right).

3.3 Impute Cell-Type-Specific Gene Expression with CIBERSORTx

In this section, we review how to use CIBERSORTx for imputing cell type-specific transcriptomes from bulk tissue GEPs without the need for physical cell sorting. To impute cell type-specific gene expression profiles, select the analysis module, "*3. Impute gene expression*" and "*Custom*" analysis mode (Fig. 4a). Two distinct modes are available:

Group-mode: This approach infers a representative transcriptome profile for each cell type in the signature matrix from a group of bulk tissue transcriptomes (Fig. 4b). Group mode is useful for learning context-dependent changes in a cell's expression profile when a biological class (or grouping) is already known. For example, one can apply group-mode to study cell-type specific transcriptional differences between responders and nonresponders to a therapy, or between tumor samples and the adjacent normal tissues.

High-resolution: This approach imputes cell-type specific GEPs from bulk tissues at sample-level resolution. As a result, the output of high-resolution mode is an expression matrix for each cell type with the same number of samples as the input mixtures (Fig. 4c). High-resolution mode is useful for exploring cell type expression variation without prior knowledge of biological or functional groupings. It can be used, for example, to investigate cell type specific gene expression in tumors in relation to patient survival or for identifying novel transcriptional states.

3.3.1 Group-Mode Expression Imputation

CIBERSORTx group-mode imputes cell type-specific gene expression profiles from a group of bulk tissue GEPs and will learn a single representative transcriptome profile for each cell type in the signature matrix across the set of mixture samples being considered. The required inputs for group-mode are a signature matrix and a mixture file in the same format as described in Subheadings 3.1.1 and 3.2. Optionally, users may wish to merge the cell types in the signature matrix into broader phenotypic classes in order to reduce the number of distinct cell subsets to deconvolve (*see* **Note 12**). To do this, CIBERSORTx supports the use of a "*Merged class file*," which is a simple text file consisting of one row with the new class labels separated by tabs. The "*Merged class file*" described allows users to group cell types in the signature matrix into a broader set of phenotypic classes. This makes it easy to reduce the number of evaluable cell types in cases where the number of mixture samples is limiting. The order of the labels must match the ordering of cell types in the signature matrix, and a label must be provided for every cell type in the signature matrix. If a merged class file is not provided, all cell types in the signature matrix will be used.

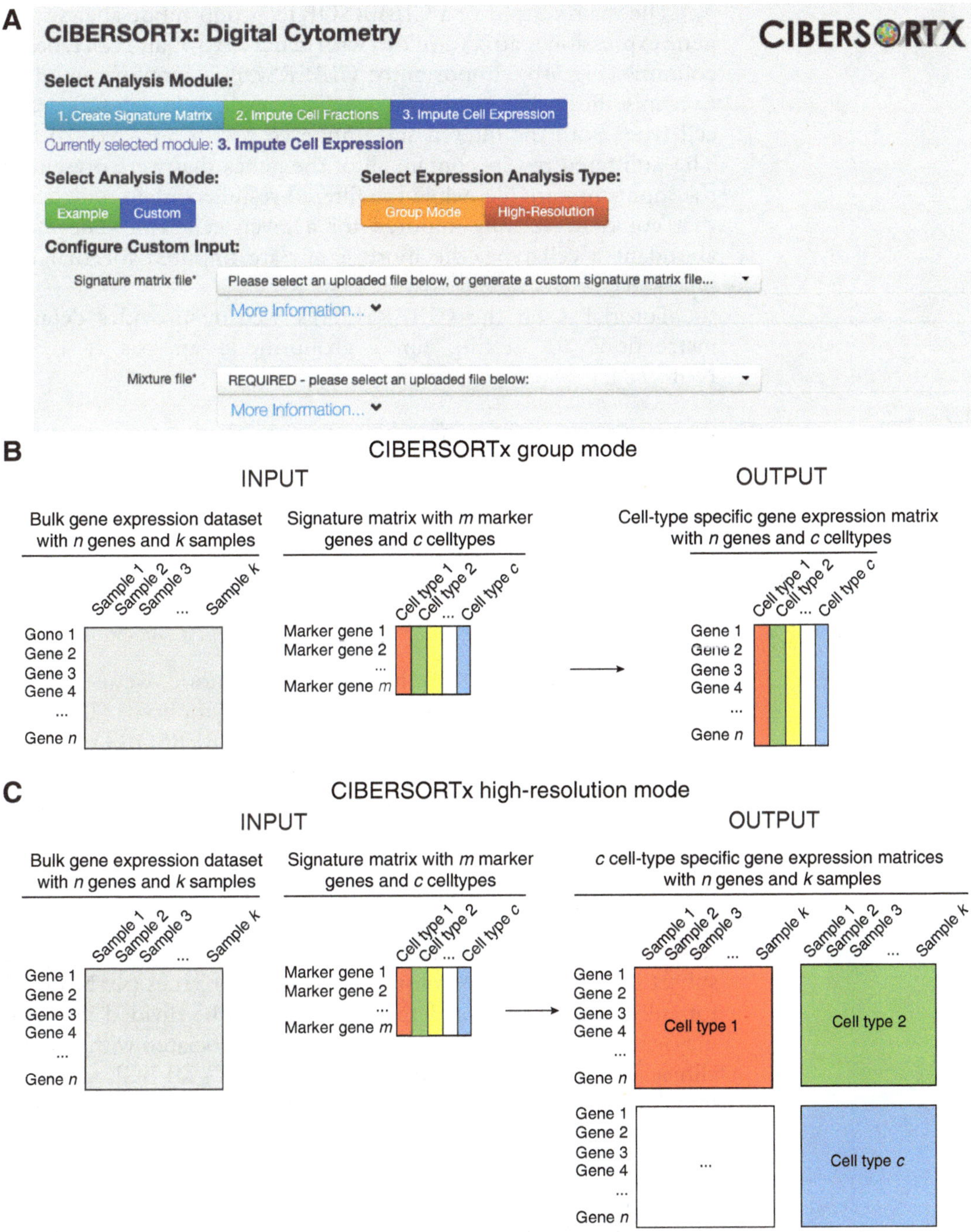

Fig. 4 Cell type-specific gene expression imputation. (**a**) Screenshot of the CIBERSORTx website showing the selection of the "*3. Impute Cell Expression*" module. (**b**, **c**) Schematic overview of group mode (**b**) and high-resolution mode (**c**) expression imputation

The main output of a CIBERSORTx group-mode analysis is a gene expression matrix (nonlog) with genes as rows and cell types as columns (Fig. 4b). Importantly, CIBERSORTx employs an adaptive noise filter that eliminates unreliably estimated genes for each cell type. Both the filtered and unfiltered results are saved to file. The unfiltered results contain all of the genes that were present in the input mixture file, while the filtered results contain only genes that could be reliably imputed for a given cell type. The more abundant a cell type, the more genes are imputed for that cell type [23].

Tutorial 4 on the CIBERSORTx website provides detailed instructions for setting up a group-mode analysis (also, *see* **Note 13**).

3.3.2 High-Resolution Expression Imputation

CIBERSORTx high-resolution mode derives cell type-specific gene expression profiles at the sample-level. Although the input requirements are the same as group-mode, the output is an expression matrix for each cell type rather than a single representative transcriptome profile per cell type (Fig. 4c). Tutorial 7 on the CIBERSORTx website provides detailed instructions for running high-resolution mode. We summarize key steps below.

3.3.3 Input Files

To illustrate CIBERSORTx high-resolution mode, we use a gene expression matrix of 150 diffuse large B-cell lymphoma (DLBCL) samples [39]. The file is available from the CIBERSORTx website by selecting "*Menu*" → "*Download*" and "*High Resolution GEPs – DLBCL (Supp. Fig. 11) (zip)*" ("*SuppFig11-DLBCL_CHOP_Lenz-arrays-bulktumors.MAS5.txt*"). Users may upload the file by selecting "*Menu*" → "*Upload Files*" and "*Mixture*" as the file type (for details on the file upload process, *see* Subheading 3.1.2).

High-resolution mode is computationally intensive. Therefore, to reduce running times, we recommend that users provide a "gene subset file" consisting of a list of at most 1000 genes (*see* **Note 14** for full-transcriptome analyses). DLBCL can be divided into two major molecular subtypes that are strongly associated with distinct clinical outcomes [40]: Activated B-cell (ABC) DLBCL and Germinal Center B-cell (GCB) DLBCL. Using the file type, "*Gene subset – High-Resolution Mode*," we upload a list of genes that are known to be differentially expressed between the two classes and that are primarily associated with B cells. The corresponding "gene subset file" is available in "*High Resolution GEPs - DLBCL (Supp. Fig. 11) (zip)*" ("*SuppFig11-DLBCL-GCBABC-genes.txt*").

3.3.4 Configure a High-Resolution Job

Once all required input files are uploaded, select "*Menu*" → "*Run CIBERSORTx*," then "*3. Impute Cell Expression*," "*Custom*," and "*High-Resolution*."

In this example, we will use "*LM22 (22 immune cell types)*" [16], a signature matrix that is available by default in the signature matrix drop-down menu. Select "*LM10 merged into 10 major cell subsets*," which is available in the "*Merged classes file*" drop-down menu. LM10 groups the LM22 cell types into ten broader phenotypes, thereby increasing the ratio between the number of samples and cell types, which will improve deconvolution performance (*see* **Note 12** and Newman et al. [23]). For the mixture file, select the DLBCL GEP dataset uploaded in the previous section. Then select the ABC/GCB gene list uploaded under "*Gene Subset file.*" Finally, check the box, "*Group genes by hierarchical clustering in output heat map*," and press "*Run.*" Note that in some instances, it may be appropriate to apply batch correction when performing expression imputation. For details, *see* **Note 15**.

3.3.5 Output of High-Resolution Mode

Expression matrices: The main output of high-resolution mode is a set of expression matrices, one for each evaluated cell subset. Each matrix contains imputed expression profiles (nonlog space) with genes as rows and samples as columns. These expression matrices can be downloaded from the CIBERSORTx job results page and used for downstream analyses elsewhere.

Heat maps: The output of high-resolution mode is also presented as a series of heat maps organized by cell type. Each heat map shows the imputed expression of the selected cell type across all of the samples (columns) and analyzed genes (rows). All samples and genes are presented in the same ordering as the original input mixture matrix. To highlight expression variation, each gene is $\log_2$ adjusted and mean-centered. Genes that have little to no expression variation across the cohort will be represented in black. If the option "*Group genes by hierarchical clustering*" has been selected, the genes are ordered by the results of the cluster analysis.

t-SNE plots: For each cell type with at least 20 imputed genes, the CIBERSORTx website will perform t-distributed stochastic neighbor embedding (t-SNE) to generate a two-dimensional projection of the imputed expression profiles. In the DLBCL example, only B-cells have enough imputed genes from the 142 gene-list to be rendered as a t-SNE plot. If users have pre-defined labels for their samples, they may upload a custom label file to highlight the samples in the t-SNE plot according to these labels. To highlight known GCB/ABC classes, select the preuploaded file "*supp. fig. 10 tSNE Class Labels*" under "*Select tSNE plot Class Labels.*"

3.3.6 When to Run CIBERSORTx Group-Mode Versus High-Resolution Mode

Because high-resolution mode performs deconvolution of cell-type-specific gene expression profiles at the sample level, it is useful for discovering novel cellular states in a dataset [23]. However, high-resolution mode requires a large number of samples for better performance (*see* **Note 12**). In settings where users have prior knowledge of sample classes (responders and nonresponders to a therapy, different molecular subtypes of a given cancer, etc.), cell type-specific differentially expressed genes can be identified in several ways:

1. Run CIBERSORTx high-resolution mode on the entire dataset, then identify differentially expressed genes between known classes using standard methods (e.g., as described in Newman et al. [23]).
2. Run CIBERSORTx high-resolution mode on the two classes separately, then combine the results by intersecting genes that are imputed in both classes. Of note, while this approach has higher power to identify differentially expressed genes than the above strategy, it requires more samples to run high-resolution mode on each class.
3. Run group-mode on each class separately, and then apply the following R script to identify statistically significant differentially expressed genes.

```
for (i in 1:ncol(geps1)){
vBetaZ <- sapply(1:nrow(geps1), function(j) (geps1[j,i]-geps2
[j,i])/sqrt(stderr1[j,i]^2+stderr2[j,i]^2))
ZPs <- 2*pnorm(-abs(vBetaZ))
Zqvals <- p.adjust(ZPs, method="BH")
}
```

Where:

- `geps1` are the cell-type-specific GEPs outputted by CIBERSORTx group mode for sample class 1.
- `geps2` are the cell-type-specific GEPs outputted by CIBERSORTx group mode for sample class 2.
- `stderr1` are the standard errors for sample class 1 outputted by CIBERSORTx Group mode and saved to file as *CIBERSORTxGEP_[...]_GEPs_StdErrs.txt.*
- `stderr2` are the standard errors for sample class 2 outputted by CIBERSORTx Group mode and saved to file as *CIBERSORTxGEP_[...]_GEPs_StdErrs.txt.*
- `i` is the cell type number for which we are calculating differentially expressed genes between class 1 and class 2.
- `BH` stands for Benjamini-Hochberg, and is the method used to adjust the p-values for multiple hypothesis testing.

The script will output a list of genes that are significantly differentially expressed between the two classes after multiple hypothesis testing.

3.4 Conclusion

CIBERSORTx is a novel computational framework for high-throughput dissection of cellular heterogeneity in bulk tissue genomic profiles. Unlike previous methods, it handles cross-platform technical variation, enabling the use of scRNA-seq reference signatures for deconvolution, and can derive cell-type-specific gene expression profiles at sample-level resolution. We anticipate that CIBERSORTx will complement existing experimental methods for studying complex tissues, with implications for the discovery of novel phenotypic states and the identification of more effective biomarkers and therapeutic targets.

4 Notes

1. While the per-sample-cost of scRNA-seq experiments remains relatively high, CIBERSORTx leverages scRNA-seq data from a small number of samples to characterize the cellular heterogeneity of bulk GEPs from a large number of tissue samples. Users can also download publicly available scRNA-seq datasets to create custom signature matrices.
2. To facilitate the identification of cell types in a single-cell RNA-seq dataset, it is possible to use deconvolution or other reference-based methods to label the cells [12, 15, 17, 22, 23, 41].
3. The scRNA-seq dataset should be in nonlog space [42] with no missing values. Data from scRNA-seq experiments summarizing expression levels using unique molecular identifiers (UMIs) are acceptable. If the maximum expression value is <50, CIBERSORTx will assume a prior logarithmic transformation, and anti-log all expression values by 2^x.
4. CIBERSORTx will automatically normalize the input data such that the sum of all normalized reads is the same for each transcriptome. If a gene length-normalized expression matrix is provided (e.g., RPKM), then the signature matrix will be adjusted to TPM (transcripts per million). If a count matrix is provided, the signature matrix will be normalized to CPM (counts per million).
5. While increasing the number of cells per phenotype and the number of biological replicates can improve the quality of the signature matrix (up to ~20 cells per phenotype and up to 2–3 donor samples), simulation experiments and empirical observations suggest that as few as three cells per phenotype and as few as one donor sample can still generate reliable results [23]. For this reason, and to limit the amount of space and time necessary to run a CIBERSORTx job, we recommend that users restrict the number of cells to at most 5000 when uploading the

scRNA-seq reference profile to the CIBERSORTx website (currently a strict upper limit of 10,000 cells is allowed).

6. When creating a signature matrix with scRNA-seq data derived from droplet-based methods (e.g., 10× Genomics [38], InDrop [5], or Drop-seq [6]), the user may wish to change the single-cell-specific parameter, "*Min. Expression*," under "*Single Cell Input Options*." This parameter refers to the minimum average $\log_2$ expression of a gene per cell phenotype (default = 0.75), and is used to filter low-level noise in plate-based scRNA-seq experiments (e.g., data generated by the SMART-Seq2 protocol) that may result from contamination, index swapping [43], or other sources. As this threshold can be too restrictive for data produced by droplet-based platforms, which capture a smaller number of genes, we generally recommend reducing this parameter to 0.50 or even 0 when processing droplet-based scRNA-seq data. Otherwise, the sparsity of the data may yield too few genes for creating a reliable signature matrix.

7. When working with solid tissues and comparing the performance of CIBERSORTx with technologies that require tissue dissociation (e.g., flow or mass cytometry), it is important to generate the gene expression profiles and perform fluorescence activated cell sorting (FACS) or scRNA-seq profiling on the same cell suspensions. If expression profiles are instead generated from intact nondissociated tissues, CIBERSORTx results will not capture dissociation-induced differences in cell composition (loss of myeloid cells due to adherence to plastics, loss of plasma cells and plasmablasts due to their fragility, etc.), leading to biases that artificially degrade CIBERSORTx performance in the comparative analysis.

8. CIBERSORTx performs a feature selection within the deconvolution step, and typically does not use all genes in the signature matrix for this step. It is therefore generally acceptable if some genes in the signature matrix are missing from the mixture file.

9. When we refer to batches in the context of expression deconvolution, we refer to two input datasets, the signature matrix and the mixture dataset. However, within a mixture dataset, one may also observe technical variation due to batch effects. Technical batches may be present between mixture datasets and within mixture datasets. For example, a dataset may consist of two experiments performed at two different time points or by two different individuals. These differences may lead to technical variation in the gene expression data. If there are known batch effects in a mixture dataset and these cannot be removed

using available methods [44–46], we recommend running CIBERSORTx on the datasets separately.

10. Batch effects between samples profiled by scRNA-seq may also occur, and these may affect the performance of the resulting signature matrix. There are available tools to remove batch effects in scRNA-seq datasets prior to construction of the signature matrix [31, 47–50].
11. When comparing CD8 T cells to flow cytometry in our example, we pooled CD8 T cells with NKTs. We did this because (a) NKTs were not separately enumerated by flow cytometry, and (b) NKT cells express both *CD8A* and *CD3D* in the scRNA-seq dataset used to build the signature matrix.
12. CIBERSORTx Gene Expression Analysis Mode works best when the number of samples to deconvolve is much larger than the number of cell types in the signature matrix. A rule of thumb is to have at least four to five times as many mixture samples as cell types [23]. The "*Merged class file*" allows users to group cell types in the signature matrix into a broader set of phenotypic classes. This makes it easy to reduce the number of evaluable cell types in cases where the number of mixture samples is limiting.
13. As detailed in the online tutorial, users may also upload a file with known gene expression profiles as a ground truth file. If a ground truth file is uploaded and selected, CIBERSORTx will generate a variety of plots that can be used for quality control purposes. For example, this will allow users to check the concordance between imputed and ground truth GEPs.
14. For full transcriptome analyses, users are encouraged to download the CIBERSORTx executable ("*Menu*" → "*Download*").
15. When performing gene expression purification, the signature matrix should represent most of the cell types in a tissue. A signature matrix that does not include all major cell types is generally not recommended for expression imputation. However, when this is the case, users may apply batch correction for expression purification to down-weight the genes that are expressed on cell types not included in the signature matrix.

Acknowledgments

We would like to thank B. Chen, B. Nabet, and M. Matusiak for assistance in beta-testing CIBERSORTx. This work was supported by grants from the 2019 AACR-AstraZeneca Lymphoma Research Fellowship (C.B.S., 19-40-12-STEE), the National Cancer Institute (A.M.N., R00CA187192; A.A.A., U01CA194389; A.A.A., R01CA188298), the Stinehart-Reed foundation (A.M.N., A.A.A.), the Stanford Bio-X Interdisciplinary Initiatives Seed Grants Program

(IIP) (A.M.N.), the Virginia and D.K. Ludwig Fund for Cancer Research (A.M.N., A.A.A.), the US Department of Defense (A.M. N., W81XWH-12-1-0498), anonymous donors (A.A.A., A.M.N.), the Shanahan and Bronzini Family Funds (A.A.A.), the V Foundation for Cancer Research (A.A.A.), the Leukemia and Lymphoma Society (A.A.A.), and the Damon Runyon Cancer Research Foundation (A.A.A.), the American Society of Hematology (A.A.A.).

References

1. Maman S, Witz IP (2018) A history of exploring cancer in context. Nat Rev Cancer 18 (6):359–376. https://doi.org/10.1038/s41568-018-0006-7
2. Valkenburg KC, de Groot AE, Pienta KJ (2018) Targeting the tumour stroma to improve cancer therapy. Nat Rev Clin Oncol 15(6):366–381. https://doi.org/10.1038/s41571-018-0007-1
3. Ribas A, Wolchok JD (2018) Cancer immunotherapy using checkpoint blockade. Science 359(6382):1350–1355. https://doi.org/10.1126/science.aar4060
4. Fan HC, Fu GK, Fodor SPA (2015) Combinatorial labeling of single cells for gene expression cytometry. Science 347(6222). https://doi.org/10.1126/science.1258367
5. Klein AM, Mazutis L, Akartuna I, Tallapragada N, Veres A, Li V, Peshkin L, Weitz DA, Kirschner MW (2015) Droplet barcoding for single-cell transcriptomics applied to embryonic stem cells. Cell 161 (5):1187–1201. https://doi.org/10.1016/j.cell.2015.04.044
6. Macosko EZ, Basu A, Satija R, Nemesh J, Shekhar K, Goldman M, Tirosh I, Bialas AR, Kamitaki N, Martersteck EM, Trombetta JJ, Weitz DA, Sanes JR, Shalek AK, Regev A, McCarroll SA (2015) Highly parallel genome-wide expression profiling of individual cells using nanoliter droplets. Cell 161 (5):1202–1214. https://doi.org/10.1016/j.cell.2015.05.002
7. Treutlein B, Brownfield DG, Wu AR, Neff NF, Mantalas GL, Espinoza FH, Desai TJ, Krasnow MA, Quake SR (2014) Reconstructing lineage hierarchies of the distal lung epithelium using single-cell RNA-seq. Nature 509 (7500):371–375. https://doi.org/10.1038/nature13173
8. Abbas AR, Wolslegel K, Seshasayee D, Modrusan Z, Clark HF (2009) Deconvolution of blood microarray data identifies cellular activation patterns in systemic lupus erythematosus. PLoS One 4(7):e6098. https://doi.org/10.1371/journal.pone.0006098
9. Ahn J, Yuan Y, Parmigiani G, Suraokar MB, Diao L, Wistuba II, Wang W (2013) DeMix: deconvolution for mixed cancer transcriptomes using raw measured data. Bioinformatics 29 (15):1865–1871
10. Angelova M, Charoentong P, Hackl H, Fischer ML, Snajder R, Krogsdam AM, Waldner MJ, Bindea G, Mlecnik B, Galon J, Trajanoski Z (2015) Characterization of the immunophenotypes and antigenomes of colorectal cancers reveals distinct tumor escape mechanisms and novel targets for immunotherapy. Genome Biol 16:64. https://doi.org/10.1186/s13059-015-0620-6
11. Aran D, Hu Z, Butte AJ (2017) xCell: digitally portraying the tissue cellular heterogeneity landscape. Genome Biol 18(1):220. https://doi.org/10.1186/s13059-017-1349-1
12. Gong T, Hartmann N, Kohane IS, Brinkmann V, Staedtler F, Letzkus M, Bongiovanni S, Szustakowski JD (2011) Optimal deconvolution of transcriptional profiling data using quadratic programming with application to complex clinical blood samples. PLoS One 6(11):e27156. https://doi.org/10.1371/journal.pone.0027156
13. Kuhn A, Thu D, Waldvogel HJ, Faull RL, Luthi-Carter R (2011) Population-specific expression analysis (PSEA) reveals molecular changes in diseased brain. Nat Methods 8 (11):945–947. https://doi.org/10.1038/nmeth.1710
14. Li B, Severson E, Pignon JC, Zhao H, Li T, Novak J, Jiang P, Shen H, Aster JC, Rodig S, Signoretti S, Liu JS, Liu XS (2016) Comprehensive analyses of tumor immunity: implications for cancer immunotherapy. Genome Biol 17(1):174. https://doi.org/10.1186/s13059-016-1028-7
15. Liebner DA, Huang K, Parvin JD (2014) MMAD: microarray microdissection with analysis of differences is a computational tool for

deconvoluting cell type-specific contributions from tissue samples. Bioinformatics 30 (5):682–689. https://doi.org/10.1093/bioinformatics/btt566

16. Newman AM, Liu CL, Green MR, Gentles AJ, Feng W, Xu Y, Hoang CD, Diehn M, Alizadeh AA (2015) Robust enumeration of cell subsets from tissue expression profiles. Nat Methods 12(5):453–457. https://doi.org/10.1038/nmeth.3337
17. Qiao W, Quon G, Csaszar E, Yu M, Morris Q, Zandstra PW (2012) PERT: a method for expression deconvolution of human blood samples from varied microenvironmental and developmental conditions. PLoS Comput Biol 8(12):e1002838. https://doi.org/10.1371/journal.pcbi.1002838
18. Quon G, Haider S, Deshwar AG, Cui A, Boutros PC, Morris Q (2013) Computational purification of individual tumor gene expression profiles leads to significant improvements in prognostic prediction. Genome Med 5(3):29
19. Shen-Orr SS, Tibshirani R, Khatri P, Bodian DL, Staedtler F, Perry NM, Hastie T, Sarwal MM, Davis MM, Butte AJ (2010) Cell type-specific gene expression differences in complex tissues. Nat Methods 7(4):287
20. Zhong Y, Wan Y-W, Pang K, Chow LM, Liu Z (2013) Digital sorting of complex tissues for cell type-specific gene expression profiles. BMC bioinformatics 14(1):89
21. Newman AM, Alizadeh AA (2016) High-throughput genomic profiling of tumor-infiltrating leukocytes. Curr Opin Immunol 41:77–84. https://doi.org/10.1016/j.coi.2016.06.006
22. Baron M, Veres A, Wolock SL, Faust AL, Gaujoux R, Vetere A, Ryu JH, Wagner BK, Shen-Orr SS, Klein AM, Melton DA, Yanai I (2016) A single-cell transcriptomic map of the human and mouse pancreas reveals inter- and intra-cell population structure. Cell Syst 3(4): e344–e360. https://doi.org/10.1016/j.cels.2016.08.011
23. Newman AM, Steen CB, Liu CL, Gentles AJ, Chaudhuri AA, Scherer F, Khodadoust MS, Esfahani MS, Luca BA, Steiner D, Diehn M, Alizadeh AA (2019) Determining cell type abundance and expression from bulk tissues with digital cytometry. Nat Biotechnol. https://doi.org/10.1038/s41587-019-0114-2
24. Jerby-Arnon L, Shah P, Cuoco MS, Rodman C, Su MJ, Melms JC, Leeson R, Kanodia A, Mei S, Lin JR, Wang S, Rabasha B, Liu D, Zhang G, Margolais C, Ashenberg O, Ott PA, Buchbinder EI, Haq R, Hodi FS, Boland GM, Sullivan RJ, Frederick DT, Miao B, Moll T, Flaherty KT, Herlyn M, Jenkins RW, Thummalapalli R, Kowalczyk MS, Canadas I, Schilling B, Cartwright ANR, Luoma AM, Malu S, Hwu P, Bernatchez C, Forget MA, Barbie DA, Shalek AK, Tirosh I, Sorger PK, Wucherpfennig K, Van Allen EM, Schadendorf D, Johnson BE, Rotem A, Rozenblatt-Rosen O, Garraway LA, Yoon CH, Izar B, Regev A (2018) A cancer cell program promotes T cell exclusion and resistance to checkpoint blockade. Cell 175 (4):984–997. e924. doi:https://doi.org/10.1016/j.cell.2018.09.006
25. Li H, van der Leun AM, Yofe I, Lubling Y, Gelbard-Solodkin D, van Akkooi ACJ, van den Braber M, Rozeman EA, Haanen J, Blank CU, Horlings HM, David E, Baran Y, Bercovich A, Lifshitz A, Schumacher TN, Tanay A, Amit I (2019) Dysfunctional CD8 T cells form a proliferative, dynamically regulated compartment within human melanoma. Cell 176 (4):775–789. e718. doi:https://doi.org/10.1016/j.cell.2018.11.043
26. Tirosh I, Izar B, Prakadan SM, Wadsworth MH 2nd, Treacy D, Trombetta JJ, Rotem A, Rodman C, Lian C, Murphy G, Fallahi-Sichani M, Dutton-Regester K, Lin JR, Cohen O, Shah P, Lu D, Genshaft AS, Hughes TK, Ziegler CG, Kazer SW, Gaillard A, Kolb KE, Villani AC, Johannessen CM, Andreev AY, Van Allen EM, Bertagnolli M, Sorger PK, Sullivan RJ, Flaherty KT, Frederick DT, Jane-Valbuena J, Yoon CH, Rozenblatt-Rosen O, Shalek AK, Regev A, Garraway LA (2016) Dissecting the multicellular ecosystem of metastatic melanoma by single-cell RNA-seq. Science 352(6282):189–196. https://doi.org/10.1126/science.aad0501
27. Puram SV, Tirosh I, Parikh AS, Patel AP, Yizhak K, Gillespie S, Rodman C, Luo CL, Mroz EA, Emerick KS, Deschler DG, Varvares MA, Mylvaganam R, Rozenblatt-Rosen O, Rocco JW, Faquin WC, Lin DT, Regev A, Bernstein BE (2017) Single-cell transcriptomic analysis of primary and metastatic tumor ecosystems in head and neck cancer. Cell 171 (7):1611–1624. e1624. https://doi.org/10.1016/j.cell.2017.10.044
28. Andor N, Simonds EF, Czerwinski DK, Chen J, Grimes SM, Wood-Bouwens C, Zheng GXY, Kubit MA, Greer S, Weiss WA, Levy R, Ji HP (2019) Single-cell RNA-seq of follicular lymphoma reveals malignant B-cell types and coexpression of T-cell immune checkpoints. Blood 133(10):1119–1129. https://doi.org/10.1182/blood-2018-08-862292
29. Karaayvaz M, Cristea S, Gillespie SM, Patel AP, Mylvaganam R, Luo CC, Specht MC,

Bernstein BE, Michor F, Ellisen LW (2018) Unravelling subclonal heterogeneity and aggressive disease states in TNBC through single-cell RNA-seq. Nat Commun 9 (1):3588. https://doi.org/10.1038/s41467-018-06052-0

30. Giraddi RR, Chung CY, Heinz RE, Balcioglu O, Novotny M, Trejo CL, Dravis C, Hagos BM, Mehrabad EM, Rodewald LW, Hwang JY, Fan C, Lasken R, Varley KE, Perou CM, Wahl GM, Spike BT (2018) Single-cell transcriptomes distinguish stem cell state changes and lineage specification programs in early mammary gland development. Cell Rep 24(6):1653–1666. e1657. https://doi.org/10.1016/j.celrep.2018.07.025
31. Butler A, Hoffman P, Smibert P, Papalexi E, Satija R (2018) Integrating single-cell transcriptomic data across different conditions, technologies, and species. Nat Biotechnol 36 (5):411–420. https://doi.org/10.1038/nbt.4096
32. Ji Z, Ji H (2016) TSCAN: pseudo-time reconstruction and evaluation in single-cell RNA-seq analysis. Nucleic Acids Res 44(13):e117. https://doi.org/10.1093/nar/gkw430
33. Kiselev VY, Kirschner K, Schaub MT, Andrews T, Yiu A, Chandra T, Natarajan KN, Reik W, Barahona M, Green AR, Hemberg M (2017) SC3: consensus clustering of single-cell RNA-seq data. Nat Methods 14(5):483–486. https://doi.org/10.1038/nmeth.4236
34. Lin P, Troup M, Ho JW (2017) CIDR: ultrafast and accurate clustering through imputation for single-cell RNA-seq data. Genome Biol 18(1):59. https://doi.org/10.1186/s13059-017-1188-0
35. Qiu X, Mao Q, Tang Y, Wang L, Chawla R, Pliner HA, Trapnell C (2017) Reversed graph embedding resolves complex single-cell trajectories. Nat Methods 14(10):979–982. https://doi.org/10.1038/nmeth.4402
36. Senabouth A, Lukowski SW, Alquicira Hernandez J, Andersen S, Mei X, Nguyen QH, Powell JE (2017, 2017) ascend: R package for analysis of single cell RNA-seq data. bioRxiv:207704. https://doi.org/10.1101/207704
37. Zurauskiene J, Yau C (2016) pcaReduce: hierarchical clustering of single cell transcriptional profiles. BMC Bioinformatics 17:140. https://doi.org/10.1186/s12859-016-0984-y
38. Zheng GX, Terry JM, Belgrader P, Ryvkin P, Bent ZW, Wilson R, Ziraldo SB, Wheeler TD, McDermott GP, Zhu J, Gregory MT, Shuga J, Montesclaros L, Underwood JG, Masquelier DA, Nishimura SY, Schnall-Levin M, Wyatt PW, Hindson CM, Bharadwaj R, Wong A, Ness KD, Beppu LW, Deeg HJ, McFarland C, Loeb KR, Valente WJ, Ericson NG, Stevens EA, Radich JP, Mikkelsen TS, Hindson BJ, Bielas JH (2017) Massively parallel digital transcriptional profiling of single cells. Nat Commun 8:14049. https://doi.org/10.1038/ncomms14049
39. Lenz G, Wright G, Dave SS, Xiao W, Powell J, Zhao H, Xu W, Tan B, Goldschmidt N, Iqbal J, Vose J, Bast M, Fu K, Weisenburger DD, Greiner TC, Armitage JO, Kyle A, May L, Gascoyne RD, Connors JM, Troen G, Holte H, Kvaloy S, Dierickx D, Verhoef G, Delabie J, Smeland EB, Jares P, Martinez A, Lopez-Guillermo A, Montserrat E, Campo E, Braziel RM, Miller TP, Rimsza LM, Cook JR, Pohlman B, Sweetenham J, Tubbs RR, Fisher RI, Hartmann E, Rosenwald A, Ott G, Muller-Hermelink HK, Wrench D, Lister TA, Jaffe ES, Wilson WH, Chan WC, Staudt LM, Lymphoma/Leukemia Molecular Profiling P (2008) Stromal gene signatures in large-B-cell lymphomas. N Engl J Med 359 (22):2313–2323. https://doi.org/10.1056/NEJMoa0802885
40. Alizadeh AA, Eisen MB, Davis RE, Ma C, Lossos IS, Rosenwald A, Boldrick JC, Sabet H, Tran T, Yu X, Powell JI, Yang L, Marti GE, Moore T, Hudson J Jr, Lu L, Lewis DB, Tibshirani R, Sherlock G, Chan WC, Greiner TC, Weisenburger DD, Armitage JO, Warnke R, Levy R, Wilson W, Grever MR, Byrd JC, Botstein D, Brown PO, Staudt LM (2000) Distinct types of diffuse large B-cell lymphoma identified by gene expression profiling. Nature 403(6769):503–511. https://doi.org/10.1038/35000501
41. Diaz-Mejia J, Meng E, Pico A, MacParland S, Ketela T, Pugh T, Bader G, Morris J (2019) Evaluation of methods to assign cell type labels to cell clusters from single-cell RNA-sequencing data [version 1; peer review: 3 approved with reservations]. F1000Research 8(296). https://doi.org/10.12688/f1000research.18490.1
42. Zhong Y, Liu Z (2011) Gene expression deconvolution in linear space. Nat Methods 9 (1):8–9.; author reply 9. https://doi.org/10.1038/nmeth.1830
43. Sinha R, Stanley G, Gulati GS, Ezran C, Travaglini KJ, Wei E, Chan CKF, Nabhan AN, Su T, Morganti RM, Conley SD, Chaib H, Red-Horse K, Longaker MT, Snyder MP, Krasnow MA, Weissman IL (2017) Index switching causes "spreading-of-signal" among multiplexed samples in Illumina HiSeq 4000 DNA sequencing. bioRxiv 2017:125724. https://doi.org/10.1101/125724

44. Smyth GK, Speed T (2003) Normalization of cDNA microarray data. Methods 31 (4):265–273. https://doi.org/10.1016/S1046-2023(03)00155-5
45. Johnson WE, Li C, Rabinovic A (2007) Adjusting batch effects in microarray expression data using empirical Bayes methods. Biostatistics 8 (1):118–127. https://doi.org/10.1093/biostatistics/kxj037
46. Giordan M (2014) A two-stage procedure for the removal of batch effects in microarray studies. Stat Biosci 6(1):73–84. https://doi.org/10.1007/s12561-013-9081-1
47. Bacher R, Chu LF, Leng N, Gasch AP, Thomson JA, Stewart RM, Newton M, Kendziorski C (2017) SCnorm: robust normalization of single-cell RNA-seq data. Nat Methods 14 (6):584–586. https://doi.org/10.1038/nmeth.4263
48. Haghverdi L, Lun ATL, Morgan MD, Marioni JC (2018) Batch effects in single-cell RNA-sequencing data are corrected by matching mutual nearest neighbors. Nat Biotechnol 36 (5):421–427. https://doi.org/10.1038/nbt.4091
49. Hie B, Bryson B, Berger B (2019) Efficient integration of heterogeneous single-cell transcriptomes using Scanorama. Nat Biotechnol. https://doi.org/10.1038/s41587-019-0113-3
50. Johnson M, Purdom E (2017) Clustering of mRNA-Seq data based on alternative splicing patterns. Biostatistics 18(2):295–307. https://doi.org/10.1093/biostatistics/kxw044
51. Schelker M, Feau S, Du J, Ranu N, Klipp E, MacBeath G, Schoeberl B, Raue A (2017) Estimation of immune cell content in tumour tissue using single-cell RNA-seq data. Nat Commun 8(1):2032. https://doi.org/10.1038/s41467-017-02289-3
52. Wang Z, Cao S, Morris JS, Ahn J, Liu R, Tyekucheva S, Gao F, Li B, Lu W, Tang X, Wistuba II, Bowden M, Mucci L, Loda M, Parmigiani G, Holmes CC, Wang W (2018) Transcriptome deconvolution of heterogeneous tumor samples with immune infiltration. iScience 9:451–460. https://doi.org/10.1016/j.isci.2018.10.028
53. Wang X, Park J, Susztak K, Zhang NR, Li M (2019) Bulk tissue cell type deconvolution with multi-subject single-cell expression reference. Nat Commun 10(1):380. https://doi.org/10.1038/s41467-018-08023-x
54. Frishberg A, Peshes-Yaloz N, Cohn O, Rosentul D, Steuerman Y, Valadarsky L, Yankovitz G, Mandelboim M, Iraqi FA, Amit I, Mayo L, Bacharach E, Gat-Viks I (2019) Cell composition analysis of bulk genomics using single-cell data. Nat Methods 16 (4):327–332. https://doi.org/10.1038/s41592-019-0355-5

Chapter 8

Visualization of Single Cell RNA-Seq Data Using t-SNE in R

Bo Zhou and Wenfei Jin

Abstract

Single cell RNA sequencing (scRNA-seq) is a powerful tool to analyze cellular heterogeneity, identify new cell types, and infer developmental trajectories, which has greatly facilitated studies on development, immunity, cancer, neuroscience, and so on. Visualizing of scRNA-Seq data is fundamental and essential because it is critical to biological interpretation. Although principal component analysis (PCA) is used for visualizing scRNA-seq at early studies, t-Distributed Stochastic Neighbor embedding (t-SNE), an unsupervised nonlinear dimensionality reduction technique, is widely used nowadays due to its advantage in visualization of scRNA-seq data. Here, we detailed the process of visualization of single-cell RNA-seq data using t-SNE via Seurat, an R toolkit for single cell genomics.

Key words Single cell RNA sequencing (scRNA-seq), t-Distributed Stochastic Neighbor embedding (t-SNE), Dimension reduction, Visualization of scRNA-seq data, Seurat

1 Introduction

Single-cell RNA-sequencing (scRNA-seq), profiling genome-wide gene expression at single-cell resolution, has become an ideal approach for identifying cellular heterogeneity, searching new/rare cell type, investigating cellular microenvironment and exploring developmental process [1–3]. Dimension reduction is critical for visualization and interpretation of the high dimensional scRNA-seq data. Principal component analysis (PCA), a classical linear transformation, was used in early single cell studies [4]. However, PCA is not able to accurately capture the features of single-cell data, especially when the cell populations have complex structure [5]. t-distributed stochastic neighbor embedding (t-SNE), a nonlinear embedding technique developed by van der Maaten, L. and G. Hinton in 2008 [6], was introduced to visualize single cell data by Amir el-AD et al. in 2013 [5]. t-SNE performs much better than lots of nonparametric visualization approaches, such as Sammon mapping, Isomap, and Locally Linear Embedding, in visualization of single cell data. The algorithms for fast implementation of

Benjamin L. Kidder (ed.), *Stem Cell Transcriptional Networks: Methods and Protocols*, Methods in Molecular Biology, vol. 2117, https://doi.org/10.1007/978-1-0716-0301-7_8,

t-SNE by several studies allowed t-SNE to be applied on large-scale single cell data [7–9]. Barnes-Hut approximation is the most popular implementation of t-SNE [7], and FFT-accelerated interpolation-based t-SNE (FIt-SNE) published recently might be the fastest one [8]. Currently, t-SNE is the most commonly used approach for single-cell data visualization and has been integrated into many scRNA-seq data analysis toolkits, such as Seurat [10]. Here, we use Seurat to demonstrate usages of the fascinating data visualization technique, t-SNE.

2 Materials

All analyses in this chapter are performed in R, with required packages and sample data.

2.1 R and R Packages

R: a free software environment for statistical computing and graphics, which can be downloaded from official website of R project (https://www.r-project.org/) and be installed following the instructions.

RStudio (optional): an integrated development environment (IDE) for R (https://www.rstudio.com/).

Seurat: an R package designed for QC, data integration, analysis, exploration and biological interpretation of scRNA-seq data. Seurat is available on CRAN (https://cran.r-project.org/), with more information available at Satija Lab websites (https://satijalab.org/seurat/). Seurat can be installed directly by:

```
>install.packages("Seurat")
```

Rtsne: an R wrapper for Barnes-Hut implementation of t-SNE, which is available to install from CRAN, or github (https://github.com/jkrijthe/Rtsne). Generally, it is automatically installed with Seurat as required dependencies.

2.2 Sample Data

The scRNA-seq of peripheral blood mononuclear cells (PBMCs) from a healthy donor is used as sample data, which is accessible at official website of 10X GENOMICS (https://support.10xgenomics.com/single-cell-gene-expression/datasets/3.0.0/pbmc_1k_v3). The PBMCs were processed by 10X Genomics Chromium and sequenced on Illumina NextSeq. The reads were processed using Cellranger (version 3.0.0), and raw matrix generated by Cellranger was set as input for this demonstration.

3 Methods

3.1 Data Reading and Preprocessing

3.1.1 Data Reading

The sample data is stored in directory raw_feature_bc_matrix, which contains three files: one barcode file (barcodes.tsv.gz), one feature file (features.tsv.gz), and one expression matrix file (matrix.mtx.gz). We read the data using *Read10X* function, which return a unique molecular identified (UMI) count matrix. Then we created a Seurat object using *CreateSeuratObject* function.

```
> library(Seurat)
> pbmc1 <- Read10X(data.dir = "raw_feature_bc_matrix/")
> pbmc1 <- CreateSeuratObject(counts = pbmc1, min.cells = 3,
min.features = 30)
> pbmc1
An object of class Seurat
15262 features across 1443 samples within 1 assay
Active assay: RNA (15262 features)
```

3.1.2 Quality Control (QC) and Filter Out of Cells

QC is critical for removing the noises generated during single cell library preparation. For example, empty droplets lead to few detected genes; Cell doublets or multiplets lead to an aberrantly high gene count. Cellular debris and dying cells lead to extensive mitochondrial contamination. Considering distribution of these parameters in this dataset (*see* Fig. 1), we kept cells with more than 200 unique feature/gene counts (nFeature_RNA), less than 26,000 UMI counts (nCount_RNA), and less than 15% mitochondrial counts (*see* **Note 1**):

```
> pbmc1[["percent.mt"]] <- PercentageFeatureSet(pbmc1, pat-
tern = "^MT-")
> FeatureScatter(pbmc1, feature1 = "nCount_RNA", feature2 =
"percent.mt")
> FeatureScatter(pbmc1, feature1 = "nCount_RNA", feature2 =
"nFeature_RNA")
> pbmc1 <- subset(pbmc1, subset = nFeature_RNA > 200 &
nCount_RNA < 26000 & percent.mt < 15)
```

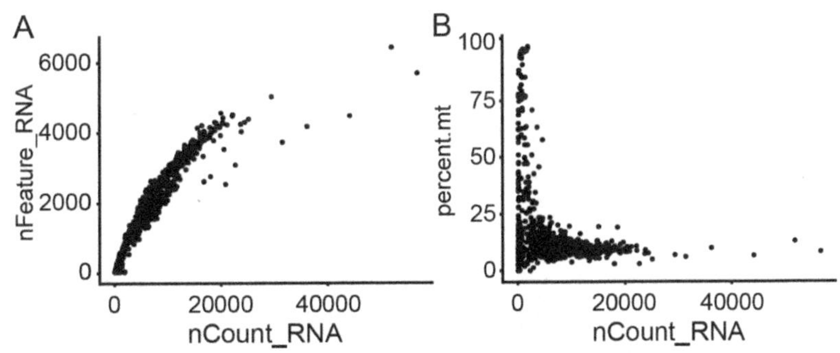

Fig. 1 Basic data information for quality control. (**a**) Scatter plot of nCount_RNA and nFeature_RNA. (**b**) Scatter plot of nCount_RNA and percent.mt.

3.1.3 Data Normalization

After QC, we normalized the data with function *LogNormalize* that normalizes the feature expression levels by the total expression for each cell.

```
> pbmc1 <- NormalizeData(pbmc1)
Performing log-normalization
0%   10   20   30   40   50   60   70   80   90   100%
[----|----|----|----|----|----|----|----|----|----|
**************************************************|
```

3.1.4 Identification of Highly Variable Features

A subset of features exhibiting high cell-to-cell variation was selected to highlight biological signal in single cell data. We implemented *FindVariableFeatures* function to calculate highly variable features, returning selected number of features (2000 by default) (*see* Fig. 2).

```
> pbmc1 <- FindVariableFeatures(pbmc1)
Calculating gene variances
0%   10   20   30   40   50   60   70   80   90   100%
[----|----|----|----|----|----|----|----|----|----|
**************************************************|
Calculating feature variances of standardized and clipped values
0%   10   20   30   40   50   60   70   80   90   100%
[----|----|----|----|----|----|----|----|----|----|
**************************************************|
> VariableFeaturePlot(pbmc1)
```

3.1.5 Data Scaling

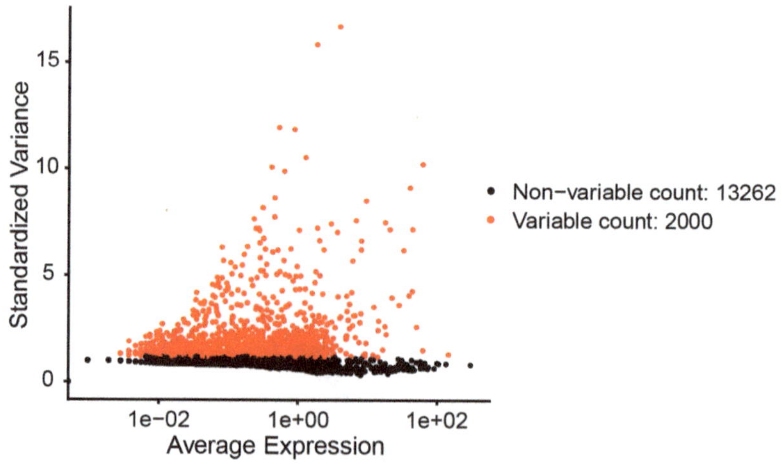

Fig. 2 Cell-to-cell variation of each feature, with each red point represents a highly variable feature

Scaling makes each gene equal weight, thus highly expressed genes do not dominate in later analyses. We implemented function *ScaleData* that scales the mean and variance of each feature across cells is 0 and 1, respectively.

```
> pbmc1 <- ScaleData(pbmc1, features = rownames(pbmc1))
Centering and scaling data matrix
|=============================================================
==| 100%
```

3.2 Visualization of scRNA-seq Data Using t-SNE

3.2.1 Principal Component Analysis (PCA)

We conducted PCA on the scaled highly variable features, returning 50 PCs:

```
> pbmc1 <- RunPCA(pbmc1, features = VariableFeatures(object =
pbmc1), npcs = 50)
```

3.2.2 Cell Clustering

To investigate cell properties, it is important to cluster cells into different groups based on their expression pattern. We clustered cells using *FindNeighbors* and *FindClusters* function based on the first 30 PCs (*see* **Note 2**) with resolution set as 1 (*see* **Note 3**):

```
> pbmc1 <- FindNeighbors(pbmc1, dims = 1:30)
Computing nearest neighbor graph
Computing SNN
> pbmc1 <- FindClusters(pbmc1, resolution = 1)
Modularity Optimizer version 1.3.0 by Ludo Waltman and Nees Jan
van Eck
Number of nodes: 1070
Number of edges: 41299
Running Louvain algorithm...
0%   10   20   30   40   50   60   70   80   90   100%
[----|----|----|----|----|----|----|-----|-----|-----|
**************************************************|
Maximum modularity in 10 random starts: 0.8079
Number of communities: 11
Elapsed time: 0 seconds
```

3.2.3 Running t-SNE

t-SNE infers the underlying manifold of the data to place similar cells together in 2-dimensional space. We visualized scRNA-seq data by running *RunTSNE* function, with the same PCs as input for the cell clustering (30 PCs) (*see* **Note 2**).

```
> pbmc1 <- RunTSNE(pbmc1, dims = 1:30, tsne.method = "Rtsne")
```

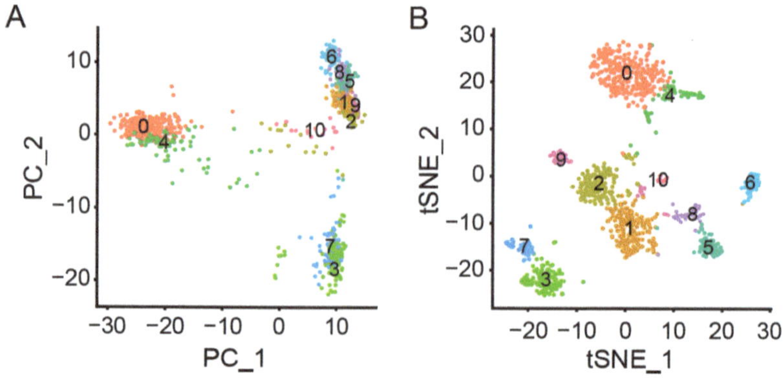

Fig. 3 Visualization of scRNA-seq data using PCA (**a**) and t-SNE (**b**). Each pair of color and number indicates an inferred cluster

3.2.4 Visualization of Single Cell RNA-seq Data Using t-SNE or PCA

Both t-SNE and PCA are used for visualization of single cell RNA-seq data, which greatly facilitate identification of cellular heterogeneity, searching new cell type, inferring cell relationship and so on. PCA is widely used for visualization of single cell data during early single cell studies. PCA visualization can be plotted using *DimPlot* with reduction as "pca" (*see* Fig. 3a):

```
> DimPlot(object = pbmc1, reduction = "pca", label = TRUE)
```

t-SNE is widely used for visualization of single cell RNA-seq data due to its advantage in showing cell clusters. Both functions *DimPlot* or *TSNEPlot* could display single cell RNA-seq data on the 2D plot (*see* Fig. 3b).

```
> DimPlot(object = pbmc1, reduction = "tsne", label = TRUE)
> TSNEPlot(object = pbmc1, label = TRUE)
```

3.2.5 Visualization of Feature Expression on t-SNE Projection

Function *FindAllMarkers* identifies cluster-specific genes (*see* **Note 4**). Seurat provides a function *FeaturePlot* to present features expression level at single cell level on t-SNE. We displayed cluster-specific expression genes on t-SNE, which provide nice visualization for distinguishing different cell clusters (*see* Fig. 4):

```
> cluster.markers <- FindAllMarkers(pbmc1)
> FeaturePlot(object = pbmc1, features = c("S100A9", "FCGR3A",
"PPBP", "CD8B", "CD3D", "CCL5", "KLRB1", "GNLY", "CD79A"),
cols = c("grey", "red"), pt.size = 0.5, reduction = "tsne")
```

3.2.6 Visualization of Other Features on t-SNE Projection

We not only could visualize cell clusters and gene expression, but also could visualize other features such as QC, counts of detected unique genes and total UMIs (*see* **Note 5**) (*see* Fig. 5). Therefore, t-SNE visualization gives us a great convenience to present cell features.

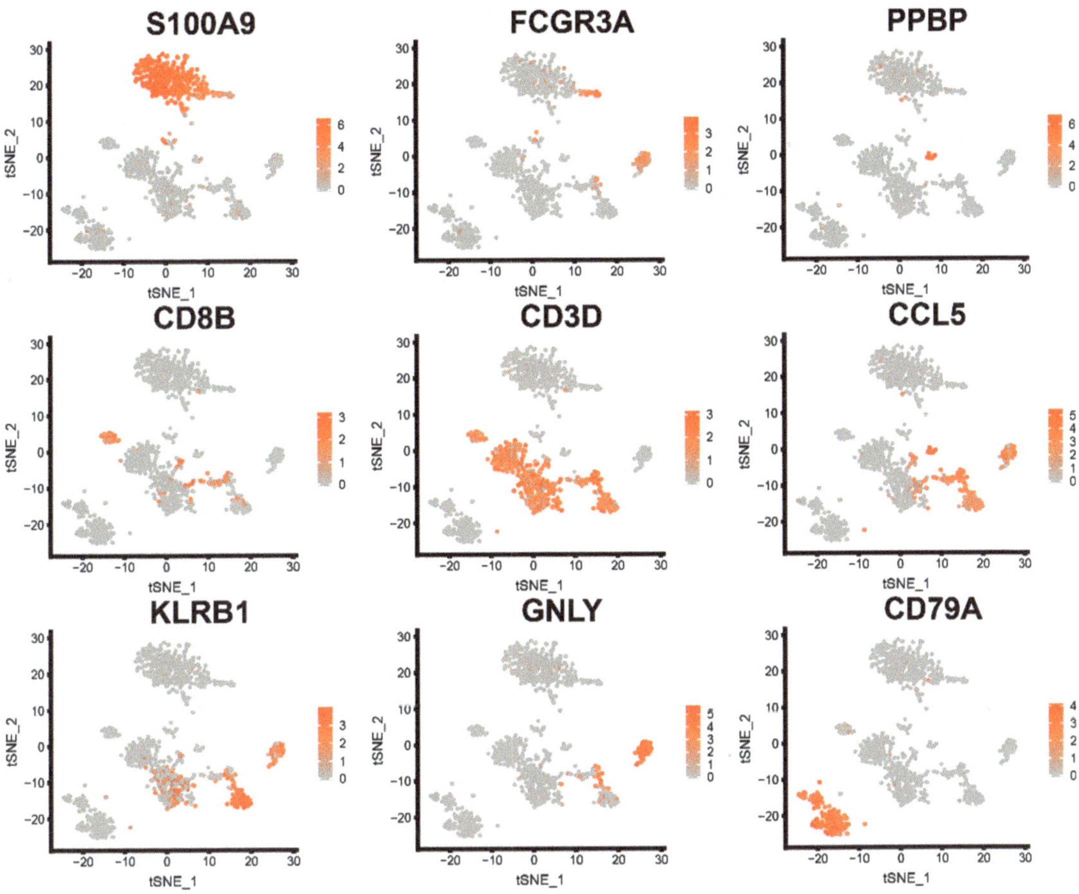

Fig. 4 Visualization of feature changes on t-SNE projection. Nine cluster-specific genes are visualized using t-SNE

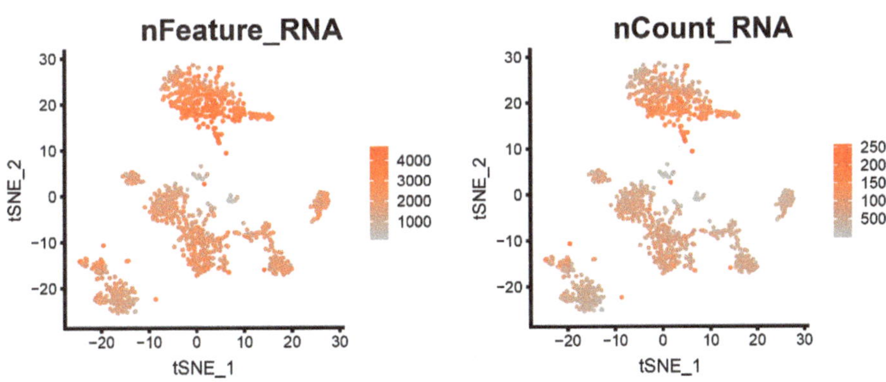

Fig. 5 Visualization of nFeature_RNA and nCount_RNA on t-SNE projection. nFeature_RNA and nCount_RNA represent number of detected genes and number of UMIs, respectively

```
> FeaturePlot(object = pbmc1, features = c("nFeature_RNA",
"nCount_RNA"), cols = c("grey", "red"), pt.size = 0.5, reduc-
tion = "tsne")
```

4 Notes

1. QC metrics include number of detected features/genes (nFeature_RNA), number of detected molecules/UMI (nCount_RNA), and percentage of reads mapped to mitochondrial genome (percent.mt). We could explore the QC metrics and customize the cell filtering criteria based on the properties of single cell data. For example, We could filter out cells that have >5% mitochondrial counts (subset = percent.mt < 5). We could filter out cells that have unique feature counts over 2500 or less than 200 (subset = nFeature_RNA > 200 & nFeature_RNA < 2500).
2. Although function *JackStrawPlot* or *ElbowPlot* was proposed to infer the dimensionality of the data, accurate determination of the number PCs for cell clustering is still a challenge. We suggest exploring different number of PCs, null models and heuristic for choosing suitable number of PCs.
3. Different resolutions would lead different number of clusters, indicating optimization of resolution is important for further analysis. Setting resolution between 0.4 and 1.2 usually returns nice results for datasets of around 3000 cells. Optimal resolution often increases for larger datasets. Sometimes, we would like to try different resolutions to optimal the results for better biological insight.
4. In many situations, we have no idea about what kinds of cells are included in our dataset. Therefore, we may define cell type based on clusters and cluster-specific genes. Using cluster-specific genes, we can observe the expression of these genes using t-SNE.
5. Any customized features of single cell can be visualized on t-SNE projection.

Acknowledgments

This work was supported by National Key R&D Program of China (2018YFC1004500), the National Science Foundation of China (81872330 and 31741077), and Basic Research Grant from Science and Technology Innovation Commission of Shenzhen Municipal Government (JCYJ20170817111841427).

References

1. Tirosh I, Venteicher AS, Hebert C, Escalante LE, Patel AP, Yizhak K, Fisher JM, Rodman C, Mount C, Filbin MG, Neftel C, Desai N, Nyman J, Izar B, Luo CC, Francis JM, Patel AA, Onozato ML, Riggi N, Livak KJ, Gennert D, Satija R, Nahed BV, Curry WT, Martuza RL, Mylvaganam R, Iafrate AJ, Frosch MP, Golub TR, Rivera MN, Getz G, Rozenblatt-Rosen O, Cahill DP, Monje M, Bernstein BE, Louis DN, Regev A, Suva ML (2016) Single-cell RNA-seq supports a developmental hierarchy in human oligodendroglioma. Nature 539(7628):309–313. https://doi.org/10.1038/nature20123
2. Yu Y, Tsang JC, Wang C, Clare S, Wang J, Chen X, Brandt C, Kane L, Campos LS, Lu L, Belz GT, McKenzie AN, Teichmann SA, Dougan G, Liu P (2016) Single-cell RNA-seq identifies a PD-1(hi) ILC progenitor and defines its development pathway. Nature 539(7627):102–106. https://doi.org/10.1038/nature20105
3. Sebe-Pedros A, Saudemont B, Chomsky E, Plessier F, Mailhe MP, Renno J, Loe-Mie Y, Lifshitz A, Mukamel Z, Schmutz S, Novault S, Steinmetz PRH, Spitz F, Tanay A, Marlow H (2018) Cnidarian cell type diversity and regulation revealed by whole-organism single-cell RNA-seq. Cell 173(6):1520–1534 e1520. https://doi.org/10.1016/j.cell.2018.05.019
4. Bendall SC, Simonds EF, Qiu P, Amir el AD, Krutzik PO, Finck R, Bruggner RV, Melamed R, Trejo A, Ornatsky OI, Balderas RS, Plevritis SK, Sachs K, Pe'er D, Tanner SD, Nolan GP (2011) Single-cell mass cytometry of differential immune and drug responses across a human hematopoietic continuum. Science 332(6030):687–696. https://doi.org/10.1126/science.1198704
5. Amirel AD, Davis KL, Tadmor MD, Simonds EF, Levine JH, Bendall SC, Shenfeld DK, Krishnaswamy S, Nolan GP, Pe'er D (2013) viSNE enables visualization of high dimensional single-cell data and reveals phenotypic heterogeneity of leukemia. Nat Biotechnol 31(6):545–552. https://doi.org/10.1038/nbt.2594
6. van der Maaten L, Hinton G (2008) Visualizing data using t-SNE. J Mach Learn Res 9:2579–2605
7. Mahfouz A, van de Giessen M, van der Maaten L, Huisman S, Reinders M, Hawrylycz MJ, Lelieveldt BP (2015) Visualizing the spatial gene expression organization in the brain through non-linear similarity embeddings. Methods 73:79–89. https://doi.org/10.1016/j.ymeth.2014.10.004
8. Linderman GC, Rachh M, Hoskins JG, Steinerberger S, Kluger Y (2019) Fast interpolation-based t-SNE for improved visualization of single-cell RNA-seq data. Nat Methods 16(3):243–245. https://doi.org/10.1038/s41592-018-0308-4
9. van der Maaten L (2014) Accelerating t-SNE using tree-based algorithms. J Mach Learn Res 15:3221–3245
10. Butler A, Hoffman P, Smibert P, Papalexi E, Satija R (2018) Integrating single-cell transcriptomic data across different conditions, technologies, and species. Nat Biotechnol 36(5):411–420. https://doi.org/10.1038/nbt.4096

Chapter 9

Use of SuperCT for Enhanced Characterization of Single-Cell Transcriptomic Profiles

Jianmei Zhong and Wei Lin

Abstract

Characterizing the cell identity in a heterogeneous tissue is essential to the in-depth understanding of this sample. Existing single-cell techniques (e.g., flow cytometry or in situ cell florescent imaging) allow us to do so using the high/low signal of a combination of multiple signature molecules or even of a single marker. Recent advance of single-cell RNA-seq technology profiles the entire transcriptome of individual cells. Using a few marker genes to characterize cell type in this new technique is less reliable due to the high noise level and the dynamic transcription behavior. Nonetheless, the "noisy" but high-throughput transcriptome profiles provide adequate information to reveal the cellular identity and to understand the detail of the molecular characteristics. In this chapter, we will demonstrate a new method that is based on the supervised learning of the single-cell transcriptome profiles of many different known cell types. We will demonstrate how this technique solves the cellular identity problem.

Key words Single-cell RNA-seq, Supervised classifier, Artificial neural network

1 Introduction

1.1 Samples of Heterogeneous Cellular Constituents

In the study of a multicellular sample of heterogeneous constituents, one of the central problems is to characterize the nontrivial cell subpopulations in this sample based on their canonical markers of the cellular lineages. Only if the cellular populations are clearly defined and the molecular signals are deconvoluted, we are able to investigate the meaningful molecular variability specific to the target cells by different conditions. Otherwise, the comparison between these samples is more like between apples and oranges, which is senseless and misleading.

1.2 Strategies of Characterizing Single-Cell Expression Profiles

The machine learning algorithms have been acknowledged and widely used in the analyses of high dimensional expression data [1], such as single-cell RNA-seq (scRNA-seq) data. In the meantime, characterization of cell types in here is a typical classification problem in machine learning. Dimensional-reduction (DR) and

Benjamin L. Kidder (ed.), *Stem Cell Transcriptional Networks: Methods and Protocols*, Methods in Molecular Biology, vol. 2117, https://doi.org/10.1007/978-1-0716-0301-7_9,

unsupervised clustering (UC) algorithms such as PCA, ICA, and tSNE have been used in the analyses of the scRNA-seq data to solve the classification problem [2–6]. Nonetheless, the major challenge is that the interpretation of the unsupervised clusters highly depends on the investigators' experience, which makes this task somehow arbitrary. A supervised classifier will make things a lot easier by memorizing the features of the known cell types in the model rather than in individual human memory. This is actually a big advantage. Besides, there is less worry about the complicated DR and UC implementation and the interpretation of the output clusters.

In the following paragraphs, we will explain how to use SuperCT, an artificial-neural-net-based (ANN-based) classifier to characterize cell identities [7].

1.3 Supervised Classifiers and Model Selections

A successful supervised-classifier consists of a mathematical model that solves the classification problem and the representative training datasets that can be used to "learn" the features of the subjects and save as the model parameters. In machine learning, there are multiple options of the models, such as decision tree, Naïve Bayesian classifier, support vector machine, and ANN by different structures. Different models have their advantages and disadvantages [8, 9].

With the recent development of deep learning methods, ANN has been successful in the application of solving the BigData problem without the assumption of the distribution fit in the linear space. In the real world, many of the signals acquired from the experiments do not explicitly fit the known distribution in the linear space or are not even necessarily distributed in the linear space. ANN model is a flexible option. Most importantly, there are some ANN structures that mimic human learning in the classification, feature extraction, and comparison over the signal profiles, which meet the multiple objectives of the single-cell data analysis. In addition, the ANN structure can be modified and the weight matrices over the layers can be transferred to the new applications in order to save the resources of the new training. We hereby proposed an ANN-based supervised classifier to characterize the known cell types, whose framework is shown as Fig. 1.

Some people had the complaint that the nonlinearity of ANN classification is hard to interpret. Nonetheless, with the development of the visualization tools that allow us to monitor the status of the hidden nodes and the evolution of the weights in the training process, the interpretability problem can be solved.

1.4 Structure of SuperCT ANN

SuperCT is a fully connected ANN structure that can be used to characterize the established cell types from the training datasets [7]. In the ANN model of the most recent version, there are 16,013 input nodes for the input gene expressions, which are corresponding to the genes that are used in the profiling of

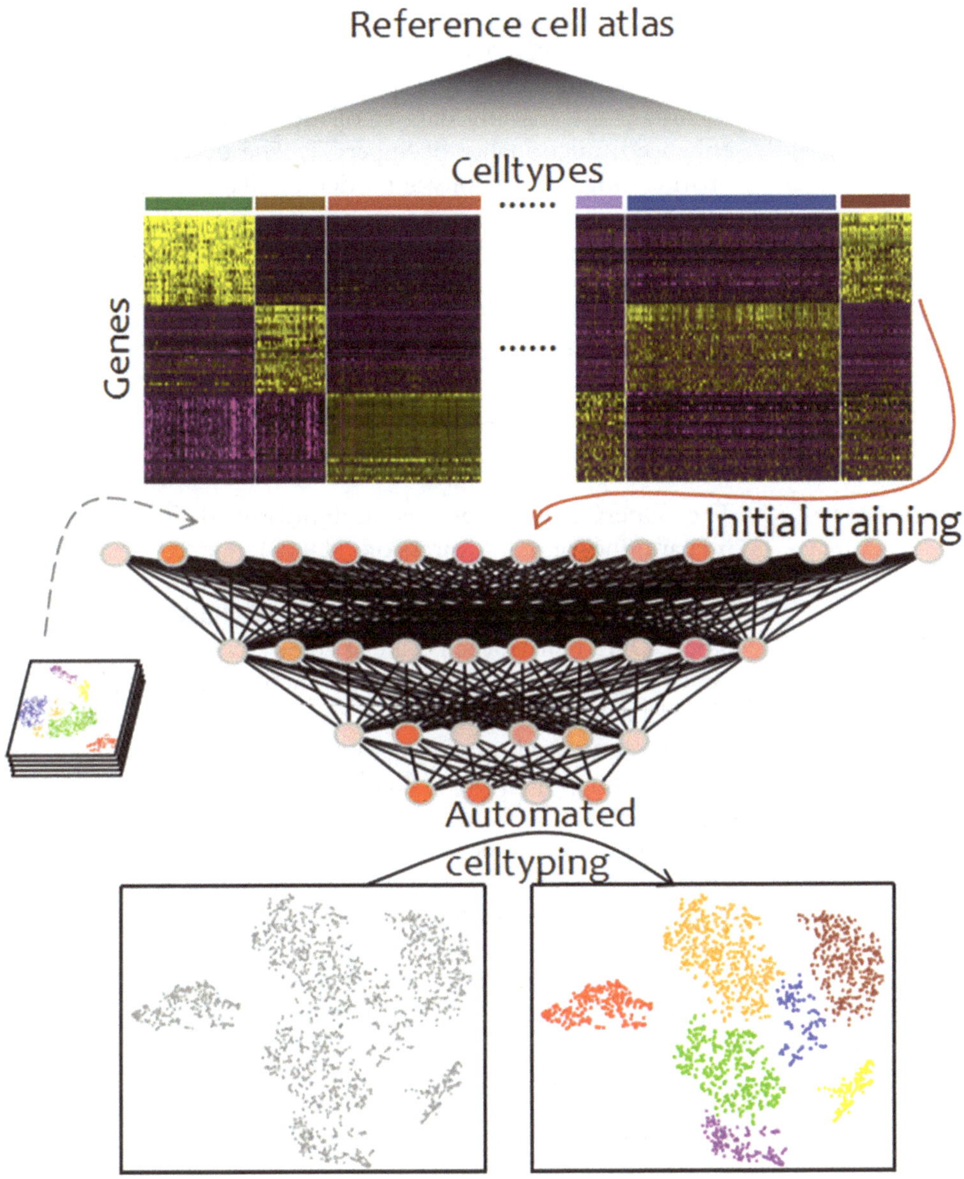

Fig. 1 The flowchart of SuperCT

single-cell transcriptome. There are two hidden layers with 200 and 100 nodes, respectively. In the output layer, there are 38 nodes, which are corresponding to the 38 defined cell types (categories) defined in the training datasets. It is notable that one of the 38 defined cell types is called unknown, which allows us to handle the unknown cell type, which indicates the possible discovery of novel cell type if "unknown" cell type comes out in the prediction.

1.5 Expansion of SuperCT Capability

Instead of a prediction tool, the SuperCT is a framework that continues to evolve to adapt the growing body of the scRNA-seq training datasets and new cell-typing knowledge to the prediction. Including the training dataset of more cell types will acquire the enhanced functionality of SuperCT. The developer will continue provide different versions of models to adapt to the specific prediction tasks. The corresponding cell-type description sheet will be available along with the prediction model. Moreover, the SuperCT developers can be contacted to discuss the training and the characterization of the specific cell-types tasks defined by the user, if the suitable training dataset is available.

2 Materials

2.1 Input File

The SuperCT prediction is straightforward. The users need to prepare the digital expression (DGE) matrix from the typical scRNA-seq assays on the prevalent platform such as 10×-Genomics [10], DropSeq [11], MicroWell-seq [12], and MARS-seq [13]. In a typical DGE matrix, each cell will give a column of copy numbers of detected gene based on the unique molecular index (UMI). Multiple cells will make a matrix with each row giving the copy numbers for an annotated gene on these cells. For example, on the 10x-Genomics platform, CellRanger software generates digital expression matrix file named "matrix.mtx" and the gene annotation file named "genes.tsv". These two files will be adequate for SuperCT to predict the cell types. DGE matrix is also the standard submission format for the scRNA-seq data for the Gene Expression Omnibus database (GEO) of NCBI.

Figure 2 shows the layout of the input matrix file that is used in the python stand-alone prediction program of SuperCT. Also, an

	cell_1	cell_2	cell_3	cell_4	cell_5	cell_6	cell_7	cell_8	cell_9	cell_10
TMSB4X	42	19	3	6	13	11	2	18	58	35
ACTB	19	10	3	2	4	5	1	1	10	4
PPBP	11	7	0	0	4	8	0	5	6	3
B2M	8	3	1	4	3	4	2	1	18	14
GNG11	6	3	0	0	1	0	0	0	2	1
SDPR	6	3	0	2	1	1	0	0	2	1
NCOA4	5	2	0	1	0	3	1	1	0	1
OAZ1	5	2	0	2	0	3	0	2	2	1
PF4	5	1	0	1	2	2	0	6	3	2
TUBA4A	4	0	0	0	0	0	0	0	0	0
CLU	3	1	0	1	1	0	0	1	1	0
......										

Fig. 2 The demo of input DGE format file

example input DGE file in CSV format can be found in the example folder of the GitHub repository of SuperCT (https://github.com/weilin-genomics/SuperCT).

2.2 SuperCT Model and Cell-Type Description Sheet at GitHub

As mentioned, we have set up a GitHub repository for the development of SuperCT and the cell-type predictions. The model and the corresponding cell-type description sheet are downloadable from the model folder. The SuperCT ANN model is implemented using keras API (*see* **Note 1**).

3 Methods

3.1 Online Prediction and Visualization Platform

For the beginner scRNA-seq user, an online application (https://sct.lifegen.com) is available for the fast and easy prediction of the cell types by uploading the 10×-Genomics output files or the DGE files. On the upload page, the user will need to input the sample descriptions, select the organism, and the input DGE format (Fig. 3).

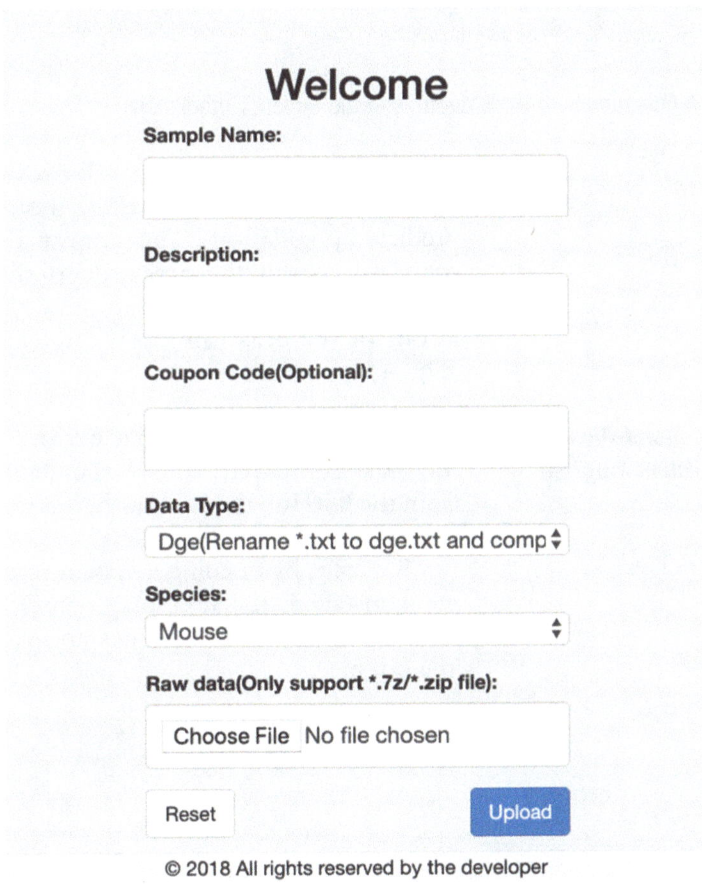

Fig. 3 The DGE matrix upload page for the SuperCT online application

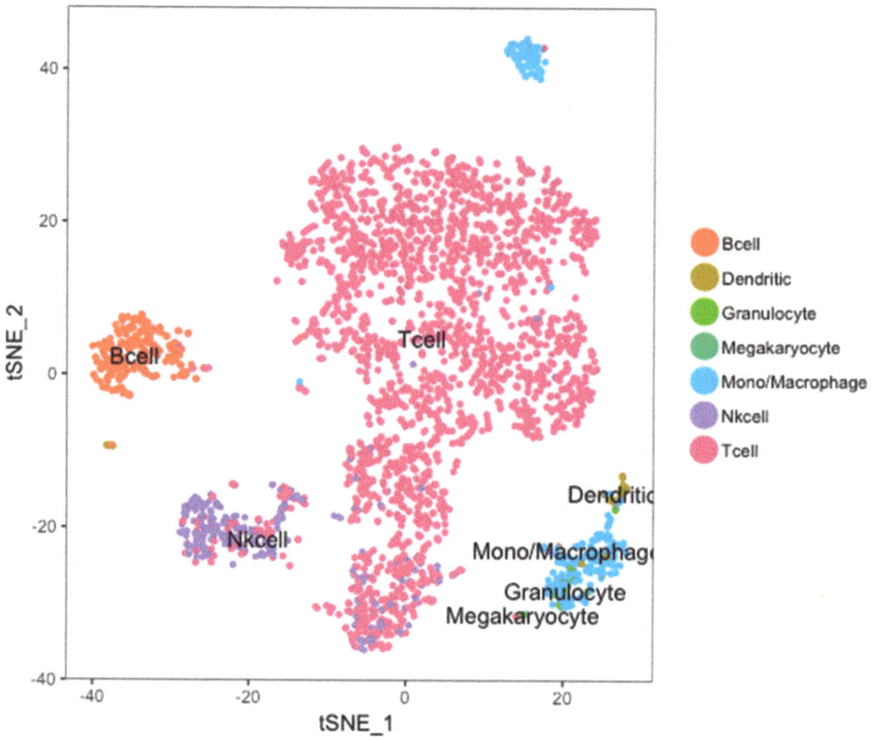

Fig. 4 An example of tSNE layout with the SuperCT predictions

This online application will provide other visualization tools that help to further explore the molecular signatures in different predicted populations. The user can also download the predictions of the cells type in the same order of the input single-cell expression profiles. The user can also browse the predictions on the tSNE layout of the cells using this SuperCT online visualization tools (Fig. 4).

3.2 Stand-Alone Prediction Program

The more advanced user may perform the SuperCT prediction on the local computer. To use the stand-alone prediction program from the GitHub repository, the user needs to download and install python 3 (*see* **Note 2**).

Basically, by opening a system terminal with python3 environment installed, the user may type a command line as below to predict the cell-type identities and retrieve the cell-type statistics in a pdf file.

```
python pred_types.py -i < input_dge.csv > -o < cell_type_output.csv > -p -m < model.h5 > -s < human/mouse> - t < id2type.csv>
```

The prediction program takes the input single-cell expression profiles, *i.e.* "input_dge.csv" and generates the output file, that is, "cell_type_output.csv" by running this command. If "-p" is used, a bar plot in PDF-format will be generated to show the cell counts of each cell type.

For the current study, other than the "input_dge.csv" file which needs to be prepared by the user, the other files in need should be found from the GitHub repository of SuperCT. The command parameters are explained as **Notes 3–6** in detail. Some minor issues when running this program are also described in **Notes 7–9**.

3.3 Prepare Input DGE File from 10× Genomics Platform

The 10xGenomics standard matrix files are very commonly seen in most of the time. The three standard files, "matrix.mtx", "genes.tsv", and "barcodes.tsv", can be transformed into the format that works for 'pred_types.py' as well. The python transformation program is available on the SuperCT GitHub repository as well, named "transform_10x_to_dge.py." The following command line shows how to run this program on a terminal.

```
python transform_10x_to_dge.py -m < matrix.mtx > -g < genes.
tsv > -b < barcodes.tsv > -o < transformed_dge.csv>
```

In here, "transformed_dge.csv" can be used as "input_dge.csv" in the stand-alone prediction program described in Subheading 3.2.

4 Notes

1. SuperCT ANN was implemented using Keras API with TensorFlow as backend. A SuperCT model is usually saved after training and loaded for future prediction. In the stand-alone prediction program, the model is one of the input arguments, shown as "-m model.h5".
2. The stand-alone prediction program is built on python 3, which is downloadable from https://www.python.org/download/releases/3.0/. The stand-alone prediction program has been tested on the mainstream operating systems (e.g., Windows, MacOS, and Unix/Linux).
3. After the clone of the repository (https://github.com/weilin-genomics/SuperCT) is downloaded on the local computer, the user can run the built-in python program "pred_types.py" easily. The python program requires a few python modules, which include keras, matplotlib, scipy, pickle, and sklearn. The

availability check of these modules can be done using the command like "python3 -c 'import module_name'". For installation of the missing one, the command line "pip3 install module_name" can be used if any module is missing by the error report.

4. A cell-type description sheet such as "id2type.csv" that covers the predictable cell-types is included corresponding to SuperCT model "model.h5" in the same folder. Both need to be included when running the prediction program.
5. The "inthomgenes.csv" lists the 16,013 genes that are homologous between human and mouse species. For the most recent SuperCT model, only these 16,013 genes are used for training. Therefore, the input DGE matrix will be rearranged within the prediction code. The redundant gene expression values will not be taken into consideration in either training or prediction. For the expression value of the gene in 16,013-list but missing in the input DGE, they will be assigned as zero.
6. The species of the sample can be specified as either "human" or "mouse." "human" is the default value.
7. Memory overflow could be reported when loading a too large matrix; rebooting the system may free some memory space and make the program go through. The other solution will be splitting a large matrix into a few small matrices with fewer columns of cells.
8. There could also be warnings due to the version conflict of the modules. In most of the time, the program will still go through and generate the prediction result.
9. The output file "cell_type_output.csv" includes two columns, one for the cell-barcodes and the other for the cell types of prediction. The resultant prediction can be used for the downstream analyses of these cells.

References

1. Hira ZM, Gillies DF (2015) A review of feature selection and feature extraction methods applied on microarray data. Adv Bioinforma 2015:13
2. Andrews TS, Hemberg M (2018) Identifying cell populations with scRNA-seq. Mol Asp Med 59:114–122
3. Butler A, Hoffman P, Smibert P et al (2018) Integrating single-cell transcriptomic data across different conditions, technologies, and species. Nat Biotechnol 36:411
4. Shekhar K, Lapan SW, Whitney IE et al (2016) Comprehensive classification of retinal bipolar neurons by single-cell transcriptomics. Cell 166(5):1308–1323.e1330
5. Wang B, Zhu J, Pierson E et al (2017) Visualization and analysis of single-cell RNA-seq data by kernel-based similarity learning. Nat Methods 14:414
6. Yao Z, Mich JK, Ku S et al (2017) A single-cell roadmap of lineage bifurcation in human ESC models of embryonic brain development. Cell Stem Cell 20(1):120–134
7. Xie P, Gao M, Wang C et al (2019) SuperCT: a supervised-learning framework for enhanced characterization of single-cell transcriptomic profiles. Nucleic Acids Res 47(8):e48–e48
8. Pranckevicius T, Marcinkevičius V (2017) Comparison of naive bayes, random forest, decision tree, support vector machines, and

logistic regression classifiers for text reviews classification. Baltic J Modern Computing 5:221–232

9. Huang, J., J. Lu,C. X. Ling (2003). Comparing naive bayes, decision trees, and SVM with AUC and accuracy. Proceedings of the Third IEEE International Conference on Data Mining, IEEE Computer Society: 553–556
10. Zheng GXJ, Terry M, Belgrader P et al (2017) Massively parallel digital transcriptional profiling of single cells. Nat Commun 8:14049
11. Macosko EZ, Basu A, Satija R et al (2015) Highly parallel genome-wide expression profiling of individual cells using nanoliter droplets. Cell 161(5):1202–1214
12. Han X, Wang R, Zhou Y et al (2018) Mapping the mouse cell atlas by microwell-seq. Cell 173 (5):1307
13. Keren-Shaul H, Kenigsberg E, Jaitin DA et al (2019) MARS-seq2.0: an experimental and analytical pipeline for indexed sorting combined with single-cell RNA sequencing. Nat Protoc 14(6):1841–1862

Chapter 10

Interactive Alternative Splicing Analysis of Human Stem Cells Using *psichomics*

Nuno Saraiva-Agostinho and Nuno Luís Barbosa-Morais

Abstract

Alternative splicing (AS) generates functionally distinct transcripts and is involved in multiple cellular processes, including stem cell differentiation. Several epithelial-to-mesenchymal transition-related splicing factors have also been associated with pluripotency. Concomitantly with the interest in studying AS in stem cell biology, the advent of next-generation sequencing of RNA (RNA-seq) has increased the public availability of transcriptomic data and enabled genome-wide AS studies. To facilitate performing such analyses in large publicly available or user-provided transcriptomics datasets, the *psichomics* R package provides an easy-to-use interface and efficient data visualization tools for AS quantification and integrative analyses of AS and gene expression data.

psichomics is employed herein to study AS changes between human stem cells and fibroblasts, based on dimensionality reduction, and median- and variance-based differential AS and gene expression analyses. Putative RNA-binding protein regulators involved in those alterations are then identified based on correlation analyses in large cohorts of human tissue transcriptomes. We identified several alterations, both novel and previously reported, in alternative splicing events and in the expression of their candidate regulators that are associated with stem cell differentiation into fibroblasts.

Key words Differential gene expression analysis, Differential splicing analysis, *psichomics*, RNA-seq, Integrative analyses, Stemness

1 Introduction

Alternative splicing (AS) is a dynamically regulated cellular process that promotes transcriptome diversity in eukaryotes [1–3]. The distinct functions of alternative transcripts that originate from a single gene are paramount for cell and tissue homeostasis, being involved in cell survival and other physiological processes [1]. However, AS regulation can also be exploited by cancer, among other diseases, as demonstrated by the key role of splicing regulators in multiple cancer hallmarks [1, 4, 5].

In the last decade, AS has been increasingly associated with pluripotency [6, 7]. Several splicing factors have been reported to be involved in it, including SRSF3, MBNL1/2, RBFOX2, and

Benjamin L. Kidder (ed.), *Stem Cell Transcriptional Networks: Methods and Protocols*, Methods in Molecular Biology, vol. 2117, https://doi.org/10.1007/978-1-0716-0301-7_10,

U2AF1 [6, 8–10]. For instance, MBNL1/2 regulates the mutually exclusive inclusion of two *FOXP1* exons, inducing a switch from its pluripotency-associated FOXP1-ES protein isoform, that promotes the expression of *OCT4*, *NANOG*, and other key pluripotency transcription factors, to the canonical differentiation-inducing FOXP1 isoform [7].

The early stage of somatic cell reprogramming, characterized by acquisition of pluripotency features, is related with mesenchymal-to-epithelial transition, a crucial development-related process affecting cell polarity and adhesion that is mediated by the aforementioned splicing regulators [6, 11]. Consistently, the AS modulation of epithelial-to-mesenchymal transition is linked with both cancer progression and metastasization and with the generation of cancer stem cells, characterized by enhanced self-renewal, proliferation, and other stemness properties [6, 11, 12].

The interest in studying the role of AS in stem cells has trailed the ever-increasing availability of RNA sequencing (RNA-seq) data. To assist scientists in performing alternative splicing analyses of publicly available or user-provided RNA-seq datasets, we have developed *psichomics*, an R package with a user-friendly graphical interface for AS quantification, analysis and visualization, especially optimized for large datasets [13]. *psichomics* integrates survival, principal component and exploratory and differential gene expression and splicing analyses. Unlike most AS analysis tools, *psichomics* takes advantage of preprocessed splice junction read counts from large publicly available RNA-seq datasets. These include data for over 30 human cancer types from The Cancer Genome Atlas (TCGA) [14], for multiple human tissues from the Genotype-Tissue Expression (GTEx) project [15] and for over 2000 studies from the Sequence Read Archive (SRA) through the recount2 project [16]. Based on these or user-provided data, *psichomics* quantifies AS by the proportion of splice junction read counts supporting inclusion of the alternative sequence (Fig. 1), the so-called percent spliced-in (PSI) metric [2].

In this chapter, we will analyze genetically (i.e., isogenic) and not genetically (nonisogenic) matched human fibroblast and (embryonic and induced-pluripotent) stem cell RNA-seq samples [17], together with GTEx and TCGA data, using *psichomics* to highlight putative differentially spliced events between such conditions and candidate RNA-binding protein regulators of those events.

We will automatically retrieve preprocessed data for the chosen dataset from recount2. We will conduct a typical workflow of filtering and normalizing gene expression and quantifying and filtering AS, followed by statistical analyses to identify differentially expressed genes and spliced events between isogenic human fibroblasts and stem cells. We will then correlate the expression of genes encoding for RNA-binding proteins with the PSI values of the most

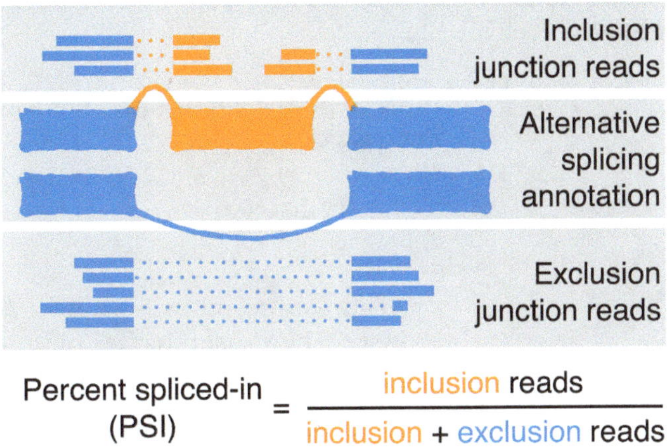

Fig. 1 In *psichomics*, alternative splicing is quantified based on the percent spliced-in (PSI) metric. Each annotated AS event is quantified based on read counts that support the inclusion of a given alternative splicing sequence (in orange) over the total sum of read counts supporting either the exclusion (in blue) or the inclusion of that sequence

interesting AS events, in the stem cell dataset as well as across multiple tissues (GTEx data) and cancer types (TCGA data), to identify among the former putative regulators of the latter. Finally, the pan-cancer prognostic value of a select AS event will be assessed.

We hereby show an example of an integrative workflow (both in terms of methods and datasets used) that can be performed using *psichomics*.

2 Materials

The protocol described in this chapter requires the latest version of *R* (https://www.r-project.org) [18], *Bioconductor* [19], and *psichomics* [13]. Code for the analysis performed herein is publicly available in https://github.com/nuno-agostinho/stem-cell-analysis-in-psichomics.

To download and install *Bioconductor* and *psichomics*, open *R* and type the following therein:

```
> if (!requireNamespace("BiocManager", quietly = TRUE))
> install.packages("BiocManager")
> BiocManager::install("psichomics")
```

3 Methods

In *R*, type the following command to load and start the visual interface of *psichomics*:

```
> library(psichomics)
> psichomics()
```

This command will launch *psichomics* in the default web browser. The visual interface of *psichomics* (Fig. 2) is composed by four sections (the main tabs in the navigation bar at the top):

- **Data** section to download, load, and manipulate TCGA, GTEx, SRA, and user-provided datasets (*see* **Note 1**).
- **Analyses** section to analyze the loaded datasets, for example, to perform principal component, differential gene expression and splicing, survival, and correlation analyses.

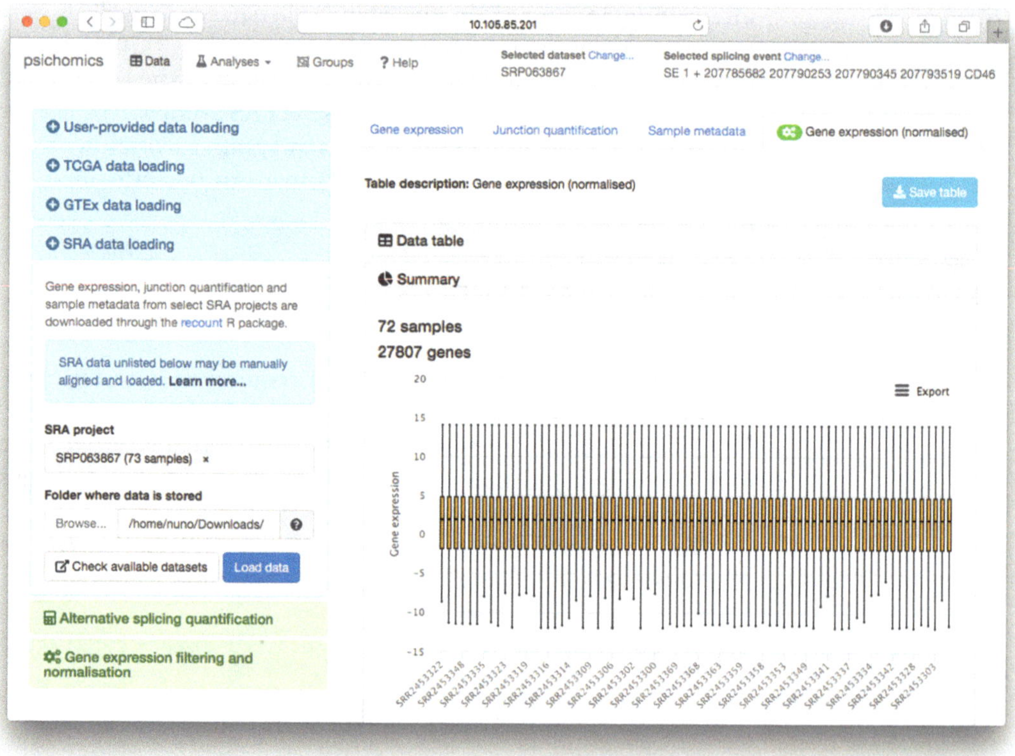

Fig. 2 Summary information on datasets. *psichomics* presents information on the loaded datasets in a dedicated tab. That information includes summary statistics, like the numbers of samples and genes profiled in each dataset, and plots, like the shown boxplot to visualize the distribution of normalized gene read counts per sample

- **Groups** section to create, manipulate (including merging, intersecting, and performing other operations between groups, as well as renaming and recoloring them), and remove groups of samples/subjects and genes/AS events that are used throughout the program.
- **Help** section to change settings and get links for tutorials, contact information and more.

3.1 Loading Data

We selected a dataset containing genetically (i.e., isogenic) and not genetically (nonisogenic) matched human induced-pluripotent stem cells (iPSC), embryonic stem cells (ESC), and fibroblasts [17] to compare changes in alternative splicing between isogenic stem cells and isogenic fibroblasts.

1. In the visual interface of *psichomics*, go to the **Data** section.
2. Click *SRA data loading* to expand the respective panel.
3. Type *SRP063867* [17] in the *SRA project* field and select the first option.
4. Click *Load data* to start downloading preprocessed RNA-seq data for all the samples in the selected SRA project. Those data include sample annotation and raw gene expression and exon–exon junction read counts. Once the files finish downloading, data will be automatically loaded into *psichomics* workspace (*see* **Note 2**).
5. The welcome screen is replaced by information summarizing each dataset (sample annotation, gene expression, and junction quantification; Fig. 2). Click on the *Sample metadata* tab and open *Data table*. This table shows information available from SRA on the samples in this project, including their respective cell types (accessible at https://www.ncbi.nlm.nih.gov/Traces/study/?acc=SRP063867).

If at any time *psichomics* is closed, the data from last session can be restored by running psichomics() again as long as the *R* session is not restarted. Data are recovered even if the program closes unexpectedly.

3.2 Gene Expression Filtering and Normalization

Gene expression filtering and normalization are performed by running functions from the *edgeR* [20] and *limma* [21] *R* packages on the loaded raw gene expression read counts.

Note that the density plot of the samples' library size (i.e.. the total number of mapped reads) features one sample in particular (SRR2453313) with relatively low read coverage (Fig. 3a). This sample may be discarded later on if shown not to be comparable with others after gene expression normalization.

1. In the **Data** section, expand the Gene expression filtering and normalization panel.

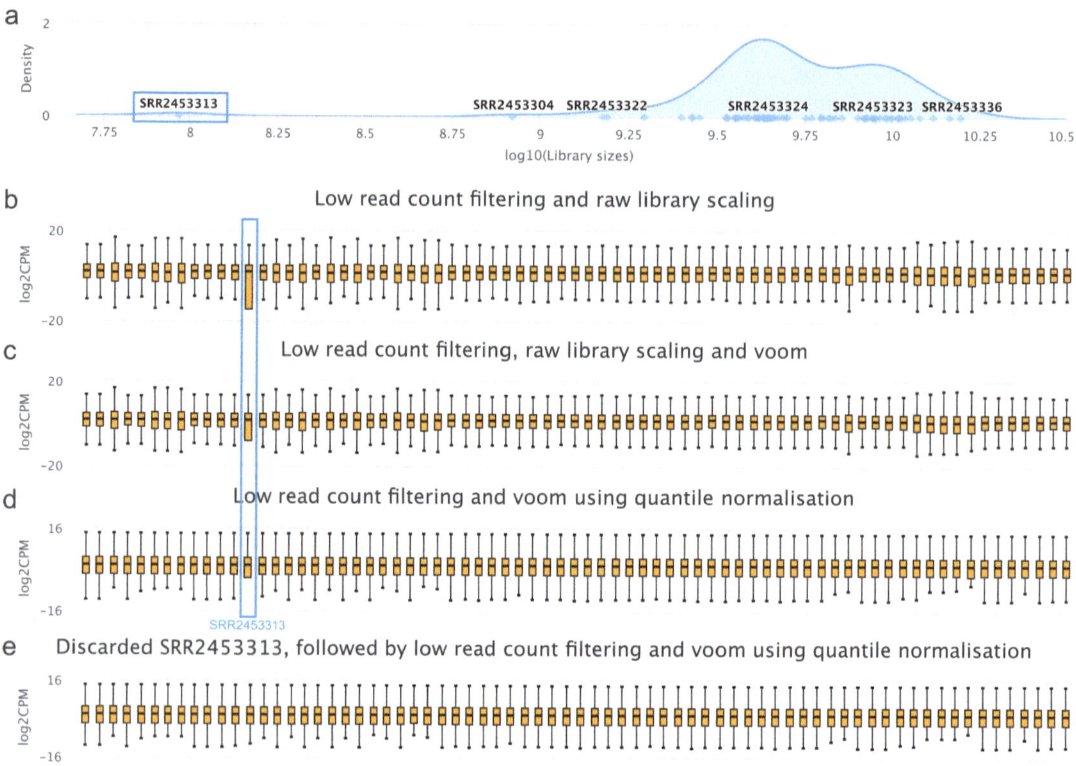

Fig. 3 Gene expression normalization. (**a**) Density plot of the distribution of raw library sizes (i.e.. total number of mapped read counts) across samples. Highlighted with the blue label is the sample with the smallest library. (**b–e**) Boxplots of distribution of gene expression in log2-transformed counts per million (log2CPM per sample after low read count filtering and raw library scaling (**b**), this procedure followed by *voom* modeling (**c**), *voom* modeling using quantile normalization instead of raw library scaling (**d**), and the latter after discarding the sample with the smallest library, highlighted in blue (**e**)

2. Select Replace unambiguous ENSEMBL gene identifiers with their gene symbols. ENSEMBL gene identifiers are kept in case of no matching gene symbols or if multiple ENSEMBL identifiers match the same gene symbol.
3. Click Filter and normalize gene expression. Genes with low numbers of read counts across samples (i.e., low expression) are discarded and gene expression data are normalized by scaling raw library sizes across samples (*see* **Note 3**).
4. The gene expression plots available within the Gene expression (normalized) tab (Fig. 3b) suggest that the default normalization is not fully effective, as very different distributions between samples are observed. We advise using voom instead, as it incorporates the mean-variance relationship of the data to normalize expression levels between samples [21]. In the Gene expression filtering and normalization panel, click Normalization, check Perform mean-variance modeling using voom and click Filter and normalize gene expression.

5. When selecting the Gene expression (normalized 1) tab, the distributions remain heterogeneous (Fig. 3c), so we replaced the default weighted trimmed mean of M-values (TMM) [20], used to normalize for library sizes, by quantile normalization [21]. In the Gene expression filtering and normalization panel, click Normalization, change Normalization method to Quantile normalization and click Filter and normalize gene expression.
6. This more vigorous normalization of gene read counts made their distributions comparable across samples, except for that already identified as having the lowest library size (Fig. 3d). To discard this sample, in the Gene expression filtering and normalization panel, go to Sample filtering, input SRR2453313 in the Samples to discard field and click Filter and normalize gene expression. No obvious outlying gene expression distribution is apparent after discarding that sample and renormalizing (Fig. 3e). The filtered and normalized gene expression dataset should now be composed of 72 samples and 27,807 genes.

3.3 AS Quantification

The percent spliced-in (PSI) metric is commonly employed to measure the relative abundance of the inclusion isoform of an AS event [2]. For each annotated event, *psichomics* quantifies AS in PSI based on the ratio of splice (exon–exon) junction read counts that support the inclusion of the alternative sequence.

1. Go to the **Data** section and expand the *Alternative splicing quantification* panel.
2. Ensure that the *Human hg38 (2018-04-30)* AS event annotation is selected (*see* **Note 4**).
3. Select to quantify the default event types: skipped exon (SE), mutually exclusive exon (MXE), alternative first and last exon (AFE and ALE, respectively), and alternative 3' and 5′ splice site (A3SS and A5SS, respectively) (*see* **Note 5**).
4. By default, only AS events with a minimum of 10 junction read counts supporting either inclusion or exclusion of the alternative sequence are considered to avoid quantifying AS events with insufficient evidence.
5. For consistency with gene expression analysis, we discarded the sample with the lowest library size. In *Sample filtering*, input *SRR2453313* in the *Samples to discard* field.
6. Click *Quantify alternative splicing*.
7. In total, 135,717 AS events were quantified. This number is very high but only events exhibiting some variance across samples will be informative when analyzing differential splicing. We therefore recommend filtering out low-variance AS events. To do so, in the *Alternative splicing quantification*

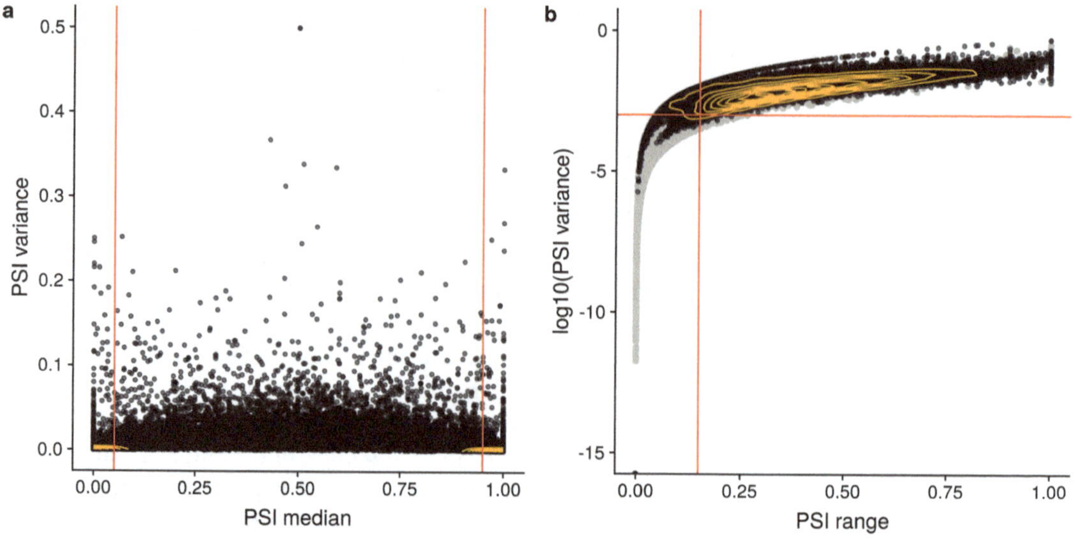

Fig. 4 Alternative splicing quantification filtering. (**a**) Selection, for further analyses, of AS events with median PSI values between 0.05 and 0.95. (**b**) Further filtering those AS events showing a PSI range above 0.15 and log10(variance) above −3. For illustration purposes, gray points represent the AS events discarded in **a**

panel, expand the *PSI Filtering* panel, click the checkbox *Filter splicing events based on their PSI values* and set the median PSI values' threshold between 0.05 and 0.95 (Fig. 4a), avoiding AS events whose median PSI is consistently near 0 and 1 (i.e. AS events that are mostly constitutive). This concomitantly filters out AS events of very low variance (Fig. 4b).

8. To further select AS events that vary across samples based on a minimum PSI variance, you can set the minimum log10(variance) to −3 (Fig. 4b).
9. AS events may be further filtered based on their PSI range (maximum—minimum PSI value across samples), as a surrogate for the minimum changes in AS that can be considered biologically meaningful. For instance, set the minimum PSI range to 0.15 (Fig. 4b).
10. Click *Quantify alternative splicing*. In the end, the number of potentially interesting AS events to analyze was reduced to 27,401.

3.4 Creating Data Groups

Data groups can be created in *psichomics* based either on sample and respective subject information (the so-called metadata) or on AS events and respective genes.

3.4.1 Grouping ESC and iPSC

1. Go to the **Groups** section.
2. In the *Subject and sample groups* tab, click *By samples* and then *Attribute*.

3. In the *Select attribute* field, type or select *cell type* and click *Create groups*. The group table should now be populated with three groups (*ESC*, *Fibroblast* and *iPSC*).
4. In the group table, select (i.e., click) the groups *ESC* and *iPSC* and click *Merge*.
5. Select the *(ESC ∩ iPSC)* group and a text field will appear in the bottom of the table allowing you to rename the group. Rename it to *Stem cell*.

3.4.2 Grouping Isogenic Stem Cells and Isogenic Fibroblasts

1. In the *Select attribute* field, type or select *data set* and click *Create groups*.
2. Select the groups *Stem cell* and *isogenic*. Click *Intersect* to create a group with the isogenic samples within ESC and iPSC. Select the newly created group, rename it to *Isogenic stem cell* and recolor the group if desired.
3. Select the groups *isogenic* and *Fibroblast*. Click *Intersect*. Select the newly created group, rename it to *Isogenic fibroblast* and recolor the group if desired.
4. Optionally, click *Save and load* and click *Save all groups*. This will allow you to save the created groups in a file that can be loaded for future sessions by clicking *Save and load* and then *Load groups*.

3.5 Principal Component Analysis (PCA)

3.5.1 Performing PCA on Normalized Gene Expression

1. Click **Analysis** and *Principal Component Analysis (PCA)*.
2. Perform PCA on AS quantification by selecting a normalized gene expression dataset in the field *Dataset to perform PCA on*.
3. Centre and scale values by checking *Center values* and *Scale values*, respectively.
4. In *Number of missing values to tolerate per event*, set to impute at most 4 (i.e., around 5% of the 72 samples) tolerated missing values per row (*see* **Note 6**).
5. Ensure that you are performing PCA on all samples and all genes and splicing events.
6. Click *Calculate PCA*.

3.5.2 PCA Plot

Once the PCA finishes, the *Plot PCA* panel will automatically expand. A variance plot (useful to compare the percentage of explained variance between principal components) can be displayed by clicking *Show variance plot* (Fig. 5a). The first two principal components explain around 50% of the observed variance in the data.

1. In the *Plot PCA* panel, assign the first principal component to the X axis and the second to the Y axis.

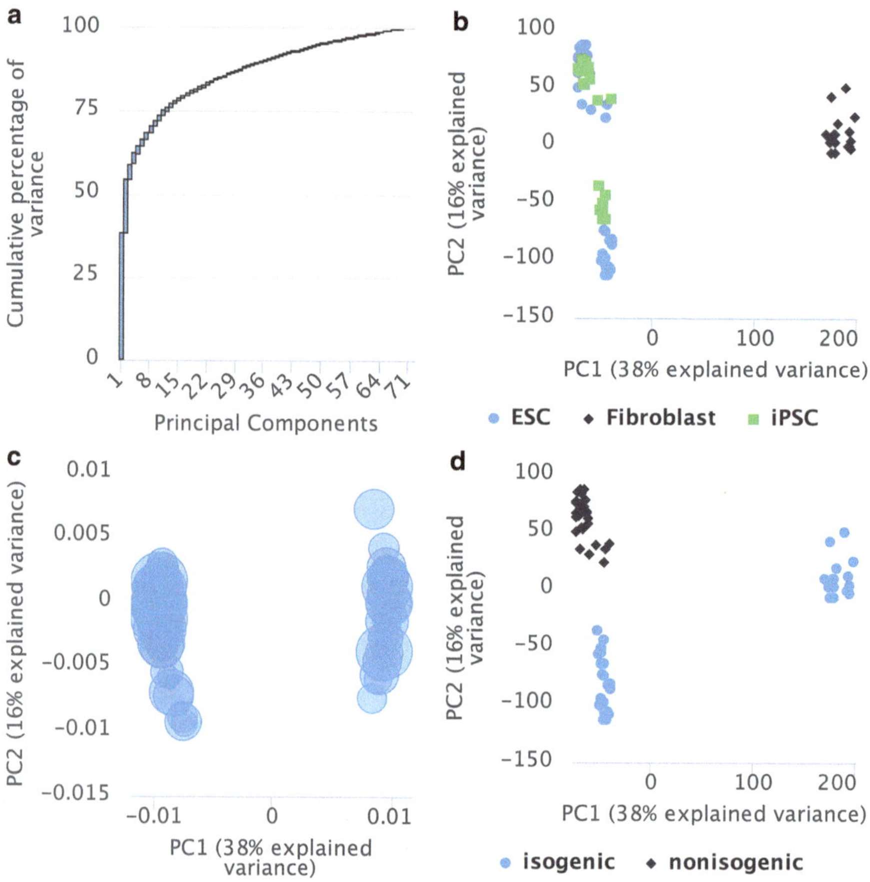

Fig. 5 Principal component analyses of normalized gene expression (**a**–**d**) and alternative splicing quantification data (**e**–**g**). (**a**) Plot displaying the cumulative percentage of total data variance explained by each principal component. (**b**, **d**) Scatter plots of scores of each sample on principal components 1 and 2, with samples colored based on cell type (**b**) and isogenicity (**d**). (**c**) Scatter plot of loadings of each gene on principal components 1 and 2. Each gene's bubble size is proportional to its relative contribution to principal components 1 and 2. (**e**, **f**) Scatter plots of scores of each sample on principal components 1 and 2, with samples colored based on cell type (**e**) and isogenicity (**f**). (**g**) Table of loadings of the 5 alternative splicing events contributing the most to principal components 1 and 2

2. In *Sample coloring*, click *Color using selected groups* and select the groups *iPSC*, *Fibroblast*, and *ESC*.
3. For performance reasons, the 100 most contributing variables (i.e., genes or AS events) to the selected principal components are plotted by default (you can choose instead to visualize *All variables* in *Variables to plot in loading plot*).
4. By clicking *Plot PCA*, two plots are rendered: a score plot (top) displaying the scores of the samples across the selected principal components (Fig. 5b) and a loading plot (below) showing the relative contribution of the variables to selected principal components (Fig. 5c).

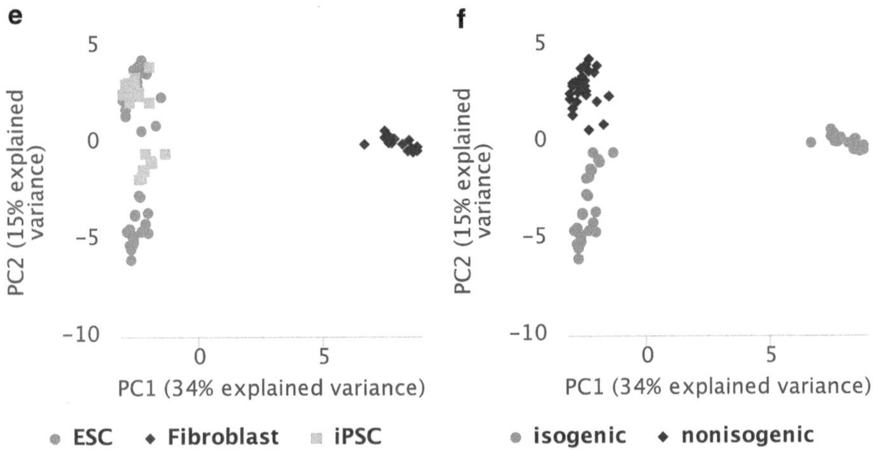

g

Rank	Gene	Event type	Chromosome	Strand	Contribution to PC1 (%)	Contribution to PC2 (%)	Contribution to PC1 and PC2 (%)
1	TMEM63B	Skipped exon	6	+	0.773936985786204	0.00406440276777961	0.537175533660727
2	SPAG9	Skipped exon	17	-	0.720386187642082	0.00255913011045217	0.499630467455487
3	MEST	Alternative first exon	7	+	0.642142302742031	0.0289812048145162	0.453574849978623
4	SCARB1	Alternative last exon	12	-	0.589350555318185	0.0134035507012602	0.412227675288103
5	SCARB1	Skipped exon	12	-	0.588841154632387	0.0119374182542734	0.41142404773143

Fig. 5 (continued)

5. To color samples in the PCA plot based on the *isogenicity*, in *Sample coloring*, select *Color using selected groups* and select the groups *Isogenic fibroblast* and *Isogenic stem cells* (Fig. 5d).

The variance observed across principal component 1 seems to be related with the cell type (fibroblast versus stem cell) (Fig. 5b), whereas principal component 2 is associated with isogenicity (i.e., isogenic vs nonisogenic; respective column named *dataset type* in the SRA metadata) (Fig. 5d). From the 15 most variance-contributing genes, as displayed in the table below the loading plot, at least *DNMT3B* and *RBPMS2* have been previously associated with pluripotency [22]. Specifically, *RBPMS2* has been reported to play a role in self-renewal following the knockdown of *ESRP1*, reported to act as a regulator of pluripotency [22].

3.5.3 PCA on AS Quantification

Perform and plot PCA on AS quantification data by selecting *Inclusion levels* in the *Dataset to perform PCA on* field and following the aforementioned instructions in Subheadings 3.5.1 and 3.5.2 (with the exception that PSI values are not required to be scaled as they are dimensionless ratios, always in the same scale ranging from 0 to 1, so make sure you uncheck *Scale values*). Similar to the observations from PCA plots on normalized gene expression, principal components 1 and 2 appear to be associated with cell type and isogenicity, respectively (Fig. 5e, f). The table below the loading plot allows you to assess which AS events contribute the most to those separations (Fig. 5g). Some of them have their cognate genes already reported to be involved in conserved splicing programs in stem cell differentiation, including *KIF13A* and *PALM* [9].

3.6 Differential Gene Expression Analysis

1. To perform differential gene expression analysis, click **Analyses** and below *Different expression analysis*, click *Exploratory (multiple genes)*.
2. Select to compare the groups *Isogenic stem cells* and *Isogenic fibroblasts*.
3. Click *Perform analyses* (*see* **Note 7**).

3.6.1 Highlight Differentially Expressed Genes

Highlight differentially expressed genes with an absolute log2-fold-change above 1 and a moderated t-test's adjusted *p*-value below 0.01 (Fig. 6a). Selected thresholds (as described below) are automatically applied to both highlight genes in the plot and to filter results in the table.

1. Expand the *Plot options and table filtering panel*.
2. Select the *Log-fold* change for the X axis. To highlight absolute log2-fold-change values above 1, expand the *Highlight points based on X value* and check *Highlight points based on X values*. Set *Lower limit* to −1, *Upper limit* to 1 and check *Invert highlighted values* (i.e.. select absolute log-fold-change values >1).
3. Click *Y axis* and set the *p-value (BH adjusted)*, that is. *p*-value after Benjamini–Hochberg (BH) adjustment for multiple testing, for the Y axis. To display -$\log_{10}$(adjusted *p*-values), in *Data transformation of Y values* select *−log10(|y|)*. To highlight values whose adjusted *p*-value is below 0.01, set *Lower limit* to *−log10(0.01)*.

3.7 Differential Splicing Analysis

1. To perform differential splicing analysis, click **Analyses** and, below *Different splicing analysis*, click *Exploratory (multiple splicing events)*.
2. In *Groups of samples to analyze*, click *Samples by selected groups* and select *Isogenic stem cells* and *Isogenic fibroblasts*.
3. Click *Perform analyses* (*see* **Note 8**).

3.7.1 Highlight Differentially Spliced Events

Highlight differentially spliced events with an absolute difference in median PSI (|Δ median PSI|) higher than 0.1 and a Wilcoxon rank-sum test's adjusted *p*-value below 0.01 (Fig. 6b). Selected thresholds are automatically applied to highlight AS events in the plot and to filter results in the table.

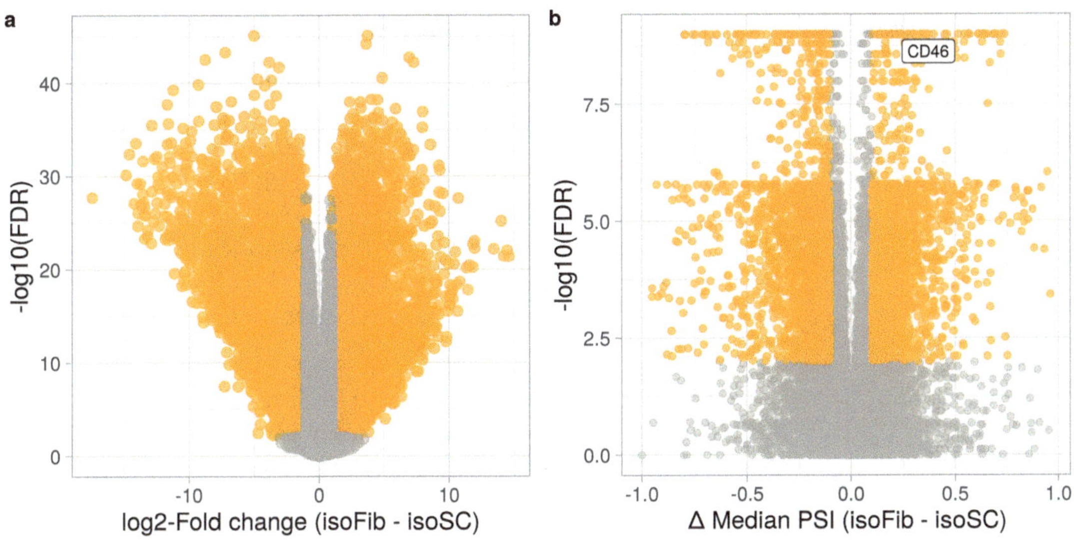

c

Event type	Chromosome	Strand	Gene	PSI distribution	Samples (isoSC)	Samples (isoFib)	Wilcoxon p-value (BH adjusted)
All	All	A	A	All	All	All	All
Skipped exon	1	+	PLEKHM2		25	17	9.36829674399094e-10
Skipped exon	1	+	NASP		25	17	9.36829674399094e-10
Skipped exon	1	+	NASP		25	17	9.36829674399094e-10
Skipped exon	1	-	C1orf43		25	17	9.36829674399094e-10
Skipped exon	1	+	CD46		25	17	9.36829674399094e-10
Skipped exon	1	+	RCOR3		25	17	9.36829674399094e-10

Fig. 6 Differential expression (**a**) and splicing (**b**, **c**) analyses. (**a**, **b**) Volcano plots where orange-highlighted genes/events correspond to adjusted *p*-value <0.01 and absolute log2-Fold change >1 (**a**) or absolute Δ median PSI > 0.1 (**b**). The *CD46* penultimate exon inclusion is labeled in **b** with the cognate gene symbol. (**c**) Table showing a subset of the differential splicing results, sorted in ascending order by the adjusted *p*-value of Wilcoxon's rank-sum test. PSI distributions are colored by groups: pink for isogenic stem cells and green for isogenic fibroblasts. The selected AS event (in blue) depicts the *CD46* penultimate exon inclusion. Legend: isogenic fibroblasts (isoFib) and isogenic stem cells (isoSC)

1. Expand the *Plot options and table filtering* panel.
2. Click the *X axis* tab and select the *Δ Median* for the X axis. To highlight Δ Median absolute values of 0.1 or more, expand the *Highlight points based on X value* and check *Highlight points based on X values*. Set *Lower limit* to −0.1, *Upper limit* to 0.1 and check *Invert highlighted values.*
3. Click the *Υ axis* tab and set the *Wilcoxon p-value (BH adjusted)* for the Y axis. To display $-\log_{10}$(adjusted *p*-values), in *Data transformation of Υ values* select *−log10(|y|)*. To highlight values whose adjusted *p*-value is below 0.01, set *Lower limit* to *−log10(0.01)*.

3.7.2 Label Specific AS Events

1. Expand the *Plot options and table filtering* panel.
2. Click the *Labels* tab and expand the *Label selected alternative splicing events* panel.
3. Check the option *Enable labeling of selected alternative splicing events* and select the event *SE 1 + 207785682 207790253 207790345 207793519 CD46*. This AS event, the penultimate exon of *CD46*, is one of the most significantly differentially spliced sequences between isogenic fibroblasts and isogenic stem cells in our analysis, as can be seen after labeling the plot (Fig. 6b).

This event is also easily identified by sorting the table based on the Wilcoxon's adjusted *p*-values: simply click the column once to sort the table in ascending (and twice for descending) order (Fig. 6c).

3.7.3 Skipping of CD46 Penultimate Exon

To evaluate the PSI distributions for the *CD46* penultimate exon in the different cell types, follow these steps:

1. In the top right corner of *psichomics*, under *Selected splicing event*, click *Change...* and input the event identifier *SE 1 + 207785682 207790253 207790345 207793519 CD46.*
2. Click **Analyses** and, under *Differential splicing analysis*, click *Individual alternative splicing event.*
3. In *Groups of samples to analyze*, click *Samples by selected groups* and select *Isogenic stem cells* and *Isogenic fibroblasts.*

Higher inclusion of the *CD46* penultimate exon is associated with fibroblasts, whereas lower inclusion is associated with stem cells, both ESC and iPSC (Fig. 7a).

3.8 Correlation Between RNA-Binding Protein Gene Expression and AS Quantification

1. To identify putative RNA-binding protein regulators of the selected AS event, click **Analyses** and *Correlation analysis.*
2. Expand the *Correlation parameters* panel and in *Genes from selected groups* type/select *RNA-binding proteins (Sebestyen et al.* [34]).
3. In *Alternative splicing events*, click *Selected event* (make sure your currently selected AS event is the penultimate exon of *CD46* inclusion).
4. Select *Pearson's product moment* as the correlation method.
5. Click *Correlate.* A table will be rendered with all the alternative splicing event and gene combinations tested, their respective correlation coefficients and associated adjusted *p*-values.
6. Expand the *Scatterplot options* panel, select which of the previously calculated correlations you wish to plot and click *Plot.* A correlation plot for each combination of selected genes and alternative splicing events will be rendered.

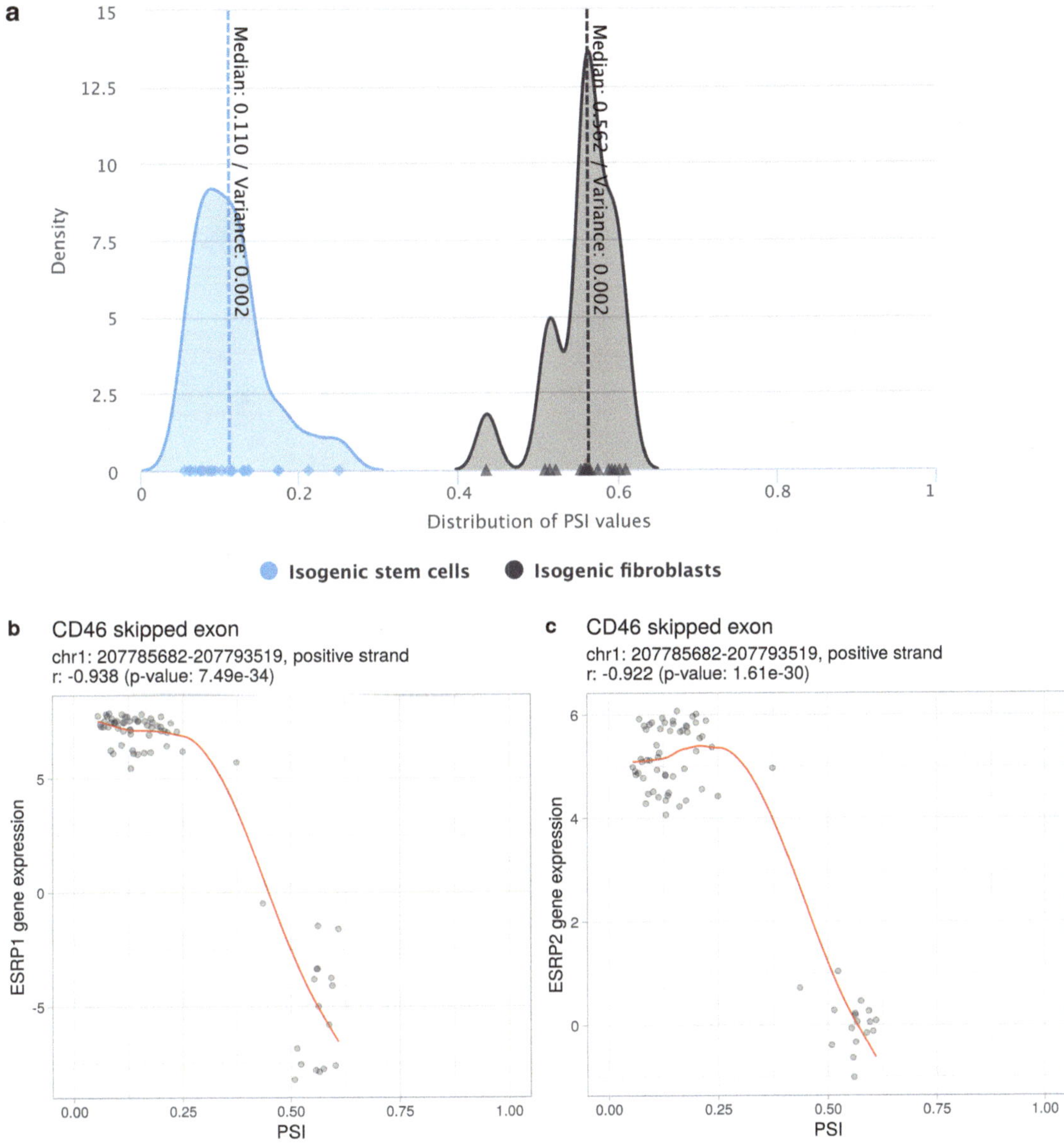

Fig. 7 Alternative splicing of *CD46* penultimate exon. (**a**) Density plots of distributions of *CD46* penultimate exon PSI values across samples, colored by isogenic stem cells (pink) and isogenic fibroblasts (green). (**b-c**) Scatterplots of PSI values for *CD46* penultimate exon inclusion *versus* normalized *ESRP1* (**b**) and *ESRP2* (**c**) expression across samples. The red line illustrates the fitted Loess regression curve. The Pearson's correlation coefficients (*r*) and associated *p*-values are shown. (**d**) Genomic alignment of *CD46* transcript isoforms with penultimate exon highlighted in an orange shade (the gray shade includes the neighboring constitutive exons to define the entire alternative splicing event)

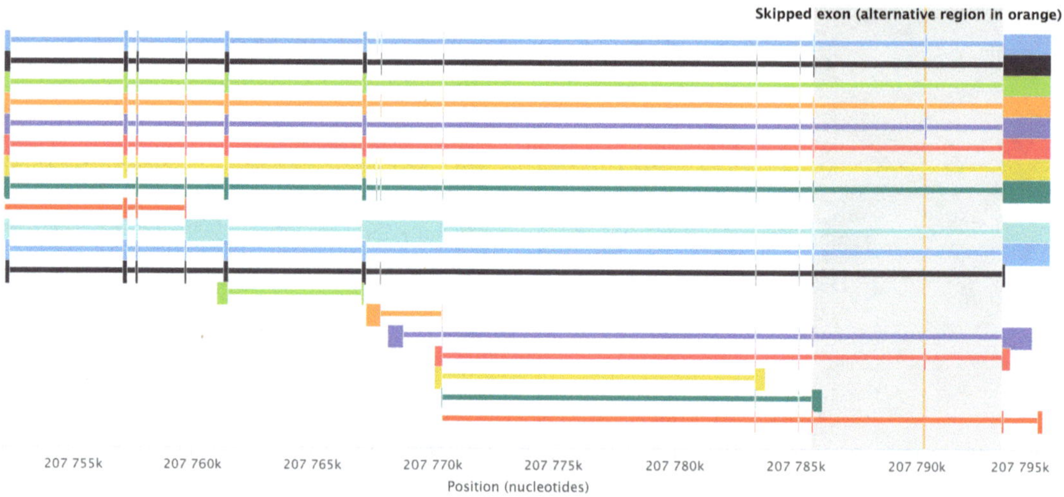

Fig. 7 (continued)

The splicing regulators *ESRP1* and *ESRP2* are among the tested genes encoding RNA-binding proteins with expression most highly correlated with the PSI values of the penultimate exon of *CD46* (Fig. 7b, c).

3.9 Gene and Transcript Annotation

Different types of information on a given AS event and its cognate gene are available in the **Analyses** section at *Gene, transcript, and protein information*. Those include basic gene description, annotated transcript isoforms (including region of AS event) (Fig. 7d) and protein domains, relevant PubMed articles [23] and links to useful external databases (namely, Ensembl [24], UniProt [25], and VAST-DB [26]).

The inclusion of *CD46* penultimate exon leads to a premature termination codon (PTC) that may cause the respective transcript to be targeted for nonsense-mediated decay (NMD) [27]. This AS event is reported as being regulated by the RNA-binding epithelial splicing regulatory proteins 1 and 2 (ESRP1 and ESRP2) [27]. ESRP1/2 are involved in the epithelial–mesenchymal transition and associated with the generation of cancer stem cells [11, 27]. ESRP1 has been previously reported to be a regulator of differentiation-poising in ES cells [22].

3.10 Extending the Analyses to GTEx and TCGA

3.10.1 Correlation of Gene Expression and AS Quantification Across GTEx Tissues

1. To automatically download and load GTEx data into your workspace, start by going to the **Data** section and clicking *GTEx data loading*.
2. Select all data types (*Sample attributes*, *Subject phenotypes*, *Gene expression*, and *Junction quantification*).
3. Click *Load data*. The selected data will start to automatically download to the folder in *Folder where data is stored* and will be instantly loaded once finished.

4. Filter and normalize gene expression using *voom* with quantile normalization. Ensure that ENSEMBL gene identifiers are converted to gene symbols by selecting the *Replace unambiguous ENSEMBL gene identifiers with their gene symbols.*
5. Quantify AS based on the Human hg19 alternative splicing annotation, as the preprocessed GTEx data provided were created following alignment against the hg19 human genome assembly and using a matching gene annotation (*see* **Note 4**).
6. Follow the instructions in Subheading 3.9 to perform correlation analysis between ESRP1/2 normalized gene expression and the AS quantification of the skipping of the penultimate exon of *CD46*, using the respective hg19 coordinates for this event: *SE 1 + 207959027 207963598 207963690 207966864 CD46.*

In tissues where *ESRP2* expression substantially varies across individuals (e.g., breast, testis, vagina, small intestine, stomach, and prostate), it is, as expected, negatively correlated with *CD46* penultimate exon inclusion (Fig. 8). Correlation with *ESRP1* expression cannot be performed, as the gene was filtered out for having low read counts across samples, suggesting its low expression in differentiated tissues.

3.10.2 Correlation of Gene Expression and AS Quantification Across TCGA Cancers

1. Load TCGA data by going to the **Data** section and clicking *Automatically load TCGA data* (*see* **Note 9**).
2. Select a tumor type (e.g., *Breast invasive carcinoma*).
3. In *Data type*, select *Clinical data*, *Junction quantification* and *Gene expression (RSEM)*.
4. Click *Load data*. If the selected TCGA data is not in the folder indicated in the *Folder where data is stored* field, a message will appear informing you that the required data will be downloaded. Click *Download data*. Once the downloads are completely finished, click *Load data* to load the selected TCGA data.
5. Filter and normalize gene expression using *voom* with quantile normalization.
6. Quantify AS based on the Human hg19 annotation, as the preprocessed TCGA data provided were created following alignment against the hg19 human genome assembly and using a matching gene annotation (*see* **Notes 4 and 10**).
7. Follow the instructions in Subheading 3.9 to perform correlation analysis between *ESRP1/2* normalized gene expression and the AS quantification for the penultimate exon of *CD46*.

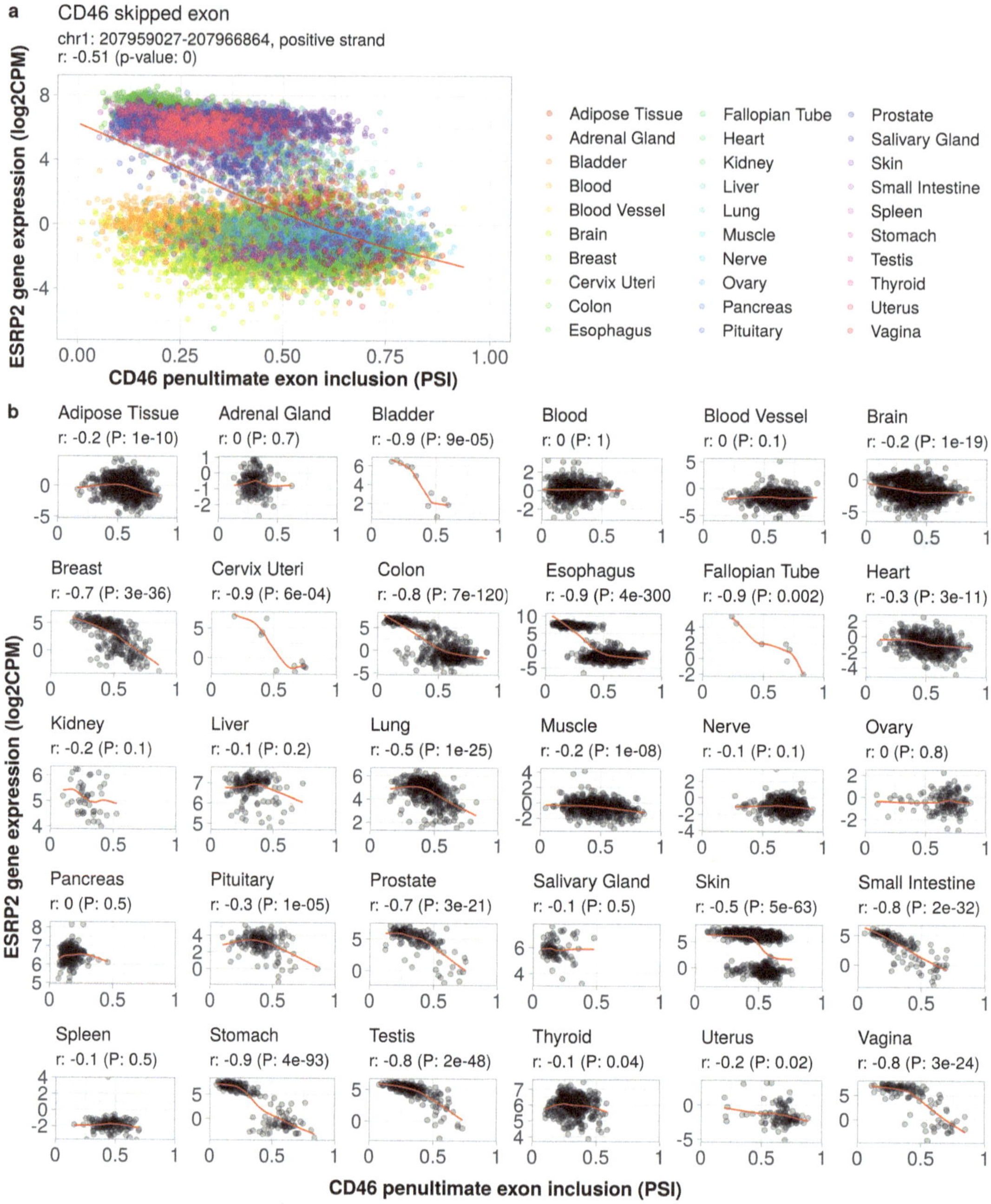

Fig. 8 Scatterplots of normalized *ESRP2* expression *versus* PSI values for *CD46* penultimate exon inclusion across GTEx tissues, altogether (**a**) and by tissue (**b**). For each plot, the red line illustrates the fitted Loess regression curve. The Pearson's correlation coefficients (*r*) and associated *p*-values (*P*) are shown

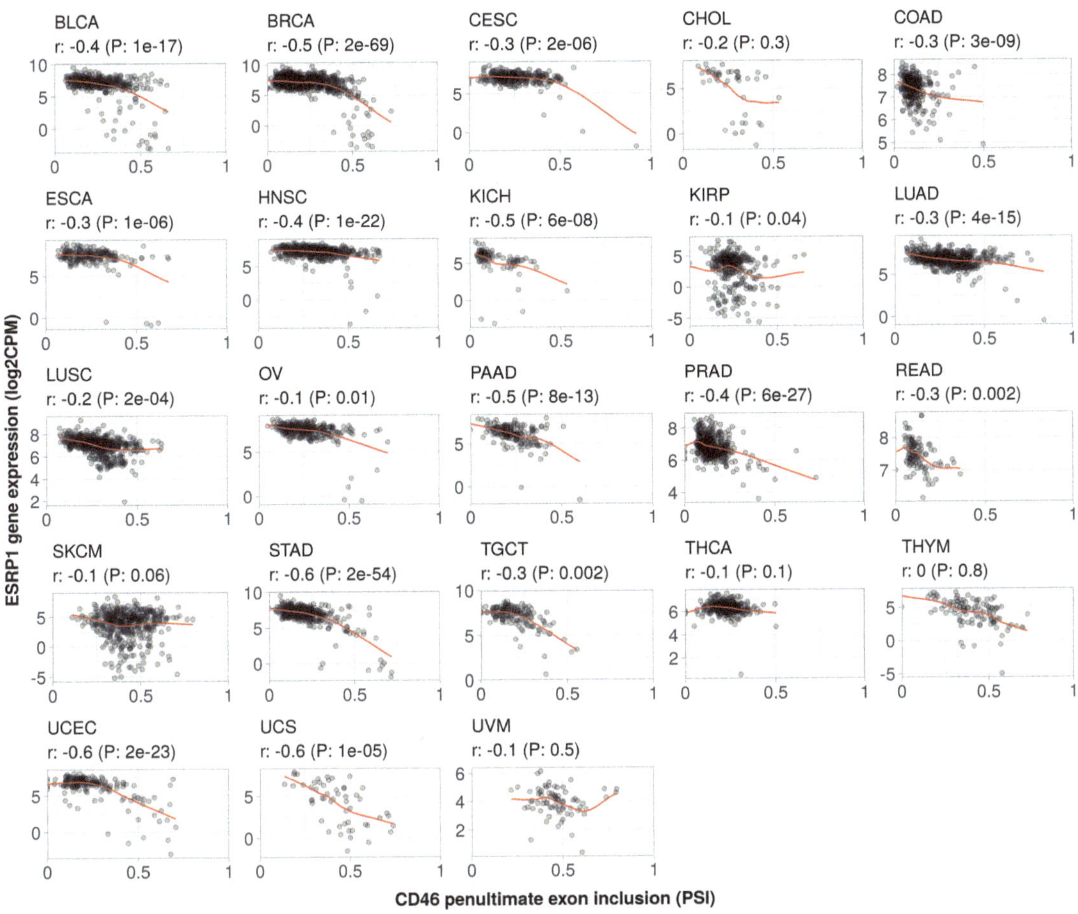

Fig. 9 Scatterplots of normalized *ESRP1* expression *versus* PSI values for *CD46* penultimate exon inclusion across TCGA tumor types. For each plot, the red line illustrates the fitted Loess regression curve. The Pearson's correlation coefficients (*r*) and associated *p*-values (*P*) are shown. *BCLA* urothelial bladder carcinoma, *BRCA* breast invasive carcinoma, *CESC* cervical squamous cell carcinoma and endocervical adenocarcinoma, *CHOL* cholangiocarcinoma, *COAD* colon adenocarcinoma, *ESCA* esophageal carcinoma, *HNSC* head and neck squamous cell carcinoma, *KICH* kidney chromophobe, *KIRP* kidney renal papillary cell carcinoma, *LUAD* lung adenocarcinoma, *LUSC* lung squamous cell carcinoma, *OV* ovarian serous cystadenocarcinoma, *PAAD* pancreatic adenocarcinoma, *PRAD* prostate adenocarcinoma, *READ* rectum adenocarcinoma, *SKCM* skin cutaneous melanoma, *STAD* stomach adenocarcinoma, *TGCT* testicular germ cell tumors, *THCA* thyroid carcinoma, *THYM* thymoma, *UCEC* uterine corpus endometrial carcinoma, *UCS* uterine carcinosarcoma, *UVM* uveal melanoma

The negative correlation between *ESRP1/2* expression and *CD46* penultimate exon inclusion is observed in most cancer types, including breast cancer (Figs. 9 and 10).

3.10.3 Pancancer Prognostic Value of AS of CD46 *Penultimate Exon*

The prognostic value of a given AS event (or gene) may be evaluated by separating subjects based on a PSI cutoff for a given AS event (or expression cutoff for a given gene). The survival

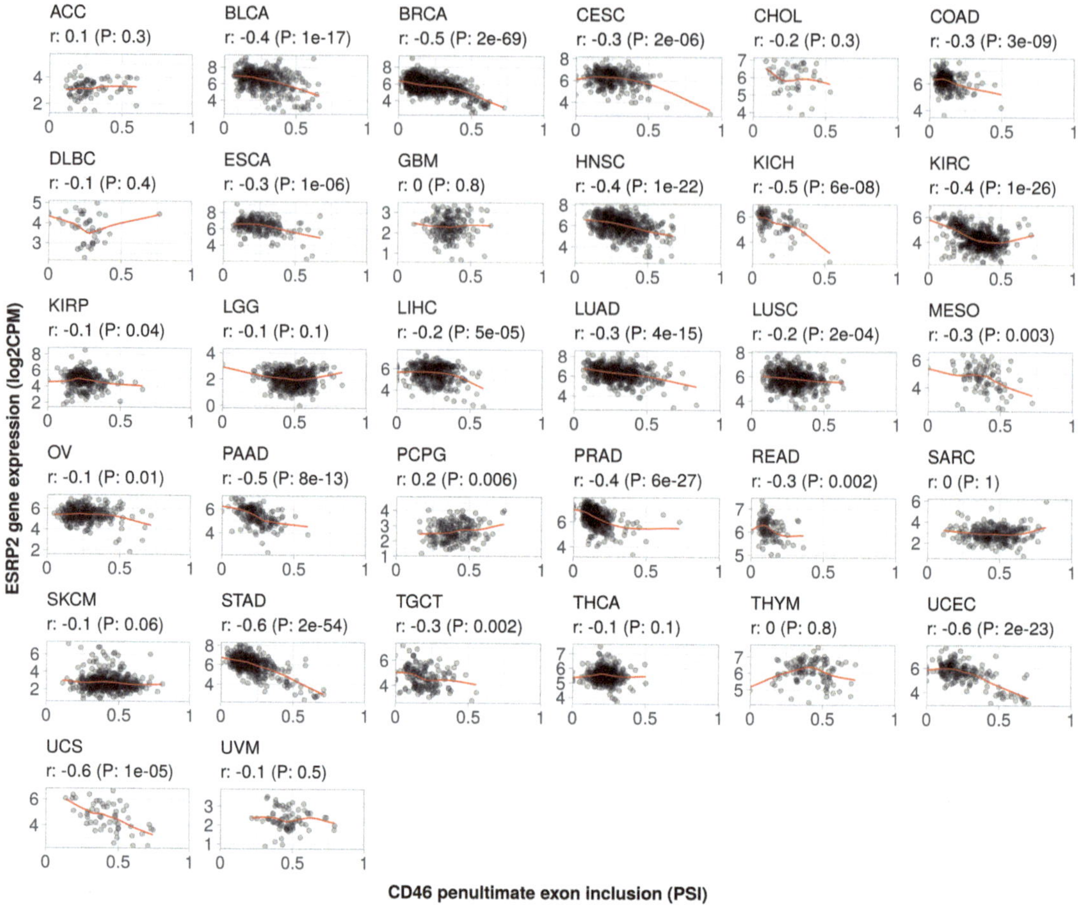

Fig. 10 Scatterplots of normalized *ESRP2* expression *versus* PSI values for *CD46* penultimate exon inclusion across TCGA tumor types. For each plot, the red line illustrates the fitted Loess regression curve. The Pearson's correlation coefficients (*r*) and associated *p*-values (*P*) are shown. Legend: *ACC* adrenocortical carcinoma, *BCLA* urothelial bladder carcinoma, *BRCA* breast invasive carcinoma, *CESC* cervical squamous cell carcinoma and endocervical adenocarcinoma, *CHOL* cholangiocarcinoma, *COAD* colon adenocarcinoma, *DLBC* lymphoid neoplasm diffuse large B-cell lymphoma, *ESCA* esophageal carcinoma, *GBM* glioblastoma multiforme, *HNSC* head and neck squamous cell carcinoma, *KICH* kidney chromophobe, *KIRC* kidney renal clear cell carcinoma, *KIRP* kidney renal papillary cell carcinoma, *LGG* brain lower grade glioma, *LIHC* liver hepatocellular carcinoma, *LUAD* lung adenocarcinoma, *LUSC* lung squamous cell carcinoma, *MESO* mesothelioma, *OV* ovarian serous cystadenocarcinoma, *PAAD* pancreatic adenocarcinoma, *PCPG* pheochromocytoma and paraganglioma, *PRAD* prostate adenocarcinoma, *READ* rectum adenocarcinoma, *SARC* sarcoma, *SKCM* skin cutaneous melanoma, *STAD* stomach adenocarcinoma, *TGCT* testicular germ cell tumors, *THCA* thyroid carcinoma, *THYM* thymoma, *UCEC* uterine corpus endometrial carcinoma, *UCS* uterine carcinosarcoma, *UVM* uveal melanoma

differences between groups are then assessed based on Kaplan-Meier estimators using the log-rank test.

In this context, we are going to make use of the clinical and transcriptomic data from TCGA to evaluate the prognostic value of the AS of *CD46* penultimate exon based on this event's PSI values in TCGA tumor samples.

1. Click **Analyses** and *Survival analysis.*
2. In *Data censoring*, select right data censoring (*see* **Note 11**).
3. The follow-up time should be *days to death* and the event of interest should be *death* (*see* **Note 12**).
4. Set to display time in *years.*
5. In *Groups for survival analyses*, select *Inclusion levels cutoff from the selected splicing event.*
6. Select the skipping of the *CD46* penultimate exon as the AS event of interest, using the respective hg19 coordinates for this event: *SE 1 + 207959027 207963598 207963690 207966864 CD46.*
7. In *Sample filtering*, select only tumor samples (*see* **Note 13**).
8. A *-log₁₀(p-value) plot by cutoff* (Fig. 11c, d) displays the *p*-values of the log-rank test of survival across multiple PSI cutoffs for the selected AS event. The PSI cutoff maximizing the significance of the survival difference is automatically selected (*see* **Note 14**).

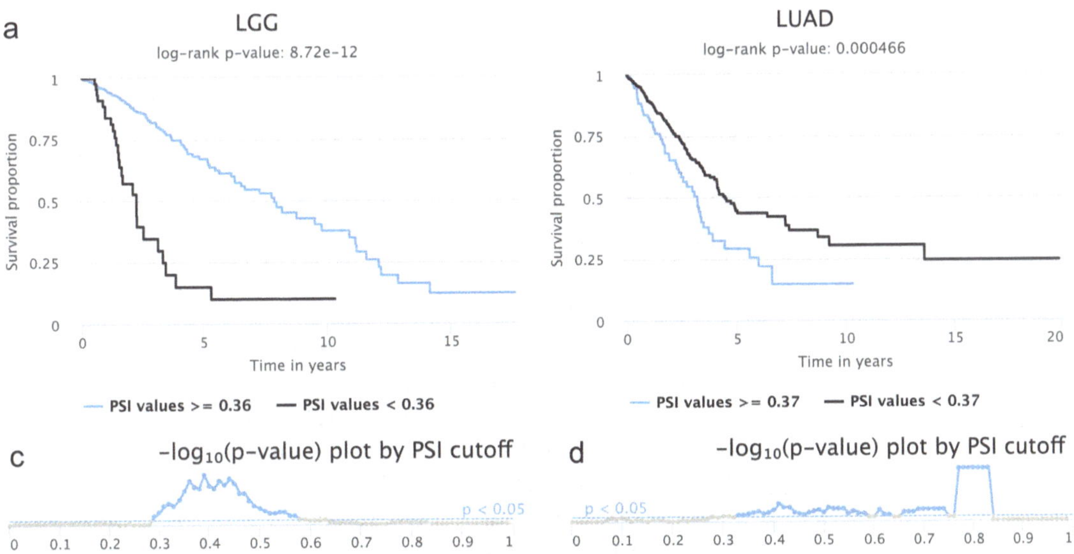

Fig. 11 Prognostic value of *CD46* penultimate exon inclusion across select TCGA cancer types. (**a**, **b**) Kaplan-Meier plots of overall survival for all patients stratified by the respective alternative splicing event's PSI cutoff that maximized the significance of differences in survival between patient groups with a reasonable number of subjects within each group. Each patient was assigned the PSI value of their tumor sample(s). (**c**, **d**) Log-rank's −log10(*p*-value) plot by PSI cutoff. Note that in **d**, for PSI values around 0.8, there are high log-rank −log10(*p*-values) although only one individual is being compared against 506 subjects. Legend: *LGG* brain lower grade glioma, *LUAD* lung adenocarcinoma

9. Click *Plot survival curves* to generate overall survival curves for patients with low and high inclusion of the *CD46* penultimate exon.
10. Assess the prognostic value of this AS event across other TCGA cancer types.

The splicing of *CD46* penultimate exon seems to have prognostic value in select cancer types, such as brain lower-grade glioma and lung adenocarcinoma (Fig. 11).

4 Notes

1. Preprocessed data from TCGA, GTEx, and select SRA projects can be automatically retrieved and loaded through panels *TCGA data loading*, *GTEx data loading* and *SRA data loading* in the **Data** section, respectively. *psichomics* uses *recount2* [16] to retrieve SRA preprocessed data and not all SRA projects are available, although unlisted SRA projects, as well as user-provided data, may be manually aligned and then loaded and analyzed in *psichomics*. For a list of SRA projects for which there are available preprocessed data, please read https://jhubiostatistics.shinyapps.io/recount. For a specific tutorial on downloading, aligning, and loading SRA or user-provided data from the raw FASTQ files, please read http://rpubs.com/nuno-agostinho/psichomics-custom-data.
2. Data will not be downloaded if available in the folder path indicated by the *Folder where data is stored* field.
3. By default, gene expression is filtered based on genes with a minimum of 10 read counts for at least one sample and with a minimum total read counts of 15. Default gene expression normalization scales for raw library sizes based on weighted trimmed mean of M-values (TMM) [20], followed by computation of log2-transformed counts per million values.
4. Currently available AS event annotations include the Human hg19 (alternativeSplicingEvents.hg19) and hg38 (alternativeSplicingEvents.hg38) genome assemblies, both available through *Bioconductor*. Selecting the AS event annotation for quantification depends on the reference gene annotation used to obtain the preprocessed data: hg19 for TCGA and GTEx data and hg38 for recount2/SRA data. Further custom annotation files can also be created and loaded. To create such files, one can parse the annotation from AS quantification tools like VAST-TOOLS [28], MISO [29], SUPPA [30], and rMATS [31]. Annotation files from MISO and VAST-TOOLS can be downloaded from their websites, whereas such annotation files can be generated by rMATS and SUPPA based on an isoform-

centered transcript annotation. More information on AS annotations is available in the following specific tutorial: http://rpubs.com/nuno-agostinho/preparing-AS-annotation.

5. AS quantification in *psichomics* is currently based on exon–exon junction read counts alone. *psichomics* is able to quantify the following types of AS events: skipped exon (SE), mutually exclusive exons (MXE), alternative 5′ splice site (A5SS), alternative 3′ splice site (A3SS), alternative first exon (AFE), and alternative last exon (ALE). Notably, intron retention events (which require (unspliced) intron–exon junction read counts for their quantification) and alternative 5′- and 3′-UTR (which require exon body read counts) are not currently quantified by *psichomics*. Nevertheless, none of the supported large dataset projects (GTEx, TCGA and recount2) currently provide intron–exon junction quantification, whereas only GTEx provides intron and exon body read quantification.
6. As PCA cannot be performed on data with missing values, these either have to be removed (thereby discarding genes or splicing events from the analysis) or imputed (i.e., replaced by estimated values). By default, if a gene (or AS event) has 5% or less missing values in *psichomics*, the median expression (or PSI) of that gene (or AS event) across samples will be used to replace them. The remaining genes or AS events (i.e.. those with more than 5% missing values, by default) will be discarded from the PCA.
7. Normalized gene expression will be linearly modeled, with explanatory variables defined based on the selected groups. Moderated t-tests and log-odds of differential expression will then be computed by empirical Bayes moderation of the standard errors toward a common value [21].
8. The volcano plot of differential splicing analysis between isogenic stem cells and fibroblasts (Fig. 6b) exhibits two strata, that is, two modes of Wilcoxon's test significance. The top significance stratum is the result of using such nonparametric test (motivated by the nonnormality of PSI distributions) when all values in one of the groups are higher than those in the other group; the number of tested groups also affects the significance of the difference. The lower significance stratum relates to a consistent number of repeated values between samples (usually occurring when one of the groups is closer to a PSI value of 0 or 1). As the Wilcoxon's test is rank-based, some ranks are not unique if there are two identical values; this occurrence (called a tie) hampers the computation of exact *p*-values. Increasing the number of identical values when performing the Wilcoxon's test decreases the significance of the comparison, which may bias the significance of differentially spliced

events when one of the groups is characterized by PSI values close to 0 or 1 (constitutive splicing) and will therefore present many 0's or many 1's.

9. Multiple files will be requested for download at once. Some web browsers (such as Google Chrome) will ask for your confirmation (please give it!) before allowing you to proceed.

   If you are using Windows, the downloaded files have names that may be longer than the Windows Maximum Path Length. A workaround is to manually rename the downloaded files to have shorter names, move all downloaded files to a single folder and load such folder by going to *Load user files*, clicking *Folder input* and selecting the newly created folder in *Folder where data is stored*.

   When working in a remote workstation and if the visual interface of *psichomics* is launched in the web browser of your local computer, *psichomics* will download TCGA data to your local computer. As *psichomics* expects files to be directly available in the workstation, please move downloaded files to a folder in your remote workstation and make sure that the *Folder where data is stored* field points to that directory.

10. Some TCGA datasets contain more than one exon–exon junction quantification file. Each junction quantification file comes from samples that were sequenced in a different platform: either Illumina HiSeq or Illumina GA. If available, use Illumina HiSeq data as it is the most recent sequencing method (using both, although not currently supported in *psichomics*, could lead to platform-related batch effects).

11. *psichomics* supports left, right, and interval data censoring for survival analysis, the type of censoring depending on the times of occurrence of events of interest. Events may occur after the last observation (right censoring, such as in the case of subjects having no event reported during the study or dropping from it altogether), before an observation is performed (left censoring, for instance when events happened in an uncertain time before the start of the study) or in-between observations (interval censoring, such as when patients require periodic follow-ups and the time of an event occurrence falls between follow-ups but is not certain) [32].

12. Analyzing *days to death* as the follow-up time and *death* as the event of interest is known as an overall survival analysis, that is, the study of the time until the subject's death following diagnosis. In cases where *days to death* is not available, the *days to last follow-up* are considered instead. Of note, in some TCGA records, the *days to last follow-up* may be a lower number than *days to death*; as we are interested in prognostic, the usage of *days to death* is recommended over *days to last follow-up*.

13. When performing survival analysis, the PSI value of a given AS event in a sample is assigned to its respective subject. In cases where multiple samples originate from the same subject, this subject is assigned the average PSI value. However, when assigning the PSI value to a given subject, not all samples should be used; for instance, PSI values from normal tissue samples should be discarded when studying disease. Such is the case with TCGA data, where only tumor samples are used to assign PSI values per subject for survival analysis, by default.
14. The survival differences are tested based on multiple PSI cutoffs for a given AS event. If the automatically selected PSI cutoff has a much lower number of subjects in one group compared to the other (e.g., in the example of Fig. 11d the lowest log-rank *p*-value corresponds to a comparison of 1 *versus* 506 subjects), try to find a cutoff with more reasonably balanced groups and for which the *p*-value is still significant. Sample size calculations can be performed based on test assumptions (e.g., probability of failure for each group during the study) using existing R packages, such as *powerSurvEpi* [33].

Acknowledgments

We would like to thank Marie Bordone, Mariana Ferreira, Sara Mendes, Marta Bica, and Arthur Schneider, for providing valuable feedback and other contributions to the chapter and *psichomics*.

This work is a result of the GenomePT project (POCI-01-0145-FEDER-022184), supported by COMPETE 2020 - Operational Programme for Competitiveness and Internationalisation (POCI), Lisboa Portugal Regional Operational Programme (Lisboa2020), Algarve Portugal Regional Operational Programme (CRESC Algarve2020), under the PORTUGAL 2020 Partnership Agreement, through the European Regional Development Fund (ERDF), and by Fundação para a Ciência e a Tecnologia (FCT). This work was also supported by: UID/BIM/50005/2019, project funded by Fundação para a Ciência e a Tecnologia (FCT)/ Ministério da Ciência, Tecnologia e Ensino Superior (MCTES) through Fundos do Orçamento de Estado; European Molecular Biology Organization [EMBO Installation Grant 3057 to N.L.B.-M.]; Fundação para a Ciência e a Tecnologia [FCT Investigator Starting Grant IF/00595/2014 to N.L.B.-M., PhD Studentship SFRH/BD/131312/2017 to N.S.-A., project PERSEIDS PTDC/EMS-SIS/0642/2014].

References

1. Paronetto MP, Passacantilli I, Sette C (2016) Alternative splicing and cell survival: from tissue homeostasis to disease. Cell Death Differ 23:1919–1929. https://doi.org/10.1038/cdd.2016.91
2. Wang ET, Sandberg R, Luo S et al (2008) Alternative isoform regulation in human tissue transcriptomes. Nature 456:470–476. https://doi.org/10.1038/nature07509
3. Barbosa-Morais NL, Irimia M, Pan Q et al (2012) The evolutionary landscape of alternative splicing in vertebrate species. Science 338:1587–1593. https://doi.org/10.1126/science.1230612
4. Gallego-Paez LM, Bordone MC, Leote AC et al (2017) Alternative splicing: the pledge, the turn, and the prestige: the key role of alternative splicing in human biological systems. Hum Genet 136:1015–1042. https://doi.org/10.1007/s00439-017-1790-y
5. Oltean S, Bates DO (2014) Hallmarks of alternative splicing in cancer. Oncogene 33:5311–5318. https://doi.org/10.1038/onc.2013.533
6. Zavolan M, Kanitz A (2018) RNA splicing and its connection with other regulatory layers in somatic cell reprogramming. Curr Opin Cell Biol 52:8–13. https://doi.org/10.1016/j.ceb.2017.12.002
7. Gabut M, Samavarchi-Tehrani P, Wang X et al (2011) An alternative splicing switch regulates embryonic stem cell pluripotency and reprogramming. Cell 147:132–146. https://doi.org/10.1016/j.cell.2011.08.023
8. Han H, Irimia M, Ross PJ et al (2013) MBNL proteins repress ES-cell-specific alternative splicing and reprogramming. Nature 498:241–245. https://doi.org/10.1038/nature12270
9. Venables JP, Lapasset L, Gadea G et al (2013) MBNL1 and RBFOX2 cooperate to establish a splicing programme involved in pluripotent stem cell differentiation. Nat Commun 4:2480. https://doi.org/10.1038/ncomms3480
10. Chen K, Dai X, Wu J (2015) Alternative splicing: an important mechanism in stem cell biology. World J Stem Cells 7(1):10. https://doi.org/10.4252/wjsc.v7.i1.1
11. Pradella D, Naro C, Sette C, Ghigna C (2017) EMT and stemness: flexible processes tuned by alternative splicing in development and cancer progression. Mol Cancer 16(8). https://doi.org/10.1186/s12943-016-0579-2
12. Aponte PM, Caicedo A (2017) Stemness in cancer: stem cells, cancer stem cells, and their microenvironment. Stem Cells Int 2017:5619472–5619417. https://doi.org/10.1155/2017/5619472
13. Saraiva-Agostinho N, Barbosa-Morais NL (2019) psichomics: graphical application for alternative splicing quantification and analysis. Nucleic Acids Res 47:e7–e7. https://doi.org/10.1093/nar/gky888
14. Cancer Genome Atlas Research Network, Weinstein JN, Collisson EA et al (2013) The cancer genome atlas pan-cancer analysis project. Nat Genet 45:1113–1120. https://doi.org/10.1038/ng.2764
15. The GTEx Consortium (2013) The genotype-tissue expression (GTEx) project. Nat Genet 45:580–585. https://doi.org/10.1038/ng.2653
16. Collado-Torres L, Nellore A, Kammers K et al (2017) Reproducible RNA-seq analysis using recount2. Nat Biotechnol 35:319–321. https://doi.org/10.1038/nbt.3838
17. Choi J, Lee S, Mallard W et al (2015) A comparison of genetically matched cell lines reveals the equivalence of human iPSCs and ESCs. Nat Biotechnol 33:1173–1181. https://doi.org/10.1038/nbt.3388
18. R Core Team (2016) R: A language and environment for statistical computing, Vienna, Austria
19. Gentleman RC, Carey VJ, Bates DM et al (2004) Bioconductor: open software development for computational biology and bioinformatics. Genome Biol 5:R80. https://doi.org/10.1186/gb-2004-5-10-r80
20. Robinson MD, McCarthy DJ, Smyth GK (2010) edgeR: a Bioconductor package for differential expression analysis of digital gene expression data. Bioinformatics 26:139–140. https://doi.org/10.1093/bioinformatics/btp616
21. Ritchie ME, Phipson B, Di W et al (2015) limma powers differential expression analyses for RNA-sequencing and microarray studies. Nucleic Acids Res 43:e47
22. Fagoonee S, Bearzi C, Di Cunto F et al (2013) The RNA binding protein ESRP1 fine-tunes the expression of pluripotency-related factors in mouse embryonic stem cells. PLoS One 8: e72300. https://doi.org/10.1371/journal.pone.0072300
23. Roberts RJ (2001) PubMed Central: The GenBank of the published literature. Proc Natl

Acad Sci U S A 98:381–382. https://doi.org/10.1073/pnas.98.2.381

24. Cunningham F, Amode MR, Barrell D et al (2015) Ensembl 2015. Nucleic Acids Res 43: D662–D669. https://doi.org/10.1093/nar/gku1010
25. Wu CH, Apweiler R, Bairoch A et al (2006) The Universal Protein Resource (UniProt): an expanding universe of protein information. Nucleic Acids Res 34:D187–D191. https://doi.org/10.1093/nar/gkj161
26. Tapial J, Ha KCH, Sterne-Weiler T et al (2017) An atlas of alternative splicing profiles and functional associations reveals new regulatory programs and genes that simultaneously express multiple major isoforms. Genome Res 27:1759–1768. https://doi.org/10.1101/gr.220962.117
27. Warzecha CC, Jiang P, Amirikian K et al (2010) An ESRP-regulated splicing programme is abrogated during the epithelial-mesenchymal transition. EMBO J 29:3286–3300. https://doi.org/10.1038/emboj.2010.195
28. Irimia M, Weatheritt RJ, Ellis JD et al (2014) A highly conserved program of neuronal microexons is misregulated in autistic brains. Cell 159:1511–1523. https://doi.org/10.1016/j.cell.2014.11.035
29. Katz Y, Wang ET, Airoldi EM, Burge CB (2010) Analysis and design of RNA sequencing experiments for identifying isoform regulation. Nat Methods 7:1009–1015. https://doi.org/10.1038/nmeth.1528
30. Alamancos GP, Pagès A, Trincado JL et al (2015) Leveraging transcript quantification for fast computation of alternative splicing profiles. RNA 21:1521–1531. https://doi.org/10.1261/rna.051557.115
31. Shen S, Park JW, Lu Z-X et al (2014) rMATS: Robust and flexible detection of differential alternative splicing from replicate RNA-Seq data. Proc Natl Acad Sci U S A 111: E5593–E5601. https://doi.org/10.1073/pnas.1419161111
32. Zhang Z, Sun J (2010) Interval censoring. Stat Methods Med Res 19:53–70. https://doi.org/10.1177/0962280209105023
33. Qiu W, Chavarro J, Lazarus R, et al power-SurvEpi: power and sample size calculation for survival analysis of epidemiological studies
34. Sebestyén E, Singh B, Miñana B, Pagès A, Mateo F, Pujana MA, Valcárcel J, Eyras E (2016) Large-scale analysis of genome and transcriptome alterations in multiple tumors unveils novel cancer-relevant splicing networks. Genome Res 26(6):732–744. https://doi.org/10.1101/gr.199935.115

Chapter 11

Gene Ontology Semantic Similarity Analysis Using GOSemSim

Guangchuang Yu

Abstract

The GOSemSim package, an R-based tool within the Bioconductor project, offers several methods based on information content and graph structure for measuring semantic similarity among GO terms, gene products and gene clusters. In this chapter, I illustrate the use of GOSemSim on a list of regulators in preimplantation embryos. A step-by-step analysis was provided as well as instructions on interpretation and visualization of the results. GOSemSim is open-source and is available from https://www.bioconductor.org/packages/GOSemSim.

Key words Semantic similarity, GOSemSim, Gene ontology, Functional prediction, Reproducible research

1 Introduction

As the accumulation of biological knowledge from various sources, gene ontology, which uses controlled vocabulary to describe genes and their products in a formal and consistent way across heterogeneous genomes, is needed to integrate the knowledge at a broad level. To understand gene ontology, we should have a basic understanding of ontology first. An ontology is the concrete form of a conceptualization of a community's knowledge of a domain. Among the many bio-ontologies including disease ontology, metabolic ontology, and pathway ontology, gene ontology is the de facto standard for the annotation of gene products. Ontologies are often organized through a formal format so that it can be accessed and interpreted by computers. The GO consortium annotates gene products with terms from three orthogonal ontologies, namely Biological Process (BP), Molecular Function (MF), and Cellular Component (CC) (*see* **Note 1**). They describe gene products in pathways, molecular activities, and locations, respectively. The three subontologies are organized as directed acyclic graphs (DAGs), in which the node represents the GO term and the edge

Benjamin L. Kidder (ed.), *Stem Cell Transcriptional Networks: Methods and Protocols*, Methods in Molecular Biology, vol. 2117, https://doi.org/10.1007/978-1-0716-0301-7_11, © Springer Science+Business Media, LLC, part of Springer Nature 2020

between the terms represents the parent-child relationship. The DAG structure lays the foundation for quantitative semantic comparisons of GO terms (*see* **Note 2**). The idea behind semantic similarity measurement is the notion that genes with similar function should have similar annotation vocabulary and have a close relationship in the GO tree structure. Since genes and gene products can be annotated by GO terms, their functional similarities can be measured by aggregating similarities of GO term sets. Measuring similarity is critical for expanding knowledge, since similar objects tend to behave similarly, which supports many bioinformatics applications to infer gene/protein functions [1] (*see* **Note 3**), miRNA function [2], genetic interaction [3], protein–protein interaction [4], miRNA–mRNA interaction [5], and cellular localization [6].

GOSemSim [7] is one of the most widely used software tools for measuring functional similarity among GO terms and gene products (*see* **Note 4**). It implemented five methods, including four information content (IC) based methods proposed by Resnik [8], Jiang [9], Lin [10], and Schlicker [11], and one graph structure based method proposed by Wang [12]. The IC-based methods depend on the frequencies of two GO terms involved and that of their closest common ancestor term in a specific corpus of GO annotations. The information content of a GO term is computed by the negative log probability of the term occurring in the GO corpus. A rarely used term contains a greater amount of information. As GO allows multiple parents for each concept, two terms can share parents and ancestors by multiple paths. The IC-based methods calculate the similarity of two GO terms based on the information content of their closest common ancestor term, which is also called the most informative common ancestor (MICA). Considering that the specificity of a GO term is usually determined by its location in the GO graph, Wang [12] proposed a graph-based approach to compute semantic similarity using the topology of the GO graph structure. Formally, a GO term can be represented as an induced DAG that contains the term and all of its ancestor terms. In Wang's method, the semantics of GO terms are encoded into a numeric format and the different semantic contributions of all terms in the induced DAG are aggregated. The similarity is determined based on the locations of the GO terms in the GO graph and their relations with their ancestor terms (*see* **Note 5**).

On the basis of semantic similarity among GO terms, GOSemSim supports computing semantic similarity among gene products and gene clusters. Since a gene product can be annotated by multiple GO terms, semantic similarity among gene products needs to be aggregated from different semantic similarity measures among GO term sets into a single value. Four combination approaches were implemented for aggregating semantic similarity scores of multiple GO terms associated with genes, including *max* which calculates the maximum similarity score over all pairs of GO terms, *avg* which uses the average value of similarity scores over all pairs of GO terms, *rcmax* which measures the maximum of *RowScore* and *ColumnScore*, where *RowScore* (*ColumnScore*) is the average of maximum similarity on each row (column) and best-match average (BMA) which measures the average of maximum similarity scores on each row and column (*see* **Note 5**).

2 Materials

1. *Operating System:* The GOSemSim package was implemented using R and C++ (via Rcpp [13]) and distributed as an R package within the Bioconductor project. It supports all platforms including Mac OS X, Linux, and Windows that can run R.
2. *Software:* For the method described in this chapter, the following software is needed:
   (a) R statistical computing environment.
   (b) Bioconducor packages, including GOSemSim [7], clusterProfiler [14], DOSE [15], ape [16], ggtree [17], and org.Hs.eg.db.

   As most of these software packages are being regularly updated, make sure to use the most recent release version. Check out the latest Bioconductor release from https://bioconductor.org/about/release-announcements/ and the latest release version of R from https://cran.r-project.org.
3. *Software Installation:* Download and install R from https://cran.r-project.org/. To download and install Bioconductor packages, open R and type the following commands:

```
> install.packages("BiocManager")
> BiocManager::install(c("GOSemSim", "clusterProfiler",
"DOSE", "ape", "ggtree", "org.Hs.eg.db"))
```

   BiocManager is an R package to install and update Bioconductor packages with the ability to resolve package dependencies in both CRAN and Bioconductor repositories.
4. *Sample Data:* We will use the candidate regulators for preimplantation development reported from an article in Nature [18]. These regulators are identified from distal ATAC-seq peaks for early development at the 2-cell, 4-cell, 8-cell; inner cell masses (ICM); and mouse embryonic stem cells (mESC).

3 Methods

3.1 Creating Semantic Data

GO term is species specific. To measure semantic similarity among genes and gene products, we need to have gene-to-GO mapping information and GO term relationships. Currently GOSemSim provides *godata* function to construct semantic data by extracting genome wide annotation from *OrgDb* (*see* **Note 6**).

Create semantic data from *org.Hs.eg.db* (*OrgDb* for human):

```
> library(GOSemSim)
> hsGO <- godata("org.Hs.eg.db", ont = "BP")
```

Users need to specify which subontology is used, that is, "BP," "CC," or "MF." By default *godata* function will also calculate information content for each of the GO terms. We can disable it by setting *computeIC* = *FALSE*, if we only want to use algorithms based on the graph structure (e.g., Wang method [12]). The semantic data, *hsGO*, contains gene to GO term mapping information, GO relationship as well as the precalculated information content of each GO term.

3.2 Measuring Semantic Similarity

GOSemSim provides several functions, including *goSim*, *mgoSim*, *geneSim*, *mgeneSim*, *clusterSim*, and *mclusterSim* (*see* **Note 5**), for measuring semantic similarity among GO terms, genes, and gene clusters. For measuring semantic similarity among a list of genes, we can use *mgeneSim* function, which accepts a vector of interesting genes.

```
> genes <- c("GABPA", "NANOG", "SOX2", "POU5F1", "TFAP2C", "GATA4", "GATA3",
 "GATA1", "TEAD4", "FOXA1", "RARG", "NR5A2", "ESRRB", "CTCF", "KLF4")
```

These genes are regulators for early development. We need to convert them to Entrez gene IDs, which can be done using *bitr* function provided by clusterProfiler package [14].

```
> library(clusterProfiler)
> genes.df < - bitr(genes, "SYMBOL", "ENTREZID", "org.Hs.eg.db")
> genes <- genes.df[,1]
> eg < - genes.df[,2]
> names(genes) < - eg
```

The *eg* variable stored the Entrez gene ID of the regulators and the named vector *genes* was created as a look-up table for converting Entrez gene ID to gene symbols.

```
> sim < - mgeneSim(eg, semData = hsGO, measure = "Wang", drop =
"IEA", combine = "BMA")
> rownames(sim) < - genes[rownames(sim)]
> colnames(sim) < - genes[colnames(sim)]
```

The *mgeneSim* function used the provided semantic data via *semData* parameter to convert Entrez gene ID to GO terms and measure semantic similarity using the specific measure method. Users can specify to ignore GO annotation from selected evidence code. In this example, *drop = "IEA"* means ignoring GO annotation inferred from electronic annotation (IEA). Semantic similarity among genes was determined by aggregating GO semantic similarity values using the specific combine method. The result was stored in *sim* variable. The column and row names were converted back to gene symbols, which are more human-friendly for interpretation.

3.3 Visualizing Semantic Similarity Matrix

```
> DOSE::simplot(sim)
```

The semantic similarity matrix can be visualized as a heatmap and there are many functions available in R for visualizing heatmap. DOSE package [15] provides a *simplot* function that displays heatmap with actual values labeled on each cell (*see* Fig. 1).

3.4 Clustering Via Similarity of Biological Knowledge

```
> library(ape)
> library(ggtree)
> hc < - hclust(as.dist(1-sim), method = "ward.D")
> ggtree(hc) + geom_tiplab() + xlim(NA, 0.5)
```

We can calculate distance based on the similarity matrix and thus perform hierarchical clustering. Data clustering is commonly used to categorize different objects into kinds. New functions can be inferred based on clustering results. The hierarchical result (Tree panel of Fig. 2) can be visualized using ggtree [17] (*see* **Note 7**).

As illustrated in Fig. 2, the hierarchical clustering tree contains two major clades. The first clade was composed of regulators for inner cell masses (ICM). POU5F1, NANOG, and ESRRB are known major regulators of ICM and they are clustered together. NR5A2 is an early regulator of ICM/trophectoderm (TE) programs and is known to upregulate the expression of POU5F1 and NANOG. TFAP2C, GABPA, and CTCF are expressed in both ICM and TE and were not regulated by NR5A2. These three genes are also clustered together. These results are consistent with experimental findings reported in an article published in Nature [18]. The second clade contains SOX2, KLF4, and FOXA1 that are

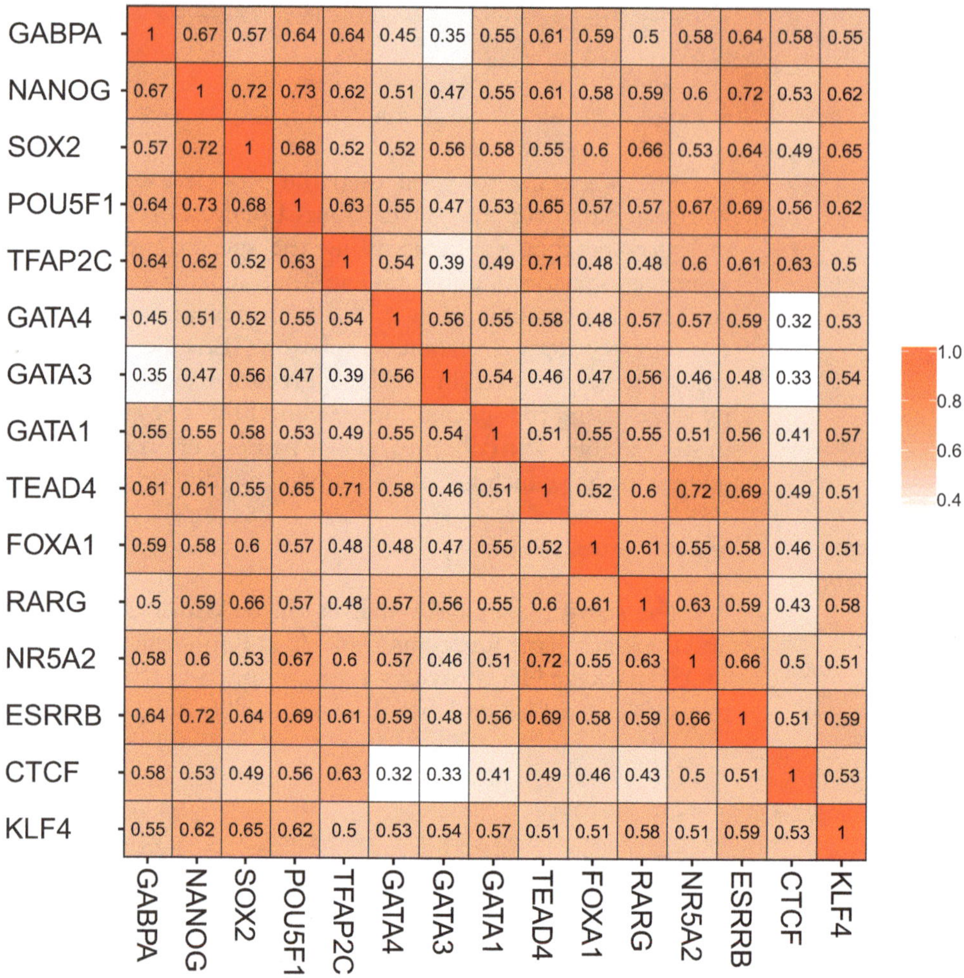

Fig. 1 Heatmap of semantic similarity matrix

known regulator for embryonic stem cells (ESC). The expression profile of these genes was visualized as a dot plot (Motif panel of Fig. 2) and aligned to the tree using *facet_plot* from ggtree [19]. The dots were colored by FPKM from RNA-seq data (GSE66582) and the dot sizes were scaled by motif enrichment obtained from ATAC-seq [18]. The source code to produce Fig. 2 is available from https://github.com/YuLab-SMU/misc/tree/master/GOSemSim_SCTN2019.

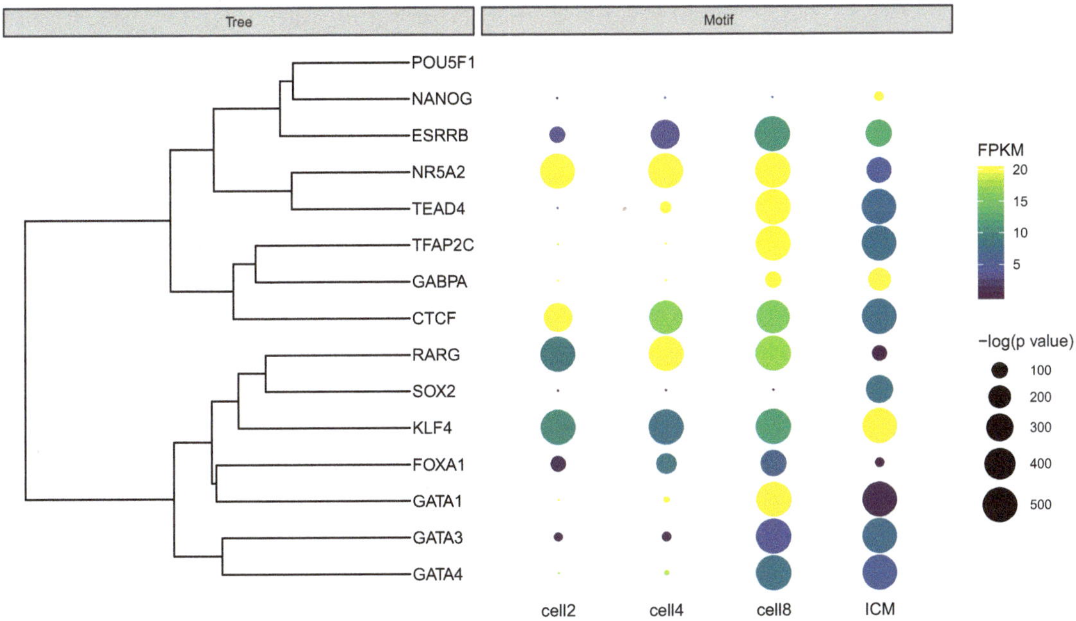

Fig. 2 Hierarchical clustering of semantic similarity matrix. The clustering result was visualized as a phylogenetic tree (Tree panel) and the expression profile of corresponding genes was displayed as dot plot and aligned to the tree (Motif panel)

4 Notes

1. Gene Ontology defines concepts/classes used to describe gene functions. Genes that annotated by a GO term can be interpreted as a specific gene set that represents a group of genes that is functionally related. Common strategies to analyze gene sets are overrepresentation analysis (ORA) and gene set enrichment analysis (GSEA). These approaches are widely used in biomedical sciences to elucidate functional associations of a gene list or genome wide expression profile. The ORA and GSEA analyses using GO and other pathway databases are supported by clusterProfiler package [14].
2. Disease ontology (DO) and Medical Subject Headings (MeSH) are also organized as directed acyclic graphs and all the methods used to measure GO semantic similarity can be applied to DO and MeSH. The DOSE [15] and meshes [20] Bioconductor packages extended GOSemSim for measuring semantic similarities of genes and gene products based on their annotations of DO and MeSH terms. Enrichment analyses including overrepresentation and gene set enrichment analysis are also available in these packages.
3. In a paper published in Journal of Proteomics [1], we found that AHNAK1 is the core protein in FMNL1 interactome in T cells by functional similarity analysis using GOSemSim. We

further verified that N-terminus of FMNL1 binds to the C-terminal part of AHNAK1. AHNAK1 has been previously reported to be involved in calcium signaling and effector functions of T cells. We further found that overexpression or downregulation of FMNL1 has an impact on ionophore-mediated calcium influx by inducing translocation of AHNAK1. In this paper, we found a core protein, AHNAK1, from functional similarities calculated by GOSemSim, which suggested a new function of FMNL1 that involved in calcium-dependent membrane processes.

4. There is a substantial number of GOSemSim users. First, the number of downloads of GOSemSim during January 2019 is more than 6000 and it is the most highly downloaded package within the Bioconductor project (rank 47 of more than 1000 packages, data from https://bioconductor.org/packages/stats). Second, GOSemSim has been cited more than 300 times in many life sciences-related fields (data from Web of Science), and it is selected as a highly cited paper (data from Essential Science Indicators).
5. For algorithm details and function examples, please refer to the online vignette of GOSemSim package, https://www.bioconductor.org/packages/GOSemSim. In future release, we plan to support more algorithms for measuring semantic similarity.
6. The Bioconductor project has already provided OrgDb for 20 species (http://bioconductor.org/packages/release/BiocViews.html#___OrgDb). There are many OrgDb for nonmodel species available online and can be queried via AnnotationHub (https://www.bioconductor.org/packages/AnnotationHub). We can also build our own OrgDb using AnnotationForge (https://www.bioconductor.org/packages/AnnotationForge). In future releases, we plan to extend GOSemSim to support building semantic data from other sources.
7. For details of ggtree visualization, users can refer to the online book "Data Integration, Manipulation and Visualization of Phylogenetic Trees," from https://yulab-smu.github.io/treedata-book/.

Acknowledgments

I thank Drs. Yin Ge and Zhongtian Xu for providing useful feedback and helpful comments on the manuscript. This work was supported by Startup funds from Southern Medical University (G618289088).

References

1. Han Y, Yu G, Sarioglu H et al (2013) Proteomic investigation of the interactome of FMNL1 in hematopoietic cells unveils a role in calcium-dependent membrane plasticity. J Proteome 78:72–82. https://doi.org/10.1016/j.jprot.2012.11.015
2. Yu G, He Q-Y (2011) Functional similarity analysis of human virus-encoded miRNAs. J Clin Bioinforma 1:15. https://doi.org/10.1186/2043-9113-1-15
3. Pirkl M, Diekmann M, van der Wees M et al (2017) Inferring modulators of genetic interactions with epistatic nested effects models. PLoS Comput Biol 13:e1005496. https://doi.org/10.1371/journal.pcbi.1005496
4. Lei C, Ruan J (2013) A novel link prediction algorithm for reconstructing protein–protein interaction networks by topological similarity. Bioinformatics 29:355–364. https://doi.org/10.1093/bioinformatics/bts688
5. Bhattacharya A, Cui Y (2015) miR2GO: comparative functional analysis for microRNAs. Bioinformatics 31:2403–2405. https://doi.org/10.1093/bioinformatics/btv140
6. Zhou H, Yang Y, Shen H-B (2017) Hum-mPLoc 3.0: prediction enhancement of human protein subcellular localization through modeling the hidden correlations of gene ontology and functional domain features. Bioinformatics 33:843–853. https://doi.org/10.1093/bioinformatics/btw723
7. Yu G, Li F, Qin Y et al (2010) GOSemSim: an R package for measuring semantic similarity among GO terms and gene products. Bioinformatics 26:976–978. https://doi.org/10.1093/bioinformatics/btq064
8. Resnik P (1999) Semantic similarity in a taxonomy: an information-based measure and its application to problems of ambiguity in natural language. J Artif Intell Res 11:95–130. https://doi.org/10.1613/jair.514
9. Jiang JJ, Conrath DW (1997) Semantic similarity based on corpus statistics and lexical taxonomy. ArXivcmp-Lg9709008
10. Lin D (1998) An information-theoretic definition of similarity. In: Proceedings of the fifteenth international conference on machine learning. Morgan Kaufmann Publishers Inc., San Francisco, pp 296–304
11. Schlicker A, Domingues FS, Rahnenführer J, Lengauer T (2006) A new measure for functional similarity of gene products based on Gene Ontology. BMC Bioinformatics 7:302. https://doi.org/10.1186/1471-2105-7-302
12. Wang JZ, Du Z, Payattakool R et al (2007) A new method to measure the semantic similarity of GO terms. Bioinforma Oxf Engl 23:1274–1281. https://doi.org/10.1093/bioinformatics/btm087
13. Eddelbuettel D, Francois R (2011) Rcpp: Seamless R and C++ Integration. J Stat Softw 40:1–18. https://doi.org/10.18637/jss.v040.i08
14. Yu G, Wang L-G, Han Y, He Q-Y (2012) clusterProfiler: an R package for comparing biological themes among gene clusters. OMICS J Integr Biol 16:284–287. https://doi.org/10.1089/omi.2011.0118
15. Yu G, Wang L-G, Yan G-R, He Q-Y (2015) DOSE: an R/Bioconductor package for disease ontology semantic and enrichment analysis. Bioinformatics 31:608–609. https://doi.org/10.1093/bioinformatics/btu684
16. Paradis E, Schliep K (2019) ape 5.0: an environment for modern phylogenetics and evolutionary analyses in R. Bioinformatics 35:526–528. https://doi.org/10.1093/bioinformatics/bty633
17. Yu G, Smith DK, Zhu H et al (2017) Ggtree: an r package for visualization and annotation of phylogenetic trees with their covariates and other associated data. Methods Ecol Evol 8:28–36. https://doi.org/10.1111/2041-210X.12628
18. Wu J, Huang B, Chen H et al (2016) The landscape of accessible chromatin in mammalian preimplantation embryos. Nature 534:652–657. https://doi.org/10.1038/nature18606
19. Yu G, Lam TT-Y, Zhu H, Guan Y (2018) Two methods for mapping and visualizing associated data on phylogeny using Ggtree. Mol Biol Evol 35:3041–3043. https://doi.org/10.1093/molbev/msy194
20. Yu G (2018) Using meshes for MeSH term enrichment and semantic analyses. Bioinformatics 34:3766–3767. https://doi.org/10.1093/bioinformatics/bty410

Part III

Derivation of Stem Cells

Chapter 12

Derivation of Maternal Epiblast Stem Cells from Haploid Embryos

Benjamin L. Kidder

Abstract

Pluripotent cells in the inner cell mass (ICM) or epiblast of mammalian embryos exhibit the capacity to differentiate into all cells represented in the three germ layers. Embryonic stem (ES) cells can be derived from the ICM of preimplantation stage blastocysts, while epiblast stem cells (EpiSCs) can be derived from the epiblast of postimplantation embryos or preimplantation stage embryos. The ability to derive distinct types of pluripotent cells from blastocyst-stage embryos suggests that optimization of culture conditions can promote self-renewal of various stem cell populations. Moreover, because mouse EpiSCs resemble human pluripotent stem (hPS) cells, EpiSCs are a useful model to study common and divergent mechanisms of self-renewal between orthologous species. In addition, studies have demonstrated that haploid embryos and ES cells can be derived from chemically activated oocytes. Here, we describe a protocol for deriving maternal (parthenogenetic/gynogenetic) EpiSCs (maEpiSCs) from haploid blastocyst-stage embryos. This protocol is suitable to establish an experimental model for the study of mechanisms of EpiSC self-renewal and differentiation.

Key words Epiblast stem cells, Haploid, Self-renewal, Epigenetics, Chromatin

1 Introduction

Pluripotent stem cells derived from the ICM of blastocyst-stage embryos or postimplantation embryos retain the capacity to self-renew in the presence of external signals or differentiate into all lineages represented in the embryo proper. ES cells can be derived from the ICM of blastocyst-stage embryos, while EpiSCs can be derived from postimplantation or preimplantation embryos [1–5]. Mouse EpiSCs resemble human pluripotent stem (hPS) cells in their morphology and ability to self-renew in the presence of activing/nodal and FGF signals [1, 4, 6]. Due to the morphological and functional similarity between mouse EpiSCs and hPS cells, mouse EpiSCs are a useful model to study common and divergent mechanisms of self-renewal and differentiation between mouse EpiSCs and hPS cells. Mouse trophoblast stem (TS) cells

Benjamin L. Kidder (ed.), *Stem Cell Transcriptional Networks: Methods and Protocols*, Methods in Molecular Biology, vol. 2117, https://doi.org/10.1007/978-1-0716-0301-7_12,

also originate from preimplantation embryos, and can maintain self-renew in the presence of external signals (activin/nodal and FGF4) [7], or differentiate into trophoblast lineages in vitro or in vivo postimplantation [8, 9]. Moreover, because trophoblast stem (TS) cells originate from preimplantation embryos, and exhibit self-renewal under similar culture conditions as mouse EpiSCs and hPS cells, it is possible that common signaling pathways may support self-renewal of multiple types of stem cells derived from preimplantation stage embryos. Here, we describe a protocol to establish long-term cultures of maternal (parthenogenetic/ gynogenetic) EpiSCs (maEpiSCs) following chemical activation of unfertilized oocytes and culture of resulting haploid preimplantation stage embryos in media containing FGF4. This protocol is sufficient to establish fully pluripotent maEpiSCs, where maEpiSCs can be differentiated in vitro and in vivo upon removal of FGF4 from the culture media.

2 Materials

2.1 Oocyte Collection, Oocyte Activation, and Embryo Culture

1. EmbryoMax M2 Medium (EMD Millipore, MR-015-D).
2. EmbryoMax KSOM Medium (EMD Millipore, MR-106-D).
3. Mineral oil for embryo culture (Sigma).
4. EGTA tetrasodium salt (Ethylene glycol-bis(β-aminoethyl ether)-*N*,*N*,*N*′,*N*′-tetraacetic acid tetrasodium salt; 0.5 M solution, Sigma). Add EGTA to media (M16 or KSOM) to obtain a concentration of 2 mM.
5. Strontium chloride hexahydrate. Add strontium chloride using a 100 mM stock solution dissolved in water to a final concentration of 5 mM in EGTA containing media (M16 or KSOM).
6. Hyaluronidase.
7. Syringe (1 cc) and 26 gauge needle.
8. Dissecting instruments (stainless steel fine forceps and fine scissors).
9. Stereomicroscope.
10. Petri dishes (sterile; 35, 60, and 100 mm).
11. PBS (Phosphate buffered saline) without calcium or magnesium (1×).
12. Aspirator mouth piece, aspirator tube assembly, and tubing (Sigma A5177).
13. Pulled capillary tube (Drummond microcaps 100-1-000-0500) or drawn Pasteur pipette.
14. Pipettes.

15. Cell culture incubator with CO_2 carbon dioxide and nitrogen gas tanks (37 °C and 5% CO_2).
16. 70% ethanol.
17. Female C57Bl6 or B6D2F1 mice (19–23 day old).
18. Penicillin–streptomycin (100×).
19. Water bath (*see* **Note 1**).

2.2 Hormones

1. PMSG (Pregnant Mare Serum Gonadotropin; Calbiochem, 367222). Resuspend PMSG to a concentration of 50 IU/mL in PBS. Aliquot and store at −80 °C. Female mice receive an injection of 5 IU of PMSG (100 μL).
2. hCG (Human Chorionic Gonadotropin, Sigma CG5). To generate a stock, resuspend lyophilized hCG to a concentration of 1000 IU/mL in PBS. Aliquot 50 IU and store at −80 °C. To generate a working solution, add PBS to make an hCG concentration of 50 IU/mL. Female mice receive an injection of 5 IU of hCG (100 μL injection).

2.3 Mouse Embryonic Fibroblast (MEF) Cell Culture

1. Dulbecco's Modified Eagle Medium (DMEM), High Glucose (Life Technologies).
2. Fetal bovine serum (FBS).
3. Penicillin–streptomycin (100×) (Life Technologies).
4. Cell culture treated dishes (24-, 12-, 6-well).
5. Polystyrene conical tubes (5, 15, and 50 mL).
6. Phosphate buffered saline (PBS 1×) without calcium or magnesium.
7. Trypsin–EDTA (0.25%, Life Technologies).

2.4 Epiblast Stem Cell (EpiSC) Cell Culture

1. Knockout DMEM media, High Glucose, L-Glutamine (Life Technologies).
2. KnockOut serum replacement (Life Technologies).
3. Penicillin–Streptomycin (100×) (Life Technologies).
4. bFGF (5 ng/mL).
5. Phosphate buffered saline (PBS 1×) without calcium or magnesium.
6. Trypsin–EDTA (0.25%, Life Technologies).
7. EmbryoMax 2-mercaptoethanol (100×; Life Technologies).
8. ROCK inhibitor (10 μM, Y27632).
9. FBS for coating cell culture dishes.
10. Cell culture treated dishes (24-, 12-, and 6-well).
11. Polystyrene conical tubes (5, 15, and 50 mL).

2.5 Maternal (Parthenogenetic/Gynogenetic) EpiSCs (maEpiSC) Cell Culture

1. RPMI (Roswell Park Memorial Institute) 1640 media with L-glutamine (200 mM).
2. FBS (Fetal bovine serum).
3. NEAA (Nonessential amino acids, 100×) (Life Technologies).
4. Penicillin–Streptomycin (100×) (Invitrogen).
5. PBS without calcium or magnesium (1×).
6. Sodium pyruvate (100 mM) (Invitrogen).
7. 2-Mercaptoethanol (100×, Invitrogen).
8. FGF4 (25 ng/mL). Resuspend lyophilized FGF4 (25 μg, R&D Systems) in PBS (0.1% BSA; *see* **Note 2**). Freeze aliquots at −80 °C.
9. Heparin (1 μg/mL). Resuspend 10,000 U of heparin (Sigma H3149) in PBS to generate 1000× stock solution (1 mg/mL). Store at −80 °C.
10. Cell culture dishes (24-, 12-, 6-well).
11. 0.1% gelatin solution in water.

3 Methods

3.1 Superovulation of Female Mice

1. Day 1. Intraperitoneally (IP) inject female mice (aged 19–23 days) with PMSG (5 IU; 100 μL of PMSG stock) between 5:00 and 7:00 pm (*see* **Note 3**):
2. Day 3. 48 h later, IP inject female mice with hCG (5 IU, 100 μL of hCG working solution).

3.2 Preparation of Culture Dishes for Oocyte Activation and Embryo Culture

1. Day 3. Multiple dishes should be prepared fresh the day before cumulus oocyte complexes (COCs) are harvested. Prepare 35 mm culture dishes with 50 μL drops of M16 or KSOM media with EGTA and strontium chloride (2 mM EGTA, 5 mM $SrCl_2$), and one 120 μL drop of M16 or KSOM media with EGTA and strontium chloride for washing. Cover the media drops carefully by overlaying 3–4 mL of mineral oil.
2. Incubate the dishes at 37 °C with 5% CO_2. On the day COCs are collected, prepare fresh 35 mm dishes containing 50–120 μL drops of M2 with and without 0.1% hyaluronidase.

3.3 Collection of Mouse Oocytes from COCs, $SrCl_2$ Induced Activation of Oocytes, and In Vitro Culture (IVC) of Embryos to the Blastocyst Stage

1. 9:00 am—Day 3. M2 media should be warmed at 37 °C in 15 mL conical tubes.
2. 2:00–3:00 pm—Day 3. Dissect oviducts from PMSG/hCG injected female mice, and place in prewarmed M2 media, then transfer to a 6 cm dish. Remove the fat surrounding the ovary, and place in a new 6 cm dish.

3. To harvest COCs use two 26 gauge needles attached to syringes (hold one in each hand) to disrupt the oviduct ampulla and release the COCs.
4. Transfer COCs from a 6 cm dish to a 35 mm dish containing M2 media. First, transfer the COCs to the drop with 120 μL M2 media to wash, and subsequently transfer COCs to the drop containing 50 μL M2 media containing 0.1% hyaluronidase. Allow the cumulus cells to disperse away from the oocyte.
5. Transfer COCs from a 35 mm dish with hyaluronidase to another 35 mm dish containing oocyte activation media. Transfer the COCs to the drop with 120 μL media to wash, and subsequently transfer COCs to the drop containing 50 μL oocyte activation media. Next, incubate for 6 h at 37 °C with 5% CO_2 (Fig. 1a, b).
6. Next, transfer activated and denuded oocytes to a fresh 35 mm dish containing 50–120 μL drops of M16 or KSOM media and incubate for 3–4 days at 37 °C with 5% CO_2. The next day, evaluate the number of embryos at the two-cell stage using standard microscopy (Fig. 1c). Continue to evaluate morulae and blastocyst formation over the course of 3–4 days.

3.4 Preparation of Cell Culture Media

1. Mouse embryonic fibroblast (MEF) media for culture of mitotically inactivated MEFs (iMEFs) derived from E13.5 to E14.5 mouse embryos. Store in liquid nitrogen or −150 °C (*see* **Note 1**). DMEM high glucose, 10% FBS, pen/strep (1×), L-glutamine (1×) at 37 °C with 5% CO_2. Mitotically inactivated MEFs (iMEFs) can be derived from mouse embryos (E13.5–14.5) [10] or purchased commercially (*see* **Note 4**).
2. maEpiSC derivation media: RPMI 1640 with L-glutamine, FBS (20%), pen/strep (1×), 1 mM sodium pyruvate, 2-mercaptoethanol (1×), 25 ng/mL FGF4, 1 μg/mL heparin [7], 37 °C with 5% CO_2.
3. EpiSC culture media: Knockout DMEM media (High Glucose, L-Glutamine), 20% KnockOut serum replacement, 5 ng/mL bFGF, 2-Mercaptoethanol (1×), pen/strep (1×), ROCK inhibitor (10 μM, Y27632), 37 °C with 5% CO_2.

3.5 Preparation of iMEF Feeder Layer

1. Twenty-four hours before deriving maEpiSCs from blastocysts generated from $SrCl_2$ induced activation of oocytes and in vitro culture (IVC), iMEFs are cultured on 24-, 12-, and 6-well gelatin-coated plates. To culture iMEFs, coat each well of a 6-well cell culture dish with 1 mL of 0.1% gelatin solution, and incubate for 20–30 min at 37 °C. For 24- and 12-well plates, scale the amount of gelatin solution accordingly.

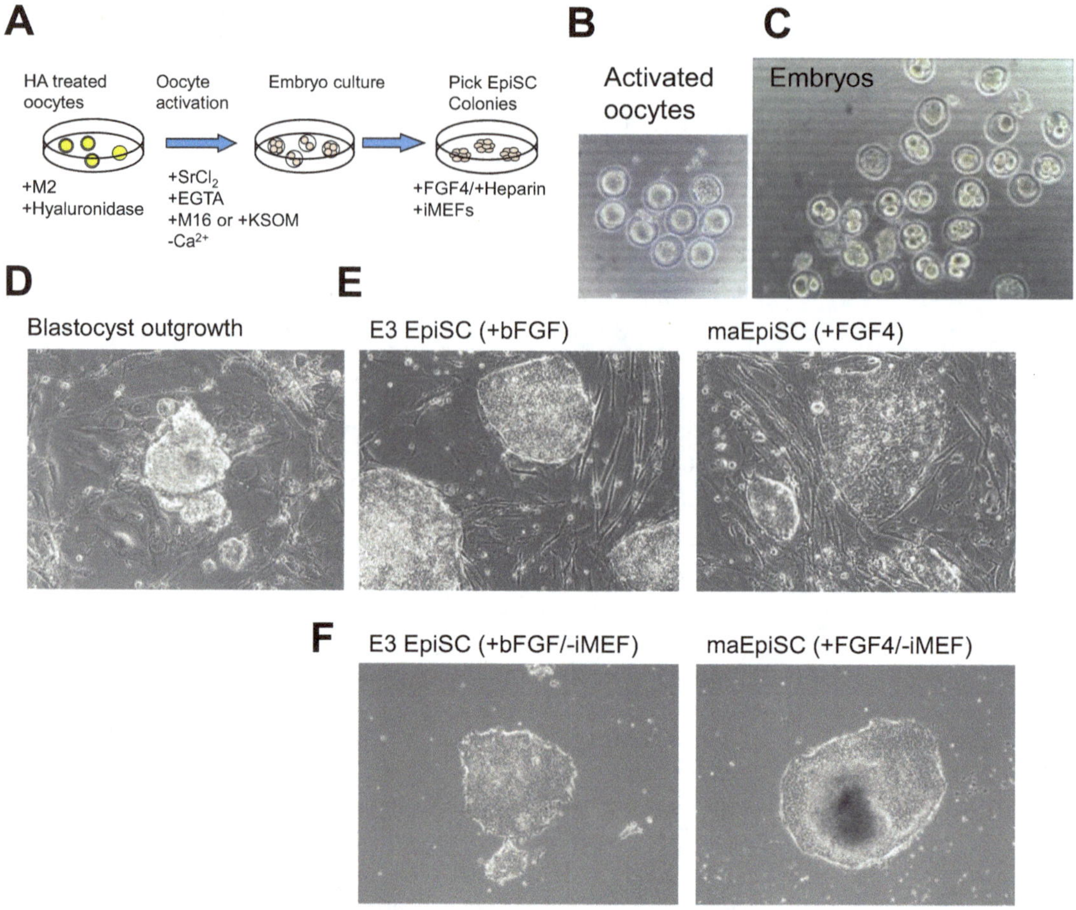

Fig. 1 Derivation of maEpiSCs from haploid embryos. (**a**) Schematic of experimental design. Superovulated ocytes were denuded using hyaluronidase and activated using $SrCl_2$-containing M16 media. (**b**) $SrCl_2$-activated oocytes were in vitro-cultured to **(c)** two-cell embryos and subsequently to the blastocyst stage. **(d)** Blastocysts were cultured in maEpiSC media containing FGF4 on iMEFs and outgrowths were picked and expanded. (**e**) Representative images of conventional E3 EpiSCs (left) cultured in EpiSC media containing bFGF and maternal (parthenogenetic/gynogenetic) EpiSCs (maEpiSCs; right) cultured in maEpiSC media containing FGF4. (**f**) Bright-field microscopy of conventional E3 EpiSCs (left) and maEpiSCs right cultured in feeder-free conditions

2. Next, warm iMEF media at 37 °C for 10 min, and subsequently thaw a vial of frozen iMEFs. Add thawed iMEFs to 3 mL of iMEF media and spin at 1500 rpm (500 × *g*) for 5 min.
3. Aspirate the gelatin solution from the culture plate, resuspend iMEFs in 2 mL of iMEF media (for 6-well plates), and incubate at 37 °C. For 12-well plates, use 1 mL media, and for 24-well plates, use 0.5 mL media.

3.6 Derivation of maEpiSCs

1. Use an inverted microscope to identify cavitated blastocysts. Use a mouth pipette apparatus and a stereomicroscope to transfer blastocysts from the 35 mm dish to a 24-well plate containing iMEFs and maEpiSC media. One blastocyst should be plated per well. Incubate at 37 °C with 5% CO_2.
2. The next day, use an inverted microscope to check attachment of the blastocysts to the iMEF layer. Attachment should occur within 2–3 days.
3. Forty-eight hours later, change media by aspirating and adding fresh maEpiSC media.
4. Culture blastocysts until the outgrowth is large enough to dissociate. Culture of blastocysts on an iMEF layer in maEpiSC media should promote outgrowth of an epiblast-like epithelial layer.
5. Dissociate the outgrowth before it becomes too large and begins to differentiate (Fig. 1d). To this end, wash with 500 μL, add 100–200 μL prewarmed 0.05% trypsin–EDTA and incubate for up to 3 min at room temperate. Alternatively, incubate for 2–3 min at 37 °C. Gently dissociate using a pipette and inactivate trypsin by adding to a conical tube containing 3 mL of prewarmed iMEF media. Spin at 1500 rpm (500 × *g*) for 5 min.
6. Resuspend the cell pellet in maEpiSC media containing FGF4, heparin, and ROCK inhibitor (10 μM, Y27632) and plate on another well of a 24-well plate containing iMEFs, and incubate at 37 °C with 5% CO_2.
7. maEpiSC colonies should become apparent within seven to 10 days following dissociation and initiation of culture on iMEFs plates. maEpiSC media should be changed every 48 h.
8. Passage the colonies once they become 50% confluent. First, wash with 500 μL PBS, add 100–200 μL prewarmed 0.05% trypsin for several minutes until clumps of cells are visible (*see* **Note 5**).
9. For established maEpiSC colonies (Fig. 1e), passaging of cells should be performed once the cultures reach 70% confluency.

3.7 Feeder-Free Culture of maEpiSCs

maEpiSCs are capable of being transitioned to feeder-free conditions to eliminate the use of iMEFs for culture and genomics analyses (e.g., RNA-Seq, ChIP-Seq) (Fig. 1f).

1. Prepare maEpiSC conditioned media. Culture iMEFs in maEpiSC media without FGF4 for 48 h, collect, and filter. maEpiSCs can be cultured on tissue culture treated dishes.
2. maEpiSCs can be cultured in 70% maEpiSC conditioned media and 30% maEpiSC media. Media should be supplemented with FGF4 (25 ng/mL) and heparin.

3. Passage feeder-free maEpiSCs in a similar way as described for culture on a layer of iMEFs. Do not single-cell dissociate, which leads to reduced viability.

3.8 Differentiation of maEpiSCs

1. maEpiSCs can be differentiated by culturing in maEpiSC media without FGF4 in feeder-free conditions. maEpiSCs should be passaged as described above in media without FGF4 or iMEFs.
2. Change media after 48 h by aspirating the old media and adding fresh maEpiSC media without FGF4.
3. Differentiation should be visible using standard microscopy after several days following culture in the absence of FGF4.
4. Directed differentiation may also be facilitated using small molecule inhibitors or growth factors as described for differentiation experiments previously described for EpiSCs.

4 Notes

1. A bead bath can be used instead of a water bath to reduce potential contamination.
2. To make a solution of 0.1% BSA in PBS, dissolve BSA in PBS, filter using a 0.45 μM syringe, aliquot, and store at −80 °C until use.
3. The timing of the PMSG injection should also depend on the time the oocytes are collected.
4. MEFs can also be obtained from various commercial vendors or prepared from E13.5 to E14.5 MEFs. Protocols for harvesting MEFs can be readily found in the literature.
5. Avoid single-cell dissociation of maEpiSCs, which leads to decreased viability.

Acknowledgments

The authors would like to thank Allegra Geller.

References

1. Tesar PJ, Chenoweth JG, Brook FA, Davies TJ, Evans EP, Mack DL, Gardner RL, McKay RD (2007) New cell lines from mouse epiblast share defining features with human embryonic stem cells. Nature 448(7150):196–199. https://doi.org/10.1038/nature05972
2. Evans MJ, Kaufman MH (1981) Establishment in culture of pluripotential cells from mouse embryos. Nature 292(5819):154–156
3. Martin GR (1981) Isolation of a pluripotent cell line from early mouse embryos cultured in medium conditioned by teratocarcinoma stem cells. Proc Natl Acad Sci U S A 78 (12):7634–7638
4. Brons IG, Smithers LE, Trotter MW, Rugg-Gunn P, Sun B, Chuva de Sousa Lopes SM, Howlett SK, Clarkson A, Ahrlund-Richter L, Pedersen RA, Vallier L (2007) Derivation of

pluripotent epiblast stem cells from mammalian embryos. Nature 448(7150):191–195. https://doi.org/10.1038/nature05950

5. Najm FJ, Chenoweth JG, Anderson PD, Nadeau JH, Redline RW, McKay RD, Tesar PJ (2011) Isolation of epiblast stem cells from preimplantation mouse embryos. Cell Stem Cell 8(3):318–325. https://doi.org/10.1016/j.stem.2011.01.016
6. Thomson JA, Itskovitz-Eldor J, Shapiro SS, Waknitz MA, Swiergiel JJ, Marshall VS, Jones JM (1998) Embryonic stem cell lines derived from human blastocysts. Science 282 (5391):1145–1147
7. Tanaka S, Kunath T, Hadjantonakis AK, Nagy A, Rossant J (1998) Promotion of trophoblast stem cell proliferation by FGF4. Science 282(5396):2072–2075
8. Kidder BL, Palmer S (2010) Examination of transcriptional networks reveals an important role for TCFAP2C, SMARCA4, and EOMES in trophoblast stem cell maintenance. Genome Res 20(4):458–472. https://doi.org/10.1101/gr.101469.109
9. Cross JC, Baczyk D, Dobric N, Hemberger M, Hughes M, Simmons DG, Yamamoto H, Kingdom JC (2003) Genes, development and evolution of the placenta. Placenta 24 (2–3):123–130
10. Kidder BL, Oseth L, Miller S, Hirsch B, Verfaillie C, Coucouvanis E (2008) Embryonic stem cells contribute to mouse chimeras in the absence of detectable cell fusion. Cloning Stem Cells 10(2):231–248

Chapter 13

Derivation of LIF-Independent Embryonic Stem Cells Using Inducible OCT4 Expression

Benjamin L. Kidder

Abstract

Pluripotent mouse embryonic stem (ES) cells, which are derived from the inner cell mass (ICM) of preimplantation stage embryos, are capable of self-renewing indefinitely in the presence of the external signal leukemia inhibitory factor (LIF), activation of Wnt signaling through inhibition of GSK3, and inhibition of MAP kinase/ERK kinase signaling. The OCT4 transcription factor is expressed highly in pluripotent cells and is a central transcriptional regulator of the pluripotent state. Here, we describe a protocol to culture ES cells in LIF-independent and serum-free media using an inducible OCT4 (iOCT4) ES cell model system. This protocol is sufficient to sustain ES cell self-renewal in vitro in defined conditions in the absence of external signals. LIF-independent iOCT4 ES cells are fully capable of differentiating following deactivation of the inducible OCT4 transgene.

Key words Embryonic stem cells, OCT4, LIF, Pluripotent, Self-renewal, Embryonic stem cells

1 Introduction

The mammalian preimplantation stage blastocyst embryo contains a pluripotent population of cells within the inner cell mass (ICM), which can be modeled in vitro by establishing embryonic stem (ES) cells. Conventional mouse ES cell self-renewal is sustained in the presence of the external signal leukemia inhibitory factor (LIF) and culture on a layer of mitotically inactivated mouse embryonic fibroblasts (MEFs). Combined inhibition of glycogen synthase kinase-3 (GSK3) using the small molecule inhibitor, CHIR99021, and MEK, using PD0325091, referred to as 2i [1], also supports ES cell self-renewal in serum- and feeder-free conditions. Activation of intrinsic OCT4-centric transcriptional networks is thought to sustain self-renewal of ES cells. OCT4 expression is elevated in ES cells and serves a critical role in maintaining ES cells identity [2, 3]. Alterations in the level of OCT4 lead to ES cell differentiation: downregulation of OCT4 has been implicated in promoting trophectoderm differentiation while

Benjamin L. Kidder (ed.), *Stem Cell Transcriptional Networks: Methods and Protocols*, Methods in Molecular Biology, vol. 2117, https://doi.org/10.1007/978-1-0716-0301-7_13,

overexpression promotes lineage commitment toward mesodermal or endodermal fates [3]. These findings suggest that OCT4 acts as a key regulator of self-renewal versus differentiation by promoting or repressing expression of target genes [4]. OCT4 interacts with transcription factors such as SOX2 and NANOG to form a core transcriptional regulatory circuit [5, 6]. Self-renewal or differentiation ES cell fate decisions are largely controlled by these intrinsic transcriptional networks and external signals. Previous studies have demonstrated that ES cell self-renewal can be maintained without STAT3 activation by overexpressing NANOG [7], ESRRB [8], GBX2 [9], KLF2 [10], KLF4, TBX3 [11], or Tfcp2l1 [12]. We recently demonstrated that inducible OCT4 expression is capable of promoting self-renewal of ES cells in LIF-independent media containing serum, or in serum- and feeder-free conditions [13]. LIF-independent inducible (iOCT4) ES cells are able to differentiate in vitro and in vivo following downregulation of the iOCT4 transgene. Here, we describe a protocol to maintain LIF-independent self-renewal using an inducible OCT4 (iOCT4) ES cell model system. This protocol is suitable for the study of ES cell pluripotency and mechanisms of self-renewal.

2 Materials

2.1 Reagents for Cell Culture

1. DMEM with L-glutamine (Dulbecco's Modified Eagle's Medium, High Glucose).
2. Fetal bovine serum (FBS, ES cell qualified).
3. Non-essential amino acids (NEAA, 100×).
4. EmbryoMax 2-mercaptoethanol (100×).
5. Sodium pyruvate (100 mM).
6. Penicillin–streptomycin (100×).
7. Leukemia inhibitory factor (LIF, 1×, Chemicon).
8. GSK3β inhibitor CHIR99021 (GSK3i).
9. MEK inhibitor PD0325091 (MEKi).
10. ESGRO Complete Basal medium (EMD Millipore).
11. Doxycycline (2 mg/mL stock solution, 2 μg/mL final concentration).
12. 0.1% gelatin solution in water
13. PBS (Phosphate buffered saline without calcium or magnesium, 1×).
14. 0.25% Trypsin–EDTA (Invitrogen).
15. Cell culture dishes (6-, 12-, and 24-well).
16. Conical tubes (polystyrene, 15 and 50 mL).
17. Water bath.
18. 70% ethanol

2.2 Equipment

1. Inverted microscope.
2. Clinical centrifuge.
3. Cell culture incubator.

2.3 Culture Media

1. MEF (Mouse embryonic fibroblast) media for culture of mitotically inactivated MEFs (iMEFs). iMEFs should be stored at −150 °C or lower. DMEM high glucose with L-glutamine, 10% FBS, penicillin–streptomycin (1×), at 37 °C with 5% CO_2. Mitotically inactivated MEFs (iMEFs) derived from E13.5 to E14.5 mouse embryos [14] can be acquired from a commercial vendor. Conventional ES cell culture involves the use of an iMEF feeder layer (*see* **Note 1**).
2. Derivation of LIF-independent iOCT4 ES cells is performed in the presence or absence of FBS and without LIF or doxycycline to induce OCT4-transgene expression. ES cell media containing DMEM high glucose, 15% ES cell qualified FBS, LIF (10 ng/mL), penicillin–streptomycin (1×), glutamine (1×), 2-mercaptoethanol (1×), nonessential amino acids (1×), and sodium pyruvate at 37 °C with 5% CO_2.

3 Methods

3.1 Prepare Feeder Layer of Mitotically Inactivated Mouse Embryonic Fibroblasts (iMEFs)

1. ES cells that have been derived on a layer of mitotically inactivated mouse embryonic fibroblasts (iMEFs) should first be cultured on iMEFs on gelatin-coated cell culture plates before transitioning them to feeder-free conditions. Add 2 mL of gelatin solution to a 6-well culture plate and incubate at 37 °C for 15–20 min.
2. Prewarm MEF media (10% FBS, DMEM, pen/strep) for 10–15 min in a water bath at 37 °C.
3. Thaw a vial of iMEFs (gamma-irradiated or mitocycin C treated) for 1–2 min in a water bath at 37 °C.
4. Add iMEFs to prewarmed MEF media (3 mL) in a conical tube (15 mL), spin for 5 min at 1500 rpm (500 × *g*). Aspirate media, resuspend iMEFs in 2 mL of MEF media, and plate in a 6-well cell culture plate.
5. Check the density of iMEFs 24 h after they are plated. iMEFs should be slightly confluent and partially overlapping but not too crowded.

3.2 Derivation of LIF-Independent iOCT4 ES Cells

1. Thaw a frozen vial of iOCT4 (ZHTc6 [3]) ES cells (Fig. 1a, b) in a water bath, add cells to 3 mL of prewarmed MEF media in a 15 mL conical tube, and spin at 1500 rpm (500 × *g*) for 3–5 min. Aspirate media, resuspend ES cells in ES cell media containing serum, LIF, 1.5 μM CHIR99021, and 2 μg/mL doxycycline, and plate on a 6-well dish containing iMEFs.

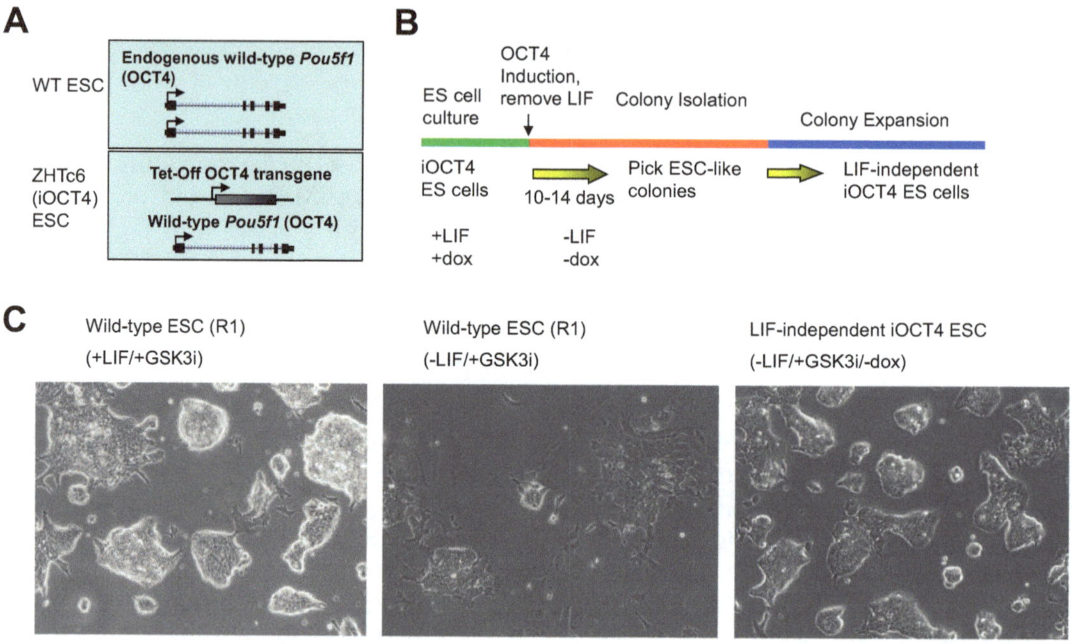

Fig. 1 Derivation of LIF-independent iOCT4 cells. (**a**) Schematic of two endogenous wild-type OCT4 (Pou5f1) alleles (top) and the Tet-Off OCT4 transgene in ZHTc6 ES cells (bottom). Wild-type (WT) ES cells have two endogenous WT OCT4 (Pou5f1) alleles, while inducible-OCT4 ZHTc6 (iOCT4) ES cells possess one WT OCT4 allele and one Tet-Off OCT4 transgene. The Tet-Off OCT4 transgene is activated in the absence of doxycycline. (**b**) Experimental design schematic for deriving LIF-independent iOCT4 ES cells. ZHTc6 iOCT4 ES cells [3] were cultured in ES cell media containing serum (FBS) in the absence of LIF and without doxycycline to induce OCT4 transgene overexpression. Colonies were pick and expended for further characterization. (**c**) Bright-field microscopy of WT ES cells (R1) cultured in ES cell media containing serum and LIF (left), or without LIF (middle). LIF-independent iOCT4 ES cells were cultured in ES cell media without LIF or doxycycline to induce OCT4 transgene expression

2. Media should be changed every 48 h. When ES cell colonies become relatively dense (30–40% confluent), they should be passaged.
3. iOCT4 ES cells should be transitioned to feeder-free conditions to eliminate contamination of fibroblasts. Prepare gelatin-coated 6-well culture plates in advance. Add 2 mL of gelatin solution to a 6-well culture plate and incubate at 37 °C for 15–20 min. Next, passage ES cells by aspirating media, washing with PBS, and adding 1 mL of 0.25% trypsin–EDTA solution and incubating at room temperature or 37 °C for 2–3 min or until cells become attached. Use a pipette and a 1 mL tip to dissociate into a single-cell suspension. Add 1 mL trypsin–cell solution to 3 mL of MEF media, spin at 1500 rpm (500 × *g*) for 5 min, aspirate media, and resuspend in ES cell media containing doxycycline, LIF, and CHIR99021.

4. After several passages, iOCT4 ES cells can be transitioned to LIF-independent conditions by culturing in ES cell media without LIF and in the absence of doxycycline. During the transition to LIF-independence, iOCT4 ES cell colony morphology should resemble a typical 3D phenotype with a tight-colony border (Fig. 1c).
5. LIF-independent iOCT4 ES cells can also be transitioned to serum- and feeder-free conditions without LIF over several passages. To transition to serum-free conditions, passage ES cells as described above with minor modifications by washing with PBS, dissociating with trypsin, and resuspending the pellet in 2 mL of a 4:1 ratio of ES cell media (without LIF and with FBS) to ESGRO complete Basal media without LIF and with doxycycline (*see* **Note 2**). For the second passage, remove the ESC/ESGRO complete basal media, wash with PBS, add trypsin (0.25%) and incubate for 2–3 min. Pipette to achieve a single-cell suspension, add to 3 mL of prewarmed MEF media, spin at 1500 rpm (500 × *g*) for 5 min, and resuspend in 2 mL of a 3:2 ratio of ES cell media (without LIF and with FBS) to ESGRO complete basal media with doxycycline. For the third passage, repeat the above conditions, except resuspend the cells in 2 mL of a 2:3 ratio of ESC–ESGRO media. For the fourth passage, resuspend the cells in 2 mL of a 1:4 ratio of ESC–ESGRO media. Following the fifth passage, ES cells should be fully transitioned to feeder-free and serum-free conditions.

3.3 Differentiation of LIF-Independent iOCT4 Cells

1. LIF-independent iOCT4 cells can be spontaneously differentiated by culturing in MEF media in the presence of doxycycline and absence of LIF in feeder-free conditions. LIF-independent iOCT4 can be passaged as described above, and replated in 6-well plates in MEF media with doxycycline and without LIF. Media should be changed every 48 h by aspirating spent media and adding 2 mL of fresh MEF media.
2. Differentiated LIF-independent iOCT4 cells should become visible after several days. LIF-independent iOCT4 cell differentiation can be evaluated using an inverted microscope and standard bright-field microscopy. The cytoplasmic to nuclear ratio should increase during ES cell differentiation. Lineage-specific cells should become visible after several days.

4 Notes

1. Culture of transduced MEFs on a feeder layer of iMEFs increases that rate of reprogramming.
2. ES cells can be cultured under the original conditions using N2B27 media [1].

References

1. Ying QL, Wray J, Nichols J, Batlle-Morera L, Doble B, Woodgett J, Cohen P, Smith A (2008) The ground state of embryonic stem cell self-renewal. Nature 453(7194):519–523. https://doi.org/10.1038/nature06968
2. Nichols J, Zevnik B, Anastassiadis K, Niwa H, Klewe-Nebenius D, Chambers I, Scholer H, Smith A (1998) Formation of pluripotent stem cells in the mammalian embryo depends on the POU transcription factor Oct4. Cell 95 (3):379–391
3. Niwa H, Miyazaki J, Smith AG (2000) Quantitative expression of Oct-3/4 defines differentiation, dedifferentiation or self-renewal of ES cells. Nat Genet 24(4):372–376
4. Pesce M, Scholer HR (2001) Oct-4: gatekeeper in the beginnings of mammalian development. Stem Cells 19(4):271–278. https://doi.org/10.1634/stemcells.19-4-271
5. Boyer LA, Lee TI, Cole MF, Johnstone SE, Levine SS, Zucker JP, Guenther MG, Kumar RM, Murray HL, Jenner RG, Gifford DK, Melton DA, Jaenisch R, Young RA (2005) Core transcriptional regulatory circuitry in human embryonic stem cells. Cell 122 (6):947–956
6. Chen X, Xu H, Yuan P, Fang F, Huss M, Vega VB, Wong E, Orlov YL, Zhang W, Jiang J, Loh YH, Yeo HC, Yeo ZX, Narang V, Govindarajan KR, Leong B, Shahab A, Ruan Y, Bourque G, Sung WK, Clarke ND, Wei CL, Ng HH (2008) Integration of external signaling pathways with the core transcriptional network in embryonic stem cells. Cell 133(6):1106–1117
7. Chambers I, Colby D, Robertson M, Nichols J, Lee S, Tweedie S, Smith A (2003) Functional expression cloning of Nanog, a pluripotency sustaining factor in embryonic stem cells. Cell 113(5):643–655
8. Martello G, Sugimoto T, Diamanti E, Joshi A, Hannah R, Ohtsuka S, Gottgens B, Niwa H, Smith A (2012) Esrrb is a pivotal target of the Gsk3/Tcf3 axis regulating embryonic stem cell self-renewal. Cell Stem Cell 11(4):491–504. https://doi.org/10.1016/j.stem.2012.06.008
9. Tai CI, Ying QL (2013) Gbx2, a LIF/Stat3 target, promotes reprogramming to and retention of the pluripotent ground state. J Cell Sci 126(Pt 5):1093–1098. https://doi.org/10.1242/jcs.118273
10. Hall J, Guo G, Wray J, Eyres I, Nichols J, Grotewold L, Morfopoulou S, Humphreys P, Mansfield W, Walker R, Tomlinson S, Smith A (2009) Oct4 and LIF/Stat3 additively induce Kruppel factors to sustain embryonic stem cell self-renewal. Cell Stem Cell 5(6):597–609. https://doi.org/10.1016/j.stem.2009.11.003
11. Niwa H, Ogawa K, Shimosato D, Adachi K (2009) A parallel circuit of LIF signalling pathways maintains pluripotency of mouse ES cells. Nature 460(7251):118–122. https://doi.org/10.1038/nature08113
12. Ye S, Li P, Tong C, Ying QL (2013) Embryonic stem cell self-renewal pathways converge on the transcription factor Tfcp2l1. EMBO J 32 (19):2548–2560. https://doi.org/10.1038/emboj.2013.175
13. He R, Xhabija B, Al-Qanber B, Kidder BL (2017) OCT4 supports extended LIF-independent self-renewal and maintenance of transcriptional and epigenetic networks in embryonic stem cells. Sci Rep 7 (1):16360. https://doi.org/10.1038/s41598-017-16611-y
14. Kidder BL, Oseth L, Miller S, Hirsch B, Verfaillie C, Coucouvanis E (2008) Embryonic stem cells contribute to mouse chimeras in the absence of detectable cell fusion. Cloning Stem Cells 10(2):231–248

Chapter 14

Simultaneous Derivation of Embryonic and Trophoblast Stem Cells from Mouse Blastocysts

Benjamin L. Kidder

Abstract

The formation of the blastocyst during mammalian development involves the segregation of two populations of cells with unequal potential: pluripotent cells of the inner cell mass (ICM) and multipotent cells of the trophectoderm (TE). ICM cells maintain the capacity to give rise to all cells represented in the organism, while TE cells, which represent the first lineage to emerge during development, are capable of differentiating into trophoblast lineages of the placenta. The ICM and TE are both essential for development. The ICM is genetically programmed to generate all cells of the embryo proper, while the TE forms extraembryonic trophoblast lineages and is required for implantation of the embryo and maternal-fetal exchange of nutrients and waste. Embryonic stem (ES) cells, which can be derived from the ICM of blastocysts in the presence of external signals such as LIF, can self-renewal indefinitely, and because they can differentiate into all cells of the organism, ES cells are a widely used in vitro model to study genetics and development. Trophoblast stem (TS) cells can be derived from the TE of blastocyst stage embryos in the presence of FGF4, and like ES cells, TS cells are also able to self-renew indefinitely. Because TS cells can differentiate into epithelial lineages of the trophoblast, TS cells are an ideal in vitro model to study the biology of the trophoblast. In this chapter, we describe protocols for simultaneous derivation of ES cells and TS cells from mouse blastocysts and culture conditions that promote self-renewal of hybrid ESC/TSC colonies. These protocols are sufficient for efficient derivation of hybrid ESC/TSC colonies.

Key words Embryonic stem cells, Trophoblast stem cells, Trophectoderm, Inner cell mass, Derivation, Blastocyst, Differentiation, Pluripotent, Multipotent

1 Introduction

Blastocyst formation in mammals yields the generation of two distinct cellular layers including the inner cell mass (ICM) and the trophectoderm (TE). Three days after fertilization of the oocyte in mice (four to five in humans), the TE forms a spherical layer of epithelial cells around the ICM of the blastocyst. In contrast to the ICM, which generates all somatic and germline lineages, the TE gives rise to epithelial trophoblast lineages of the placenta including trophoblast giant cells, spongiotrophoblasts, glycogen trophoblast cells, and syncytiotrophoblasts [1]. The ICM and TE both contain

Benjamin L. Kidder (ed.), *Stem Cell Transcriptional Networks: Methods and Protocols*, Methods in Molecular Biology, vol. 2117, https://doi.org/10.1007/978-1-0716-0301-7_14, © Springer Science+Business Media, LLC, part of Springer Nature 2020

stem cell compartments that are capable of self-renewing in the presence of appropriate external signals. The ICM and TE can be modeled in vitro through the derivation of embryonic stem (ES) cells [2] and trophoblast stem (TS) cells [3], respectively. ES cells can be maintained in vitro indefinitely in the presence of LIF, BMP4, and WNT3a, while TS cells can be propagated indefinitely in the presence of FGF4, INHBA, NODAL, and TGFB1 [3]. TS cells can be differentiated by removal of the self-renewal factors, which is accompanied by decreased proliferation and trophoblast differentiation. In contrast, ES cells can be differentiated into all cellular phenotypes of the adult organism that are programmed into the mammalian genome.

TS cells are a useful model to study the development, gene expression, and the epigenomic landscape of placental lineages. Previous findings have demonstrated that TS cells and ES cells share common and distinct expression programs. ES cell self-renewal and pluripotency require the core transcription factors Oct4 [4], Sox2 [5], and Nanog [6, 7], while TS cell multipotency requires transcription factors such as Cdx2 [8], Eomes [9], Esrrb [10], and Tead4 [11]. Genome-wide transcriptomic and epigenomic analyses have provided global views of molecular networks that regulate transcription and pattern the epigenetic landscape of TS cells [12, 13]. Previously, we found that ES cells and TS cells coexpress a number of transcriptional regulators including Sox2, Stat3, Tbx3, Esrrb, Klf5, Lin28, Rest, Rex1, Sall4, Smarca4, and Utf1 [12]. Using an RNAi knockdown approach, we also found that Tfap2c, Smarca4 (Brg1), and Eomes are important for TS cell self-renewal [12]. These findings provide additional insight into epigenetic mechanisms of TS cell self-renewal. Here, we describe protocols for simultaneously deriving ES cells and TS cells from mouse blastocysts.

2 Materials

2.1 Cell Culture Reagents

1. Dulbecco's Modified Eagle's Medium (DMEM), High Glucose.
2. Penicillin–streptomycin (100×) (Invitrogen).
3. L-glutamine (200 mM).
4. Fetal bovine serum (FBS, ES cell qualified).
5. Nonessential amino acids (NEAA, 100×) (Invitrogen).
6. Phosphate buffered saline without calcium or magnesium (PBS, 1×).
7. Leukemia inhibitory factor (LIF, 1000×, Chemicon).

8. FGF4 (25 μg, R&D Systems). Resuspend lyophilized FGF4 in PBS with 0.1% BSA. Aliquot tubes and freeze at −80 °C until use (*see* **Note 1**). To make a solution of 0.1% BSA in PBS, dissolve BSA in PBS, filter using a 0.45 μM syringe, aliquot, and store at −80 °C until use.
9. Heparin (10,000 U, Sigma H3149). Prepare a 1000× stock tube of heparin by resuspending in PBS to 1 mg/mL (1000×). Store at −80 °C until use.
10. 0.25% Trypsin–EDTA (Invitrogen).
11. 0.05% Trypsin–EDTA (Invitrogen).
12. Sodium pyruvate (100 mM) (Invitrogen).
13. 2-Mercaptoethanol (100×, Invitrogen).
14. 24-, 12-, 6-Well culture dish.
15. 0.1 % gelatin solution in water.
16. M2 Medium (EmbryoMax 1×, Millipore).
17. 5, 15, and 50 mL polystyrene conical tubes.
18. Lab Armor bead bath.
19. 70% ethanol.

2.2 Equipment

1. Pulled capillary tube or drawn Pasteur pipette.
2. Aspirator tube assembly: aspirator mouth piece, tubing (Sigma A5177).
3. Stereomicroscope.
4. Clinical centrifuge.
5. Cell culture incubator.

2.3 Culture Media

1. Mouse embryonic fibroblast (MEF) media for culture of mitotically inactivated MEFs (iMEFs) derived from E13.5 to E14.5 mouse embryos. Store in liquid nitrogen until use. DMEM high glucose, 10% FBS, penicillin–streptomycin (1×), L-glutamine (1×) at 37 °C with 5% CO_2. Mitotically inactivated MEFs (iMEFs) can be harvested from mouse embryos (E13.5–14.5) [14] or purchased from a commercial vendor.
2. ESC/TSC derivation media: DMEM, 20% FBS (ESC qualified), L-glutamine (1×), penicillin-streptomycin (1×), 2-mercaptoethanol (1×), nonessential amino acids (NEAA), Sodium pyruvate (1 mM), LIF (1×), 25 ng/mL FGF4, 1 μg/mL heparin [3]. Store at 4 °C until use.
3. Feeder-conditioned ESC/TSC cell media: Conditioned media (CM) is prepared by culturing iMEFs in ESC/TSC media without LIF or FGF4, and harvesting the conditioned media two to three days later. CM can be stored at −20 °C until use, or at 4 °C for a short period. Avoid multiple freeze–thaw cycles.

3 Methods

3.1 Preparation of Feeder Layer

1. For culture of mitotically inactivated MEFs (iMEFs) on 24-, 12-, and 6-well gelatin-coated plates, add 0.5, 1, or 2 mL, respectively of gelatin solution to culture plates and incubate at 37 °C for 10 min.
2. Prewarm MEF media at 37 °C in a Lab Armor bead bath or water bath.
3. Thaw a vial of MEFs in a Lab Armor bead bath or water bath, add cells to 3 mL of prewarmed MEF media in a 15 mL conical tube, and spin at 1500 rpm for 3–5 min.
4. Meanwhile, aspirate gelatin from culture plates. After spinning has completed, resuspend iMEFs in MEF medium, and plate the iMEFs in the culture dishes. The total volume of MEF media for a 24-, 12-, and 6-well plates should be 0.5, 1, and 2, respectively. Before placing the plate in the incubator, gently move the plate vertically then horizontally to evenly distribute the MEFs.

3.2 Blastocyst Isolation

1. Day 0—afternoon. Set up mouse matings by placing a female mouse in estrus into a cage housing a male in the afternoon. Check for the presence of a vaginal plug the following morning (E0.5). Place the female into a new cage.
2. Day 3—morning. Collection of blastocysts. Prepare reagents for harvesting blastocysts from the female mouse. Prewarm M2 medium at 37 °C.
   (a) Sterilize dissection area by spraying with 70% ethanol (*see* **Note 2**).
   (b) Sacrifice female mouse by cervical dislocation or CO_2 asphyxiation (*see* **Note 3**).
   (c) Spray abdomen with 70% ethanol to sterilize the fur, then use scissors to cut through skin and peritoneum.
   (d) Dissect uterus by cutting below oviduct and above the cervix.
   (e) Wash uterus briefly in PBS in a 10 cm Petri dish.
   (f) Place the uterus in a 35 mm dish containing M2 medium.
   (g) Use a 1 cc syringe with a 32 gauge needle to flush the E3.5 blastocysts from the oviduct side of the uterus horn through the cervix. Use forceps to hold the uterus near the oviduct side while flushing the uterus. Flush each horn twice with M2 medium. When finished flushing, place uterus in another dish.

3.3 Simultaneous Derivation of ES Cells and TS Cells

1. Prewarm ESC/TSC media containing LIF and FGF4 at 37 °C.
2. Remove 24-well plate containing iMEFs from incubator, aspirate MEF media, and add fresh ESC/TSC derivation media. Incubate the 24-well dish at 37 °C with 5% CO_2 for 1 h to equilibrate the media.
3. Use a mouth pipette and stereomicroscope to transfer E3.5 blastocysts harvested from pregnant mice to 24-well plate containing iMEFs and ESC/TSC derivation media. Place one blastocyst per well and incubate at 37 °C with 5% CO_2.
4. The following day, use an inverted microscope to check the attachment of the blastocysts to the iMEFs. Blastocysts should attach to the iMEFs within 2–3 days.
5. After 48 h, check for attached blastocysts to the iMEFs. For wells that contain attached blastocysts, aspirate the medium and add 500 μL of fresh ESC/TSC derivation media. For wells that contain unattached blastocysts, add fresh medium without aspirating the old medium.
6. Culture the blastocysts until the outgrowth has reached a size that is appropriate for disaggregation. Culture of blastocysts in ESC/TSC media should promote growth of the inner cell mass (ICM) and expansion of the trophectoderm (TE).
7. It is important to disaggregate the blastocyst outgrowth before it becomes too large in size (*see* **Note 4**). Prolonged culture of the blastocyst outgrowth may result in differentiation of ESCs and TSCs, or expansion of endoderm progenitors. To dissociate the outgrowth, wash once with 500 μL PBS, then add 100 μL prewarmed 0.05 trypsin–EDTA. Incubate at room temperature for 2–3 min. Briefly pipette up and down several times, then add to a 15 mL conical tube containing 3 mL of prewarmed MEF media. Centrifuge for 3 min at 1500 rpm (500 × *g*).
8. Aspirate the media and resuspend the cells in 500 μL ESC/TSC media. Next, plate the cells in a fresh well of a 24-well plate containing iMEFs. Incubate at 37 °C with 5% CO_2.
9. ES cell and TS cell colonies should become visible within 3–10 days after disaggregation and replating. At this stage the morphology of the early passage disaggregated blastocyst outgrowth should resemble a heterogeneous mixture of ES cells and TS cells (Fig. 1). Change the ESC/TSC cell medium every 2 days.
10. When the colonies reach semiconfluence (e.g., 50% confluent), passage the ESCs/TSCs onto a larger dish (e.g., 12-well). To this end, wash the 24-well with 500 μL PBS, add 100 μL of prewarmed 0.05% trypsin and incubate for 2–3 min at room

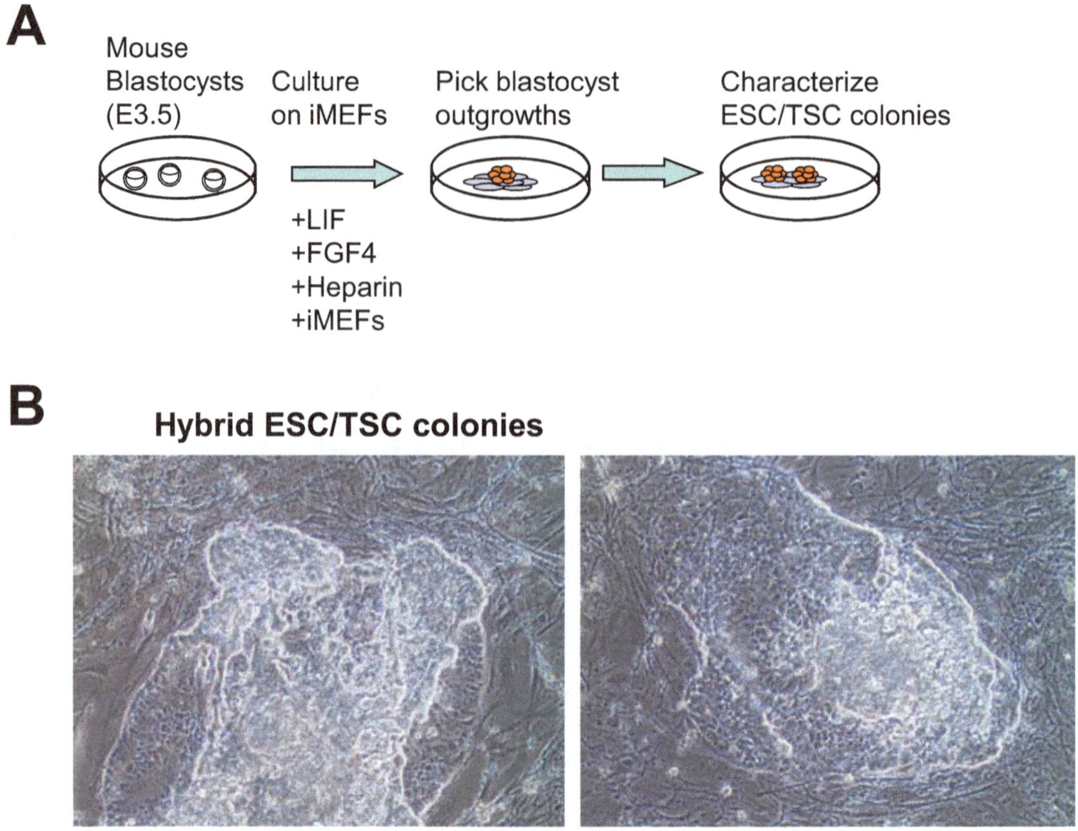

Fig. 1 Simultaneous derivation and culture of ES cells and TS cells from mouse blastocysts. (**a**) Schematic of experimental design for deriving ES cells and TS cells from mouse blastocysts. (**b**) Bright field microscopy of early passage ES cells and TS cells cultured on iMEFs. Hybrid ESC/TSC colonies represent an in vitro expansion of two predominant compartments of the blastocyst, the ICM and the TE

temperature. Pipet several times to minimally dissociate the ESCs/TSCs. It is important to avoid pipetting to a single cell suspension because TS cells exhibit reduced self-renewal when dissociated to a single cell suspension. Thus, maintain small clumps of TS cells while passaging during the early stages of derivation to avoid differentiation of TS cells.

11. Once the colonies reach 50% confluence, passage the ESCs/TSCs onto a 6-well dish. Repeat the steps for dissociating ESCs/TSCs as described in **step 10**.
12. Continue to culture ESCs/TSCs on 6-well plates, and passage the cells whenever they reach 50–70% confluence. Established hybrid ESC/TSC colonies can be passaged onto larger dishes (e.g., 10 cm) using a high split ratio (1:10). It is important to empirically determine the optimal split ratio for passaging newly derived cocultures of ESCs/TSCs. Hybrid colonies can also be cultured in conditioned ESC/TSC media supplemented with LIF and FGF4.

4 Notes

1. Avoid multiple freeze–thaw cycles.
2. Autoclaving and spraying with 70% ethanol prior to use can be used to sterilize dissection instruments.
3. Mice should be euthanized in accordance with an approved Institutional Animal Care and Use Committee (IACUC) protocol.
4. Disaggregate blastocysts when an appropriate outgrowth perimeter is visible.

References

1. Cross JC, Baczyk D, Dobric N, Hemberger M, Hughes M, Simmons DG, Yamamoto H, Kingdom JC (2003) Genes, development and evolution of the placenta. Placenta 24 (2-3):123–130
2. Martin GR (1981) Isolation of a pluripotent cell line from early mouse embryos cultured in medium conditioned by teratocarcinoma stem cells. Proc Natl Acad Sci U S A 78 (12):7634–7638
3. Tanaka S, Kunath T, Hadjantonakis AK, Nagy A, Rossant J (1998) Promotion of trophoblast stem cell proliferation by FGF4. Science 282(5396):2072–2075
4. Nichols J, Zevnik B, Anastassiadis K, Niwa H, Klewe-Nebenius D, Chambers I, Scholer H, Smith A (1998) Formation of pluripotent stem cells in the mammalian embryo depends on the POU transcription factor Oct4. Cell 95 (3):379–391
5. Avilion AA, Nicolis SK, Pevny LH, Perez L, Vivian N, Lovell-Badge R (2003) Multipotent cell lineages in early mouse development depend on SOX2 function. Genes Dev 17 (1):126–140
6. Mitsui K, Tokuzawa Y, Itoh H, Segawa K, Murakami M, Takahashi K, Maruyama M, Maeda M, Yamanaka S (2003) The homeoprotein Nanog is required for maintenance of pluripotency in mouse epiblast and ES cells. Cell 113(5):631–642
7. Chambers I, Colby D, Robertson M, Nichols J, Lee S, Tweedie S, Smith A (2003) Functional expression cloning of Nanog, a pluripotency sustaining factor in embryonic stem cells. Cell 113(5):643–655
8. Strumpf D, Mao CA, Yamanaka Y, Ralston A, Chawengsaksophak K, Beck F, Rossant J (2005) Cdx2 is required for correct cell fate specification and differentiation of trophectoderm in the mouse blastocyst. Development 132(9):2093–2102. https://doi.org/10.1242/dev.01801
9. Russ AP, Wattler S, Colledge WH, Aparicio SA, Carlton MB, Pearce JJ, Barton SC, Surani MA, Ryan K, Nehls MC, Wilson V, Evans MJ (2000) Eomesodermin is required for mouse trophoblast development and mesoderm formation. Nature 404(6773):95–99
10. Luo J, Sladek R, Bader JA, Matthyssen A, Rossant J, Giguere V (1997) Placental abnormalities in mouse embryos lacking the orphan nuclear receptor ERR-beta. Nature 388(6644):778–782. https://doi.org/10.1038/42022
11. Nishioka N, Yamamoto S, Kiyonari H, Sato H, Sawada A, Ota M, Nakao K, Sasaki H (2008) Tead4 is required for specification of trophectoderm in pre-implantation mouse embryos. Mech Dev 125(3-4):270–283
12. Kidder BL, Palmer S (2010) Examination of transcriptional networks reveals an important role for TCFAP2C, SMARCA4, and EOMES in trophoblast stem cell maintenance. Genome Res 20(4):458–472. https://doi.org/10.1101/gr.101469.109
13. Kidder BL, Palmer S (2012) HDAC1 regulates pluripotency and lineage specific transcriptional networks in embryonic and trophoblast stem cells. Nucleic Acids Res 40 (7):2925–2939. https://doi.org/10.1093/nar/gkr1151
14. Kidder BL, Oseth L, Miller S, Hirsch B, Verfaillie C, Coucouvanis E (2008) Embryonic stem cells contribute to mouse chimeras in the absence of detectable cell fusion. Cloning Stem Cells 10(2):231–248

Part IV

Transcriptional Programs that Promote Reprogramming, Transdifferentiation, and Cancer Formation

Chapter 15

Reprogramming Fibroblasts to Neural Stem Cells by Overexpression of the Transcription Factor Ptf1a

Kangxin Jin, Min Zou, Dongchang Xiao, and Mengqing Xiang

Abstract

Neural stem cells (NSCs) have the features of both neural progenitors and stem cells, and show great potentials in translational research and regenerative medicine. Studies on NSCs have been greatly accelerated by the introduction of induced neural stem cells (iNSCs). The iNSCs are usually differentiated from induced pluripotent stem cells (iPSCs) or transdifferentiated from somatic cells such as fibroblasts or glial cells. Here, we describe a detailed protocol to reprogram human and mouse fibroblasts into iNSCs by overexpression of a transcription factor Ptf1a delivered by lentiviruses. The obtained iNSC lines have a strong self-renewal ability and are capable of differentiating into various types of neurons, astrocytes, and oligodendrocytes both in vitro and in vivo. The protocol is quite simple but powerful to produce iNSC lines.

Key words Neural stem cells, Inducible, Reprogram, Differentiation, Neuron, Astrocyte, Oligodendrocyte

1 Introduction

Neural stem cells (NSCs) are defined by two criteria. First, they are self-renewable and maintain their identity after proliferation; second, they are tripotent in neural lineages and capable of differentiating into neurons, astrocytes and oligodendrocytes. By this definition, early neural progenitors in development mostly belong to NSCs. The number of NSCs gradually decline with the maturation of tissues and organs. In the adulthood, however, there are still some, but not many, adult NSCs left in the neural tissues [1–4]. Most of the adult NSCs are maintained in a quiescent status within a special microenvironment called stem cell niche [5, 6]. In case of a minor tissue injury and stimulated by growth factors released from nearby damaged cells, adult NSCs enter the cell cycle and divide asymmetrically. One of the daughter cells begins to differentiate and commits to a fate of a neuron, astrocyte, or oligodendrocyte.

Benjamin L. Kidder (ed.), *Stem Cell Transcriptional Networks: Methods and Protocols*, Methods in Molecular Biology, vol. 2117, https://doi.org/10.1007/978-1-0716-0301-7_15, © Springer Science+Business Media, LLC, part of Springer Nature 2020

In cell replacement therapies to treat degenerative neural diseases, NSCs have several favorable features when compared to embryonic stem cells (ESCs) or induced pluripotent stem cells (iPSCs). Firstly, NSCs still possess a strong self-renewal and proliferative ability, and are able to provide sufficient number of cells. Secondly, NSCs can only differentiate into neural-restricted lineages but no other cell types when repairing neural tissues. Thirdly, NSCs are capable of generating neuronal or glial growth factors which promote the survival and growth of neighboring cells by paracrine effect. Lastly but most importantly, unlike ESCs or iPSCs, NSCs are not tumorigenic when transplanted in vivo. These features make NSCs an ideal source for cell transplantation and other regenerative purposes [7–11]. In addition, NSCs can be used as in vitro disease models to study pathogenesis and screen small molecule drugs [12–15].

Two roadblocks barricaded the early studies and applications of NSCs. The first is that it is difficult to maintain the NSCs in vitro and they have the tendency to differentiate when cultivated in cell culture medium. This roadblock was removed by the Weiss laboratory. They found that some growth factors, such as EGF and bFGF, could stimulate the NSCs to enter the proliferative cell cycle and keep self-renewal [1, 3]. The second roadblock is that separation of NSCs from neural tissues, and their cultivation and identification are tricky and time-consuming [16]. This barrier was cleared with the advent of in vitro reprogramming and emergence of induced neural stem cells (iNSCs). Compared to NSCs separated from neural tissues, iNSCs are generated from autologous cells and have low or no immunogenicity in cell transplantation.

There are at least two successful routes to establish iNSC lines. The first technical route is to differentiate iPSCs into iNSCs. Through overexpression of transcription factors delivered by retrovirus, lentivirus, adeno-associated virus (AAV), plasmid, or episome into somatic cells, the somatic cell fates can be reversed and induced into iPSCs [17–20]. Alternatively, small molecule compounds plus growth factors can also induce somatic cells into iPSCs. Then iNSCs can be derived from iPSCs in various culture medium containing growth factors, small molecule compounds, and so on [15, 21]. One problem of this route is that iNSCs must be confirmed to be free of iPSC contamination before they can be used for other purposes, especially for cell transplantation since iPSCs are highly tumorigenic in vivo. The second route is direct reprogramming of somatic cells into iNSCs with combination of transcription factors, growth factors, and/or chemical compounds. This has been confirmed so far in fibroblasts [22–25], astrocytes [26], urine cells [27, 28], liver cells [29], peripheral blood cells [30], and so on. Theoretically all somatic cells could be converted into iNSCs with the correct combination of transcription factors and/or small molecules.

Previously we have found that the basic helix–loop–helix transcription factor Ptf1a alone is sufficient to reprogram mouse and human fibroblasts into iNSCs [31, 32]. Ptf1a is mostly expressed in postmitotic cells and specifies terminal cell fates in neural tissues. During retinal development, Ptf1a is only transiently expressed in postmitotic precursors and specifies the amacrine and horizontal cell fates [33, 34]. Its function is mediated by downstream effectors such as Prdm13 [35] and Tfap2a/b [36]. Here, we describe in detail the protocol in which Ptf1a is used to reprogram fibroblasts into iNSCs, including lentivirus packaging, MEF (mouse embryonic fibroblast) preparation, induction process, iNSC identification, and in vitro differentiation. Though we used the *tet-on* inducible lentivirus to deliver Ptf1a into fibroblasts, other types of delivery methods should serve the same purpose.

2 Materials

2.1 Instruments

1. Dissection microscope. It is used to dissect mouse embryos and prepare MEFs.
2. Beckman high-speed centrifuge with SW-28 rotor. We use this piece of equipment to concentrate lentiviruses. Any brand should serve the same purpose.
3. HERA cell 150i CO_2 incubator. For cell culture.
4. Thermo Scientific MSC-Advantage Class II Biological Safety Cabinet. For cell culture.
5. Nikon eclipse TS100 microscope. For cell culture.
6. Real time PCR machine. It is used to determine virus titers and examine gene expressions in iNSCs.

2.2 Dissecting Tools

They are used to dissect mouse embryos and should be kept autoclaved.

1. 10 cm straight forceps.
2. 10 cm fine straight and iris scissors.
3. Dumont #5 or #55 fine dissecting forceps.

2.3 Cell Culture for HEK293T, MEF and iNSC

2.3.1 Materials and Reagents

70% Ethanol.

Steritop-GP 0.22 μm, polyethersulfone, 500 ml 33 mm, radio-sterilized funnel (Millipore).

10 cm sterile tissue culture dishes.

6 cm sterile tissue culture dishes.

5 ml sterile serological pipettes.

10 ml sterile serological pipettes.

25 ml sterile serological pipettes.

15 ml Polypropylene conical tube.

50 ml Polypropylene conical tube.

12-well cell culture plates.

24-well cell culture plates.

50 ml Reagent Reservoir, polystyrene.

1× PBS, pH 7.4.

1× HBSS.

Fetal bovine serum (FBS).

Dimethyl Sulfoxide Hybrid-Max (DMSO), sterile filtered (Sigma).

2-Mercaptoethanol (β-Mercaptoethanol) (Sigma).

Dulbecco's Modified Eagle Medium (DMEM) (Gibco).

DMEM/F-12 (Gibco).

Penicillin/Streptomycin 100× (Gibco).

MEM nonessential amino acids 100× (Gibco).

0.25% trypsin–EDTA.

Accutase.

Sodium Selenite, 1 mg (Sigma).

Progesterone, 1 g (Sigma).

Putrescine dihydrochloride, 1 mg (Sigma).

Insulin, human, 50 mg (Sigma).

Apo-Transferrin, human, 500 mg (Sigma).

Doxycycline hyclate, 1 g (Sigma).

Recombinant mouse EGF, 200 μg (R&D systems).

Recombinant human FGF-basic, 25 μg (R&D systems).

Hexadimethrine bromide (Polybrene) (Sigma).

Poly-L-ornithine hydrobromide, 10 mg (Sigma).

Laminin, 1 mg (Sigma).

2.3.2 Reagent Setup

100× Poly-L-Ornithine stock: To make 100× stock, dilute 100 mg in 67 ml water. Filter with 0.45 μm syringe filter and store at −20 °C in 1 ml aliquots.

1000× Doxycycline stock: 2 g doxycycline dissolved in 100 ml water. Sterilize with a 0.45 μm syringe filter and store at −20 °C. Protect from light.

1000× Hexadimethrine bromide (Polybrene) stock: 10 mg/ml dissolved in water. Filter through a 0.45 μm syringe filter. Store at −20 °C.

500× Apo-Transferrin stock: Dissolve 500 mg in 20 ml water. Sterilize with a 0.45 μm syringe filter and store at −80 °C in 1 ml aliquots.

250× Insulin stock: Dissolve 500 mg in 80 ml 10 mM NaOH. Sterilize with a 0.45 μm syringe filter and store at −80 °C in 2 ml aliquots.

Sodium Selenite stock: Dissolve 86.5 mg in 10 ml water. Sterilize with a 0.45 μm syringe filter and store at −80 °C in 1 ml aliquots.

Progesterone stock: Dissolve 32 mg in 50 ml ethanol, dilute 1:100 in ethanol, and store at −80 °C in 0.5 ml aliquots.

Putrescine stock: Dissolve 0.16 g in 10 ml water. Sterilize with a 0.45 μm syringe filter and store at −80 °C in 0.5 ml aliquots.

1000× EGF: Dissolve 200 μg in 20 ml sterile PBS (10 μg/ml or 1000× stock). Store at −20 °C in 100 μl aliquots.

1000× bFGF: Dissolve 25 μg in 2.5 ml sterile PBS with 0.1% BSA (10 μg/ml or 1000× stock). Store at −20 °C in 100 μl aliquots.

MEF Medium: To make 500 ml medium: 440 ml DMEM, 50 ml FBS, 5 ml 100× Pen/Strep, 5 ml 100× MEM NEAA, 4 μl 2-Mercaptoethanol. Sterile filter with 0.22 μm Steritop filter into sterile medium bottle. Store at 4 °C for up to one month.

N3 medium: To make ~500 ml medium: 500 ml DMEM/F-12, 5 ml Pen/Strep, 2 ml insulin stock, 1 ml Apo-transferrin stock, 500 μl progesterone stock, 500 μl putrescine stock, 30 μl sodium selenite stock.

NSC medium: N3 medium (see above), 10 ng/ml bFGF, 10 ng/ml EGF, and 2 mg/ml Doxycycline Hyclate.

Cell storage medium: To make 100 ml: 80 ml MEF medium (or N3 medium), 10 ml FBS, 10 ml DMSO. Filtered to be sterile and store at 4 °C for up to one month.

Poly-L-Ornithine and Laminin coated plates: Thaw the polyornithine and laminin slowly at 4 °C. Dilute 100× polyornithine stock to 1× with cold, sterile water. Dispense polyornithine solution to completely cover the culturing surface. Incubate at 37 °C for at least 1 h. Remove solution and rinse once with sterile water. Dilute laminin to 5 μg/ml in cold, sterile water. Dispense laminin solution to completely cover the culturing surface. Incubate at 37 °C for at least 1 h. Remove solution and rinse once with sterile PBS before use.

2.4 Lentivirus Preparation and Transfection

2.4.1 Materials and Reagents

Ultra-Clear ultracentrifuge tubes (25 × 89 mm) (Beckman Coulter).

Sterile syringe filter w/0.45 μm cellulose acetate membrane (VWR).

10 ml sterile syringes.

BES (Sigma).

$CaCl_2$ (Sigma).

FUW-tetO-Ptf1a (mouse Ptf1a ORF was cloned into the FUW-tetO vector).

FUW-tetO-GFP (GFP gene was cloned into the FUW-tetO vector).

FUW-M2rtTA (Addgene, plasmid 20342).

pRSV-rev (Addgene, plasmid 12253).

pMDLg/pRRE (Addgene, plasmid 12251).

pMD2.G (Addgene, plasmid 12259).

HEK293T cells (ATCC, CRL-11268).

2.4.2 Reagent Setup

2× BBS solution (1000 ml): 50 mM BES, 280 mM NaCl, 1.5 mM Na_2HPO_4. Add H_2O up to 900 ml. Dissolve, titrate to pH 6.95 with 1 M NaOH and bring volume to 1 liter. Sterile filter, aliquot and store at 4 °C or −20 °C.

2.5 M $CaCl_2$: Dissolve in sterile water. Aliquot into 1.5 ml Eppendorf tubes and store at −20 °C.

2.5 iNSC Identification

2.5.1 Materials and Reagents

VWR micro slides, Superfrost plus (VWR).

VWR micro cover glass (VWR).

PAP pen.

Tween 20 or Triton X-100 (Sigma).

Normalized donkey serum (Jackson Immuno Research).

Mouse anti-Nestin (Millipore, MAB353), 1:100.

Mouse anti-Pax6 (DHSB, PAX6), 1:1000.

Mouse anti-Sox2 (Santa Cruz, sc365823), 1:200.

Alexa Fluor-labeled secondary antibodies (Host: Donkey) are from Life Technologies.

DAPI (0.2 μg/ml).

Paraformaldehyde (PFA).

1× and 10× PBS, pH 7.4.

30% sucrose in 1× PBS.

Tissue-Tek O.C.T. Compound.

Aqueous Mounting Medium (Vectorlabs).

TRIzol Reagent (Invitrogen).

AMV First-strand cDNA Synthesis Kit.

SYBR Fast universal qPCR kit.

2.5.2 Reagent Setup

4% PFA: Add 90 ml H_2O into a 250–500 ml flask, heat to 50–70 °C in a microwave; add a couple of drops of 1–5 M NaOH; add 4 g paraformaldehyde and shake by hand until dissolved; cool down immediately in ice water; add 10 ml 10× PBS (pH 7.4), add H_2O to total volume of 100 ml. Store at 4 °C and use it as soon as possible. Discard it if stored for more than 3 days (*see* **Note 1**).

PBST: Add 1 ml Tween 20 or Triton X-100 to 100 ml PBS (pH 7.4) and mix well. Store at 4 °C (*see* **Note 2**).

Blocking buffer: Dissolve 5 g normalized donkey serum into 100 ml PBST, aliquot and store at −20 °C (*see* **Note 3**).

2.6 In Vitro Differentiation of iNSC

2.6.1 Materials and Reagents

Neurobasal medium (Life Technologies).

B27 serum free supplement (50×) (Life Technologies).

N2 supplement (Gibco).

GlutaMAX (Gibco).

DMEM/F12 (Gibco).

Penicillin/Streptomycin 100× (Gibco).

Fetal bovine serum (Gibco).

Recombinant mouse FGF-basic (R&D systems).

Recombinant human PDGF (Peprotech).

Recombinant human BDNF (R&D systems).

Recombinant human CNTF (R&D systems).

Forskolin (Sigma).

L-glutamic acid (Sigma).

L-glutamine (Sigma).

Mouse anti-Tuj1 (Sigma, T8660), 1:250.

Mouse anti-Tau (Santa Cruz, sc-2176), 1:50.

Rabbit anti-TH (Millipore, AB152), 1:2000.

Goat anti-Doublecortin (Santa Cruz, sc-8066), 1:500.

Mouse anti-Map2 (Millipore, BM1243), 1:200.

Mouse anti-NeuN (Millipore, MAB377), 1:200.

Rabbit anti-GFAP (DAKO, Z0334), 1:2000.

Rabbit anti-BLBP (Millipore, ABN14), 1:500.

Mouse anti-CNP (Sigma, c5922), 1:500.

Mouse anti-MBP (Biolegend, 836,502), 1:500.

Mouse anti-O1 (Millipore, MAB344), 1:1000.

2.6.2 Reagent Setup

Neuron differentiation medium (50 ml): 1% B27, 1% N2, 5 μg/ml insulin, 20 ng/ml BDNF, 20 ng/ml CNTF, 10 μm Forskolin, 25 mM L-glutamic acid, 200 mM L-glutamine, 500 μl 100× Pen/Step, 50 ml Neurobasal medium.

Astrocyte differentiation medium (50 ml): 1% N2, 2 mM GlutaMAX, 1% FBS, 50 ml DMEM.

Oligodendrocyte differentiation medium (50 ml): 10 ng/ml FGF-basic, 10 ng/ml PDGF, 10 nM Forskolin, 50 ml DMEM/F12.

3 Methods

3.1 Lentivirus Preparation Using HEK293T Cells (See Note 4)

1. Twenty-four hour prior to transfection, plate 4.0×10^6 HEK293T cells with 6 ml MEF medium in a 10 cm tissue culture dish. For preparation of each virus, plate 3 dishes of cells (*see* **Note 5**).
2. one hour before transfection, replace the culture medium with 9 ml of fresh, prewarmed MEF medium, and put the dish back into the incubator.
3. For every 10 cm plate of HEK293T cells, prepare 10 μg of Tet-O-FUW-Ptf1a (lentiviral plasmid), 5 μg of pMDLg/pRRE, 2.5 μg of pRSV-rev, and 2.5 μg of pMD2.G in sterile water, for a total volume of 400 μl in a 1.5 ml Eppendorf tube.

   The Tet-O-FUW-Ptf1a plasmid can be replaced by FUW-M2rtTA plasmid to produce the transactivator virus for TetO promoter, or by Tet-O-FUW-GFP plasmid to produce the control GFP virus.
4. Add 100 μl of 2.5 M $CaCl_2$ dropwise to water–DNA mixture. The final volume of this solution should be 500 μl. Vortex for 10 s.
5. Immediately add 500 μl of 2× BBS to the transfection mixture. Vortex for 10 s.
6. Leave the mixture at room temperature for 30 min.
7. Gently mix solution by pipetting up and down and add 1 ml transfection mixture dropwise to each plate of HEK293T cells.
8. 16–20 h after transfection, replace medium with 4 ml of fresh MEF medium.
9. Collect viral supernatant 24 h later (40–44 h after transfection) and replace it with another 4 ml of fresh MEF medium. Collect the supernatant next day and combine with those collected before for each plate.

10. Filter viral supernatant into a falcon tube using a 0.45 μm syringe-driven filter to remove any HEK293T cells or cell debris.
11. Concentrate the filtered viral supernatant in a polyallomer ultracentrifuge tube with an ultracentrifuge at 25,000 rcf (or g-force) for 90 min at 4 °C (about 21,000 rpm for SW28 rotor).
12. Slowly pour off the supernatant from the ultracentrifuge tube and leave about one drop of liquid at the lip of the tube without falling out. Seal the tube with Parafilm and shake slowly at 4 °C overnight.
13. Gently resuspend the viruses by pipetting up and down several times and transfer them into a 1.5 ml Eppendorf tube. The volume will be 200–400 μl. Take out 5–10 μl to determine the virus titer by qRT-PCR. Viruses can be kept at 4 °C for up to 2 weeks, or aliquoted into five Eppendorf tubes and kept at −80 °C for long-term storage. Once thawed, viruses from one tube should be used completely or discarded. Do not freeze/thaw repeatedly, as this will cause significant loss of activity.

3.2 MEF Preparation

1. Collect E13.5-E14.5 C57BL/6 mouse embryos and place them in a 10 cm tissue culture dish with HBSS.
2. Wash embryos briefly with HBSS and transfer them to a fresh 10 cm tissue culture dish.
3. Remove the limbs, tail, head, spinal cord and visceral organs of the embryos using a pair of fine-tip forceps. Avoid including any tissues above the shoulder and hip joints. Wash the collected tissue pieces with HBSS and place them in a fresh 10 cm tissue culture dish.
4. Combine tissue pieces from no more than eight embryos into one dish. Add 1 ml 0.25% trypsin–EDTA onto the tissues. Use sterile surgical scissors and forceps to thoroughly mince the tissues to homogenization. Incubate for 15 min at 37 °C. Shake the dish once or twice during incubation.
5. Triturate dissociated tissues with a 10 ml pipette. Split cells obtained from 2 to 4 embryos into one 10 cm dish.
6. Culture the cells in MEF medium in 37 °C cell incubator.
7. For each 10 cm dish, when cells reach confluence (about 2 days), remove the medium, rinse once with PBS, and add 1 ml 0.25% trypsin–EDTA. Incubate at 37 °C for 3 min. Then add 5 ml MEF medium and pipet up and down several times to dissociate cells. Transfer cells to a 15 ml Falcon tube.
8. Spin down cells at 1000 rpm (200–300 rcf or g-force) for 5 min. Remove supernatant and resuspend the cells with 4 ml MEF medium. Split cells at a ratio of 1:4 into four fresh 10 cm dishes.

9. For each 10 cm dish, remove MEF medium when cells reach confluence, rinse once with PBS, and add 1 ml 0.25% trypsin–EDTA. Incubate at 37 °C for 3 min. Add 5 ml MEF medium and dissociate cells by pipetting. Spin down cells at 1000 rpm for 5 min. Remove the supernatant and resuspend cells in 5 ml cell storage medium and mix well.
10. Quickly aliquot cell suspension into cryotubes at 1 ml per tube. Place cryotubes on ice immediately after adding cell suspension. When finished, transfer all cryotubes to a −80 °C freezer and store overnight. The next day, transfer cryotubes to liquid nitrogen for long-term storage (*see* **Note 6**).

3.3 Infection of MEFs with Lentiviruses (See Note 7)

1. Retrieve a vial of passage 1 (or 2) MEFs from liquid nitrogen storage (*see* **Note 8**).
2. Thaw cells in a 37 °C water bath with constant shaking by hand. As soon as the cells are thawed, transfer cells into a Falcon tube containing 10 ml of fresh MEF medium.
3. Spin down cells at 1000 rpm (200–300rcf) for 5 min. Remove supernatant and resuspend cells in MEF medium. Plate cells onto one 10 cm tissue culture dish and cultivate in an incubator.
4. When cells reach 80–90% confluence, dissociate cells with 1 ml 0.25% trypsin–EDTA. After resuspension, count the cell number with a hemocytometer.
5. Culture MEF cells on a 12-well plate precoated with poly-L-ornithine and laminin. Plate 3–4 × 10^4 cells in 1 ml MEF medium for each well. Incubate overnight at 37 °C.
6. About 24 h later, prepare virus mixture for infection. Generally, one tube of frozen concentrated virus stock should be sufficient to infect one 12-well plate. Combine Ptf1a (or GFP) and M2rtTA viruses by a ratio of 1:1. Add 1 ml fresh MEF medium to each mixture. Add Polybrene to a final concentration of no more than 5–10 μg/μl (*see* **Note 9**).
7. Remove culture medium from MEF cells and add 1 ml virus mixture into each well.
8. After 16 h, remove medium containing viruses and replace with fresh NSC medium.
9. Change with fresh NSC medium every 2 days during the cultivation period. Monitor the morphological changes of MEF cells under microscope daily. Typically, MEF cells infected with Ptf1a viruses tend to aggregate into cell clusters compared to MEF cells without viral infection or infected with the control GFP lentiviruses (Fig. 1. Reproduced with permission from Nature Communications [32]). Usually by day 8–10 after infection, small neural spheroids should be visible in each well. The sizes of the spheroids can increase dramatically over 3–4 days.

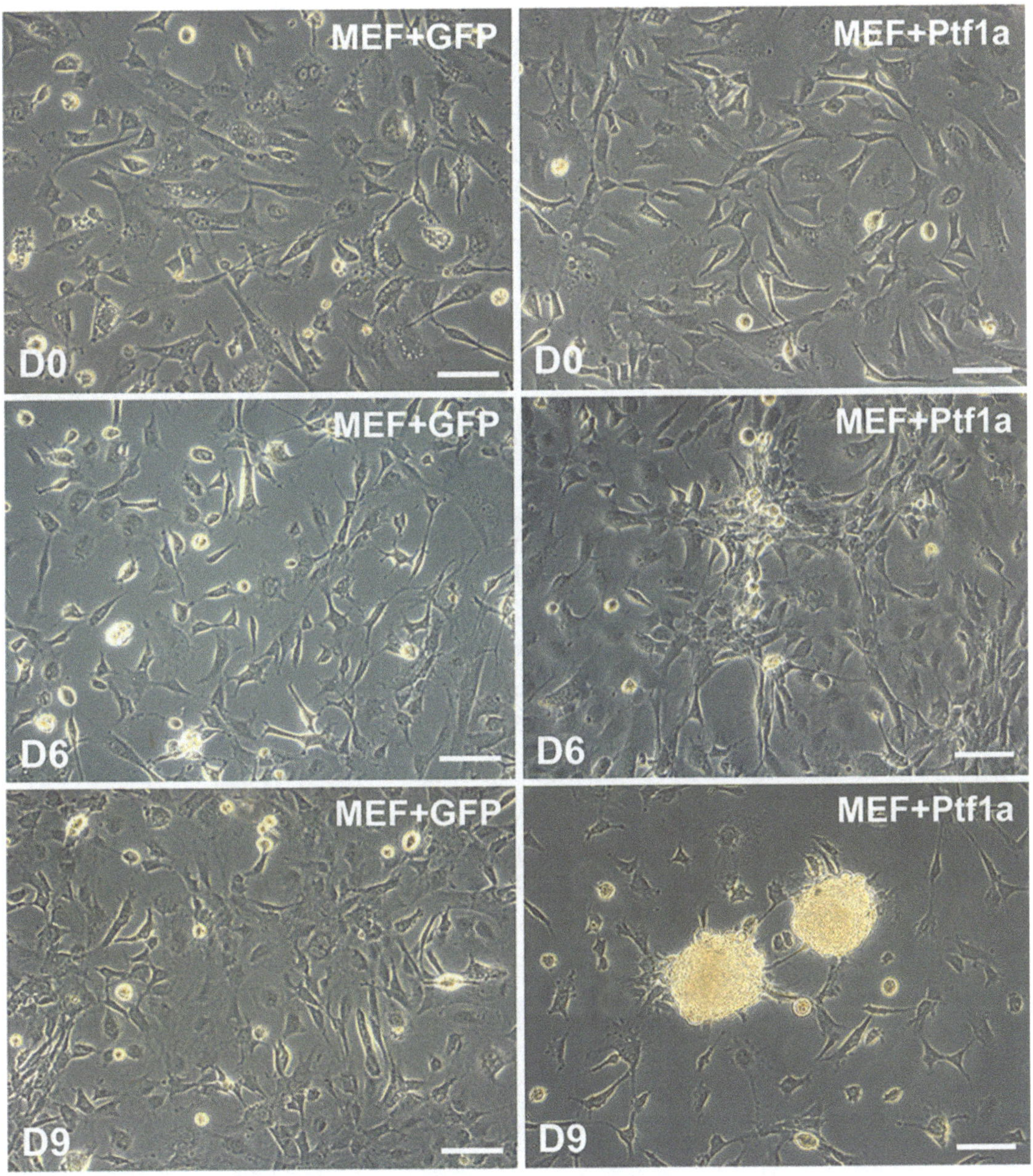

Fig. 1 Morphological changes of MEFs reprogrammed by Ptf1a. From Day 0 (D0) to Day 6 (D6) and Day 9 (D9), MEFs infected with the control GFP lentivirus (left panel) show little changes in morphology. In contrast, MEFs infected with the Ptf1a lentivirus (right panel) aggregate into cell clusters at D6, and grow into typical neural spheroids at D9. Scale bars: 80 μm

3.4 iNSC Expansion, Passage, and Storage

1. When reaching a certain size (close to the diameter of a 200 μl pipette tip), individual spheroid can be dislocated and picked up with a 200 μl pipette tip, and replated into a well of 96-well plate to be further expanded.

2. Add 100 μl NSC medium to each well and change the medium every 2 days until the spheroid nearly occupies the whole bottom of the well. Generally, spheroids with larger sizes grow better than the ones with smaller sizes after replating.
3. Select around 30 best clones to further expand. Remove the culture medium with a pipettor, and add 100 μl 0.25% trypsin–EDTA (or Accutase, *see* **Note 10**) to each well, digest for 5 min at 37 °C. Add 200 μl DMEM with 10% FBS, dissociate cells by repeated pipetting. Transfer the cells in each well to a 1.5 ml Eppendorf tube, centrifuge at 200 rcf for 5 min. Discard the supernatant. Resuspend the cells in 1 ml NSC medium and transfer cells to a well of the 24-well plate precoated with poly-L-ornithine and laminin. Continue cultivation and change NSC medium every 2 days.
4. Likewise, continue to expand each clone to a well of the 6-well plate and finally to a 6 cm culture dish. Take a few colonies from each well for identification while culturing in 24-well or 6-well plates (*see* **Note 11**).
5. Aliquot and freeze adequate numbers of iNSCs for each clone in cell storage medium.

3.5 Identification of iNSCs with qRT-PCR, Cytoimmuno-fluorescence, and Karyotyping

With qRT-PCR, we examine the expression levels of NSC biomarker genes such as Pax6, Sox2, Nestin, and so on. The expression of GAPDH is used as a reference.

1. Take 1–2 × 10^6 NSC cells, spin down and discard the supernatant. Add 1 ml TRIzol, prepare the total RNA following the manufacturer's protocol. Determine the RNA concentration with a spectrometer (*see* **Note 12**).
2. Prepare cDNA from 1 μg of total RNA using the First-strand cDNA Synthesis Kit and following the manufacturer's protocol. Elute the cDNA and dilute it to a volume of 20 μl (*see* **Note 13**).
3. Perform qRT-PCR using the KAPA Sybr Fast qPCR Kit. PCR reaction system (total 10 μl): 5 μl 2× Master Mix, 0.25 μl 10 μM Forward primer, 0.25 μl 10 μM Reverse primer, 1 μl cDNA from **step 2**, 3.5 μl H_2O. PCR cycle setup: 95 °C, 5 s (denaturation); 60 °C, 45 s (annealing and amplification), total 40 cycles. The primers are listed in Table 1 (*see* **Note 14**).

*With cytoimmunofluorescence, we examine the expression of abovementioned NSC biomarkers at the peptide/protein level in the iNSC cells. (**Steps 1–10**: collecting samples and preparation for immunostaining)*

1. Dislocate spheroids from dishes/wells by gently pipetting the culture medium with a transfer pipette. Transfer the medium containing floating spheroids into a 15 ml Falcon tube.

Table 1
Primers used for qRT-PCR analysis

Gene	Forward primer	Reverse primer
Pax6	CGCGGATCTGTGTTGCTCAT	CTTAAATCCATGGCAAATCTTGTCG
Olig2	GCAGAGCACTGCACTTGACTTCTT	CTTCTGCAGTGCGGTTCCTG
Nestin	CCTCAACCCTCACCACTCTATTTT	GCTTTTTACTGTCCCCGAGTTCTC
Sox2	CATGAGAGCAAGTACTGGCAAGA	ACCAACGATATCAACCTGCATGG
Gapdh	CGTGCCGCCTGGAGAAACCTG	GAGTGGGAGTTGCTGTTGAAGTCGC

2. Let it sit for 10–15 min. Check if all spheroids have deposited at the bottom of the tube by gravity. Carefully aspirate the supernatant from the tube without removing the spheroids.
3. Wash spheroids by adding 10 ml of PBS into the tube. Shake gently for 5 min at room temperature. Sit the tube to collect spheroids at the bottom. Remove PBS carefully and add 10 ml of 4% PFA. Shake gently at 4 °C for 20 min.
4. Collect spheroids by gravity. Remove PFA solution and add PBS to wash three times.
5. Collect spheroids by gravity. Remove PBS from the tube and add 10 ml of 30% sucrose in PBS. Shake gently at 4 °C from 1 h to overnight.
6. Collect spheroids by gravity. Using a p1000 pipettor, insert the pipette tip at the bottom of the tube, carefully pull the spheroids into the tip and transfer them to a tissue embedding mold. Let it sit at 4 °C for 2 h.
7. Carefully remove the remaining sucrose solution and add OCT compound to the mold. Gently mix it and let it sit at 4 °C for 1–2 h (*see* **Note 15**).
8. Check to make sure most of the spheroids have fallen to the bottom of the mold. Freeze the OCT compound in a dry ice–ethanol bath.
9. Prepare 14 μm sections with a Cryostat Microtome. Be sure to start collecting sections onto the *Superfrost* slides from the very beginning as spheroids are accumulated at the very bottom of the OCT blocks (*see* **Note 16**).
10. Air dry sections on slides at room temperature for 4–5 h (*see* **Note 17**).

    (Steps 11–14: immunostaining)
11. If the slides are taken out from −20 °C/−80 °C freezer, let the slides dry for 30 min in air. Delimit the section boundaries with a PAP pen. Wash three times with PBST, 5 min each. Add blocking buffer and incubate at room temperature for at least 1 h (*see* **Note 18**).

12. Remove blocking buffer from sections and add primary antibodies diluted with blocking buffer onto sections. Incubate at 4 °C overnight.
13. Remove primary antibodies and wash three times with PBST. Add secondary antibodies diluted at 1:1000 with blocking buffer onto sections. Incubate at room temperature in the dark for 1 h.
14. Remove secondary antibodies and wash three times with PBST. Add DAPI (0.5 μg/ml) onto sections and incubate for 2 min at room temperature. Wash once with PBS and mount sections with the Aqueous Mounting Medium. Fluorescent labeling is visualized with a laser scanning microscope (Fig. 2. Reproduced with permission from Nature Communications [32]).

We also examined the karyotypes of the iNSC clones. This was done by a local facility with related expertise. The detailed protocol will not be described here.

3.6 iNSC In Vitro Differentiation

The ability to differentiate into neurons, astrocytes and oligodendrocytes is a fundamental characteristic of iNSCs. We describe here the protocol for in vitro differentiation of iNSCs.

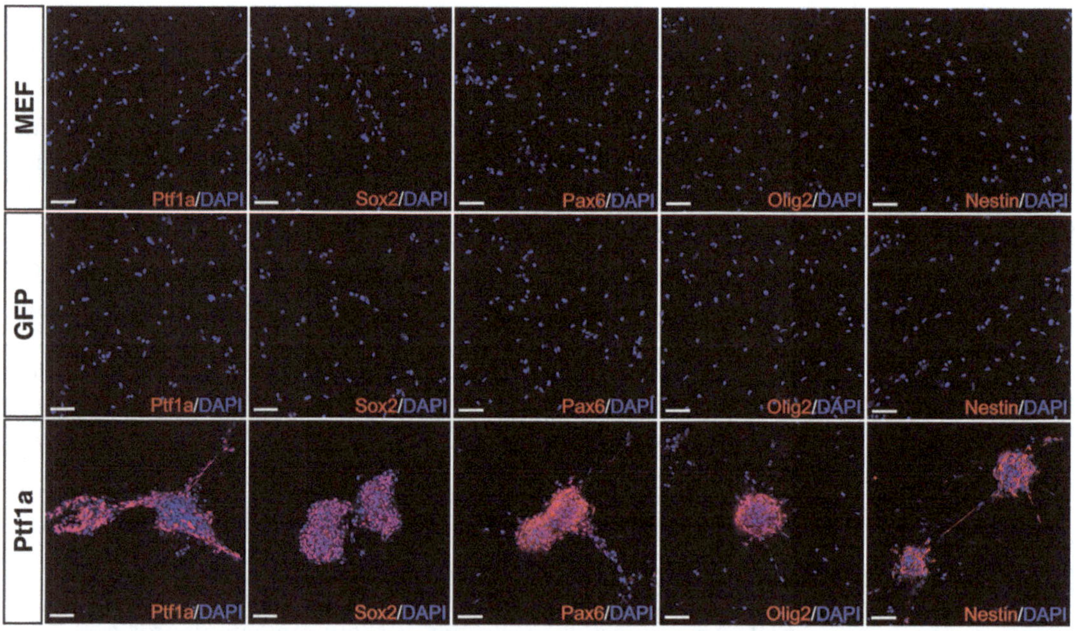

Fig. 2 Identification of iNSCs by cytoimmunofluorescence. MEFs without infection (top panel), MEFs infected with the GFP lentivirus (Middle panel), and neural spheroids induced from MEFs infected with the Ptf1a lentivirus (bottom panel) were subjected to immunostaining with antibodies against Ptf1a and NSC biomarkers Sox2, Pax6, Olig2, and Nestin. The cells in spheroids were positive (in red) for all these NSC biomarkers. In comparison, MEFs without infection or infected with the GFP virus were negative for these markers. The cells were counterstained with DAPI (in blue). Scale bars: 40 μm

1. Place precleaned, sterile cover glasses into the wells of a 24-well plate. Coat the wells with poly-L-ornithine and laminin (see above) and plate 1×10^5 iNSCs into each well. Culture in NSC medium.
2. After 24 h, remove the NSC medium and change the medium according to your experimental purposes:
   (a) To differentiate iNSCs into neurons, add the neuron differentiation medium; replace medium every 2 days, continue culturing for 2 weeks.
   (b) To differentiate iNSCs into astrocytes, add the astrocyte differentiation medium; replace medium every 2 days; culture for 2 weeks.
   (c) To differentiate iNSCs into oligodendrocytes, add the oligodendrocyte differentiation medium; replace medium every 2 days, culture for 10 days.
3. Take out cover glasses with cells for further analysis such as immunostaining (Fig. 3. Reproduced with permission from Nature Communications [32]).

4 Notes

1. PFA is a suspected carcinogen. All operations should be carried out in a fume hood. There are usually some unsolvable white small particles at the bottom, which do not affect the fixation. Those particles can be filtered out with a paper filter. If possible, use freshly prepared 4% PFA for fixation since PFA tends to decompose into formaldehyde when stored in water.
2. Triton X-100 is more powerful than Tween 20 to penetrate through the cell membrane. We routinely use Tween 20 which is less disruptive to the cell morphology and has satisfactory performance in immunostaining.
3. Normalized goat, sheep and donkey serums are commonly available in the market. Sometimes the goat or sheep serum interferes with your experiment when primary antibodies are generated from goats or sheep.
4. We used the $CaCl_2$ method to transfect the HEK293T cells in the protocol. Lipofectamine 2000 or other transfection reagent can be used instead, and the protocol should be adjusted accordingly. The HEK293T cells should be examined to be mycoplasma-free. Mycoplasma greatly decreases the titer of lentiviruses produced, which would seriously affect reprogramming efficiency later.
5. Scale the dish numbers according to your needs.

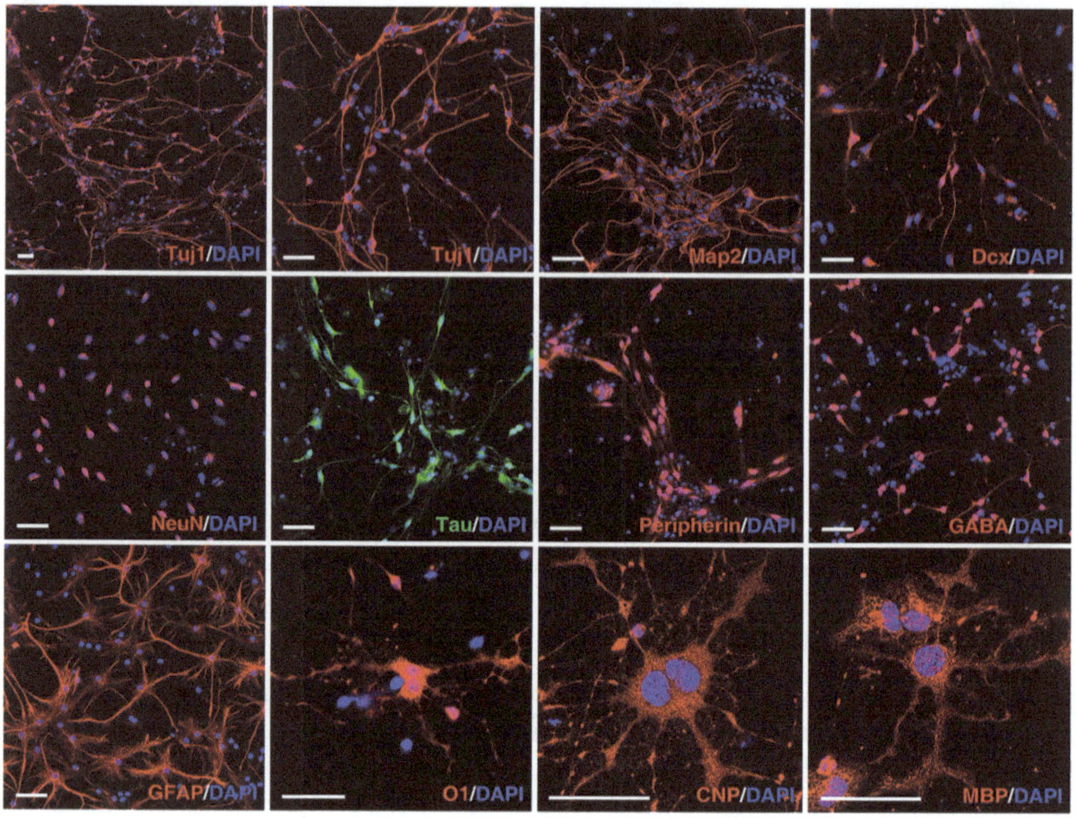

Fig. 3 Directed in vitro differentiation of iNSCs into neurons, astrocytes, and oligodendrocytes. When cultured in various differentiation media, iNSCs differentiated into many types of neurons, labeled by Tuj1, Map2, Doublecortin (Dcx), NeuN, Tau, Peripherin, and GABA; or astrocytes, labeled by GFAP; or oligodendrocytes, labeled by O1, CNP, and MBP. Cells were counterstained with DAPI (in blue). Scale bars: 40 μm

6. Make sure that the cryotube caps are tightly closed. Otherwise, the liquid nitrogen may infiltrate into the tubes and causes contamination as well as possible explosion of the tubes when taken out of the liquid nitrogen tank.
7. We also tried the protocol with P7 tail tip fibroblasts and it worked well. The protocol works with human fibroblasts as well, though it takes longer time for human fibroblasts than mouse ones to be reprogramed into iNSCs.
8. For optimal reprogramming efficiency, it is generally considered to be a thumb rule to use fibroblasts less than 3 passages, either for mouse or human cells. The transdifferentiation efficiency decreases significantly with more passaged cells. You may have to try multiple times to achieve a successful reprogramming with cell lines which have been passaged for many generations.

9. Polybrene is a small, positively charged molecule that binds to the cell surface and neutralizes surface charge. It increases virus gene transfer efficiency by enhancing receptor-independent virus adsorption on target cell membranes [37] . Target cells could be killed by high concentration of polybrene. This is an especially known problem with some B- and T-cell lines. Stop usage or lower the concentration of Polybrene if necessary.
10. Accutase is tender and causes less damage to stem cells, but more costly.
11. At about passage 10, the iNSCs become homogeneous and grow in a typical monolayer. The proliferation is not affected by the presence or absence of doxycycline.
12. Avoid taking too much aqueous fluid in the top phase to prevent DNA contamination when separating RNA from proteins and DNAs in the bottom phase. For a relative pure RNA sample, the A230:260:280 ratio should be around 1:2:1.
13. To achieve better qRT-PCR results, we use no less than 20 ng of cDNA template (calculated with the amount of total RNA used in reverse-transcription divided by final cDNA volume) for each reaction. Typically, we use 50–100 ng of cDNA template in one reaction, depending on the quantity of cDNA and the scarcity of the target genes.
14. We employed the two-temperature cycle here. Adjusting the annealing temperature and duration according to your primers.
15. Avoid putting the samples in OCT compound for too long. In our experience, over 2 h in OCT compound would somehow affect immunostaining quality.
16. Use the *Superfrost* or other specially treated slides with positive-charged surfaces which help to attach sample sections.
17. Wet sections are easy to peel off during immunostaining.
18. Make sure that the OCT compound is completely removed by PBST. The trick is to wash with as much PBST as possible on slides.

Acknowledgments

This work is supported by funds from the National Key R&D Program of China (2017YFA0104100), National Basic Research Program (973 Program) of China (2015CB964600), National Natural Science Foundation of China (81670862, 81721003, and 31871497), Science and Technology Program of Guangzhou (201904010358 and 201904020036), and the Fundamental Research Funds of the State Key Laboratory of Ophthalmology, Sun Yat-sen University.

References

1. Reynolds BA, Weiss S (1992) Generation of neurons and astrocytes from isolated cells of the adult mammalian central nervous system. Science 255(5052):1707–1710
2. Morshead CM, Reynolds BA, Craig CG, McBurney MW, Staines WA, Morassutti D et al (1994) Neural stem cells in the adult mammalian forebrain: a relatively quiescent subpopulation of subependymal cells. Neuron 13(5):1071–1082
3. Weiss S, Dunne C, Hewson J, Wohl C, Wheatley M, Peterson AC et al (1996) Multipotent CNS stem cells are present in the adult mammalian spinal cord and ventricular neuroaxis. J Neurosci 16(23):7599–7609
4. Gritti A, Bonfanti L, Doetsch F, Caille I, Alvarez-Buylla A, Lim DA et al (2002) Multipotent neural stem cells reside into the rostral extension and olfactory bulb of adult rodents. J Neurosci 22(2):437–445
5. Otsuki L, Brand AH (2018) Cell cycle heterogeneity directs the timing of neural stem cell activation from quiescence. Science 360 (6384):99–102
6. Than-Trong E, Ortica-Gatti S, Mella S, Nepal C, Alunni A, Bally-Cuif L (2018) Neural stem cell quiescence and stemness are molecularly distinct outputs of the Notch3 signalling cascade in the vertebrate adult brain. Development 145(10):dev161034
7. Kalladka D, Sinden J, Pollock K, Haig C, McLean J, Smith W et al (2016) Human neural stem cells in patients with chronic ischaemic stroke (PISCES): a phase 1, first-in-man study. Lancet (London, England) 388 (10046):787–796
8. Luan Z, Liu W, Qu S, Du K, He S, Wang Z et al (2012) Effects of neural progenitor cell transplantation in children with severe cerebral palsy. Cell Transplant 21(Suppl 1):S91–S98
9. Bagó JR, Okolie O, Dumitru R, Ewend MG, Parker JS, Werff RV et al (2017) Tumorhoming cytotoxic human induced neural stem cells for cancer therapy. Sci Transl Med 9(375): eaah6510
10. Feldman EL, Boulis NM, Hur J, Johe K, Rutkove SB, Federici T et al (2014) Intraspinal neural stem cell transplantation in amyotrophic lateral sclerosis: phase 1 trial outcomes. Ann Neurol 75(3):363–373
11. Rosenzweig ES, Brock JH, Lu P, Kumamaru H, Salegio EA, Kadoya K et al (2018) Restorative effects of human neural stem cell grafts on the primate spinal cord. Nat Med 24(4):484–490
12. Choi SH, Kim YH, Hebisch M, Sliwinski C, Lee S, D'Avanzo C et al (2014) A three-dimensional human neural cell culture model of Alzheimer's disease. Nature 515 (7526):274–278
13. Tang H, Hammack C, Ogden SC, Wen Z, Qian X, Li Y et al (2016) Zika virus infects human cortical neural progenitors and attenuates their growth. Cell Stem Cell 18 (5):587–590
14. Quinti L, Dayalan Naidu S, Trager U, Chen X, Kegel-Gleason K, Lleres D et al (2017) KEAP1-modifying small molecule reveals muted NRF2 signaling responses in neural stem cells from Huntington's disease patients. Proc Natl Acad Sci U S A 114(23): E4676–E4e85
15. Lorenz C, Lesimple P, Bukowiecki R, Zink A, Inak G, Mlody B et al (2017) Human iPSC-derived neural progenitors are an effective drug discovery model for neurological mtDNA disorders. Cell Stem Cell 20(5):659–74.e9
16. Vescovi AL, Snyder EY (1999) Establishment and properties of neural stem cell clones: plasticity in vitro and in vivo. Brain Pathol (Zurich, Switzerland) 9(3):569–598
17. Takahashi K, Yamanaka S (2006) Induction of pluripotent stem cells from mouse embryonic and adult fibroblast cultures by defined factors. Cell 126(4):663–676
18. Takahashi K, Tanabe K, Ohnuki M, Narita M, Ichisaka T, Tomoda K et al (2007) Induction of pluripotent stem cells from adult human fibroblasts by defined factors. Cell 131 (5):861–872
19. Shi Y, Do JT, Desponts C, Hahm HS, Scholer HR, Ding S (2008) A combined chemical and genetic approach for the generation of induced pluripotent stem cells. Cell Stem Cell 2 (6):525–528
20. Yu J, Hu K, Smuga-Otto K, Tian S, Stewart R, Slukvin II et al (2009) Human induced pluripotent stem cells free of vector and transgene sequences. Science 324(5928):797–801
21. D'Aiuto L, Zhi Y, Kumar Das D, Wilcox MR, Johnson JW, McClain L et al (2014) Large-scale generation of human iPSC-derived neural stem cells/early neural progenitor cells and their neuronal differentiation. Organogenesis 10(4):365–377
22. Kim J, Efe JA, Zhu S, Talantova M, Yuan X, Wang S et al (2011) Direct reprogramming of mouse fibroblasts to neural progenitors. Proc Natl Acad Sci U S A 108(19):7838–7843

23. Han DW, Tapia N, Hermann A, Hemmer K, Hoing S, Arauzo-Bravo MJ et al (2012) Direct reprogramming of fibroblasts into neural stem cells by defined factors. Cell Stem Cell 10 (4):465–472
24. Ring KL, Tong LM, Balestra ME, Javier R, Andrews-Zwilling Y, Li G et al (2012) Direct reprogramming of mouse and human fibroblasts into multipotent neural stem cells with a single factor. Cell Stem Cell 11(1):100–109
25. Thier M, Worsdorfer P, Lakes YB, Gorris R, Herms S, Opitz T et al (2012) Direct conversion of fibroblasts into stably expandable neural stem cells. Cell Stem Cell 10(4):473–479
26. Corti S, Nizzardo M, Simone C, Falcone M, Donadoni C, Salani S et al (2012) Direct reprogramming of human astrocytes into neural stem cells and neurons. Exp Cell Res 318 (13):1528–1541
27. Cheng L, Lei Q, Yin C, Wang HY, Jin K, Xiang M (2017) Generation of urine cell-derived non-integrative human iPSCs and iNSCs: a step-by-step optimized protocol. Front Mol Neurosci 10:348
28. Wang L, Wang L, Huang W, Su H, Xue Y, Su Z et al (2013) Generation of integration-free neural progenitor cells from cells in human urine. Nat Methods 10(1):84–89
29. Cassady JP, D'Alessio AC, Sarkar S, Dani VS, Fan ZP, Ganz K et al (2014) Direct lineage conversion of adult mouse liver cells and B lymphocytes to neural stem cells. Stem Cell Rep 3(6):948–956
30. Liao W, Huang N, Yu J, Jares A, Yang J, Zieve G et al (2015) Direct conversion of cord blood CD34+ cells into neural stem cells by OCT4. Stem Cells Transl Med 4(7):755–763
31. Jin K, Xiang M (2019) Transcription factor Ptf1a in development, diseases and reprogramming. Cell Mol Life Sci 76(5):921–940
32. Xiao D, Liu X, Zhang M, Zou M, Deng Q, Sun D et al (2018) Direct reprogramming of fibroblasts into neural stem cells by single non-neural progenitor transcription factor Ptf1a. Nat Commun 9(1):2865
33. Fujitani Y, Fujitani S, Luo H, Qiu F, Burlison J, Long Q et al (2006) Ptf1a determines horizontal and amacrine cell fates during mouse retinal development. Development 133 (22):4439–4450
34. Nakhai H, Sel S, Favor J, Mendoza-Torres L, Paulsen F, Duncker GI et al (2007) Ptf1a is essential for the differentiation of GABAergic and glycinergic amacrine cells and horizontal cells in the mouse retina. Development 134 (6):1151–1160
35. Watanabe S, Sanuki R, Sugita Y, Imai W, Yamazaki R, Kozuka T et al (2015) Prdm13 regulates subtype specification of retinal amacrine interneurons and modulates visual sensitivity. J Neurosci 35(20):8004–8020
36. Jin K, Jiang H, Xiao D, Zou M, Zhu J, Xiang M (2015) Tfap2a and 2b act downstream of Ptf1a to promote amacrine cell differentiation during retinogenesis. Mol Brain 8(1):28
37. Davis HE, Morgan JR, Yarmush ML (2002) Polybrene increases retrovirus gene transfer efficiency by enhancing receptor-independent virus adsorption on target cell membranes. Biophys Chem 97(2–3):159–172

Chapter 16

Reprogramming of Fibroblasts to Neural Stem Cells by a Chemical Cocktail

Chuijin Wei, Shumin Xiong, and Lin Cheng

Abstract

Chemically induced cell fate conversion, including reprogramming to pluripotent stem cells and direct reprogramming to somatic cells, has been proved to be an alternative strategy with many advantages, in comparison with conventional transcription factors- or microRNAs-enabled cell reprogramming. Many functional and desirable cells have been generated via the chemically induced reprogramming. Neural stem cells (NSCs) hold great potential in basic research and clinical application. Here, we describe a detailed protocol for converting mouse fibroblasts into NSCs by a cocktail of chemical compounds.

Key words Direct cell reprogramming, Transdifferentiation, Neural stem cells, Chemical compounds, Mouse fibroblasts

1 Introduction

Nervous system diseases are usually accompanied by unpleasant body compatibility with even lethality to patients, which do not have efficient drug treatment, physical treatment, or operation strategy. Neural stem cells (NSCs) with self-renewal property and multipotent differentiation potency to neurons, astrocytes, and oligodendrocytes have been considered as a promising cell seed for treating neural diseases and drug screening [1]. Traditional methods for generating NSCs include isolation from primary tissues, such as brain and spinal cord, differentiation from pluripotent stem cells, such as embryonic stem cells and induced pluripotent stem cells, and transdifferentiation from somatic cells. The first two methods are limited to insufficient tissue source and potential safety concern. The third method is mainly based on virus-mediated exogenous gene expression [2–4], which is often accompanied by random gene insertion into genome and low transforming efficiency. All the disadvantages limit the application of NSCs.

Recently, chemical induced direct cell reprogramming with more compelling advantages comparing with the above one has

Benjamin L. Kidder (ed.), *Stem Cell Transcriptional Networks: Methods and Protocols*, Methods in Molecular Biology, vol. 2117, https://doi.org/10.1007/978-1-0716-0301-7_16, © Springer Science+Business Media, LLC, part of Springer Nature 2020

successively generated neural stem/progenitor cells. The advantages include rapid biological effects, cost-effective, relatively easy preservation and access, and inducing procession reversibility [5]. There are three recipes about neural stem/progenitor cell generation from fibroblasts via chemical induction have been reported so far [6–8]. All these chemical recipes can highly activate endogenous Sox2 expression, which is one of the key transcription factors in maintaining NSC function. Consistent with other reports, external factors also play an important role during the induction procession expect for internal signals. In this chapter, we describe a protocol in detail about direct reprogramming of mouse fibroblasts into NSCs by a chemical cocktail in a conditional medium.

2 Materials

2.1 Reagents for Isolating and Culturing Mouse Fibroblasts

1. Dulbecco's Modified Eagle's Medium (DMEM), high glucose (Gibco). Storage at 2–8 °C. Protect from light.
2. Fetal bovine serum (FBS, Gibco). Storage at −80 °C.
3. 1× Dulbecco's phosphate-buffered saline (DPBS) without Calcium and Magnesium.
4. Trypsin–EDTA (0.25%), Phenol red (Gibco). Storage at 2–8 °C.
5. GlutaMAX-I (100×) supplement (Gibco). Storage at 2–8 °C.
6. MEM Non-Essential Amino Acids Solution (100×) (NEAA, Gibco). Storage at 2–8 °C.
7. Penicillin-Streptomycin (10,000 U/mL, Gibco). Storage at 2–8 °C.
8. 70% EtOH.
9. Dimethyl sulfoxide (DMSO).

2.2 Reagents and Plasticware for NSC Induction

1. NeuroCult™ Proliferation Kit (Stemcell Technologies), containing medium for expansion of mouse neural stem and progenitor cells, heparin solution, human recombinant basic fibroblast growth factor (bFGF), and human recombinant epidermal growth factor (EGF).
2. Recombinant murine bFGF (PeproTech). Storage at −80 °C.
3. Recombinant murine EGF (Stemcell Technologies). Storage at −80 °C.
4. MyeloCult™ M5300 (Stemcell Technologies). Storage at 2–8 °C.
5. Recombinant mouse LIF protein (R&D Systems). Storage at −80 °C.

6. Recombinant murine IL-6 (PeproTech). Storage at −80 °C.
7. Recombinant human FGF-5 Protein (R&D Systems). Storage at −80 °C.
8. Valproic acid sodium salt (VPA) (Sigma-Aldrich), 1 M in H_2O, storage at −20 °C.
9. CHIR99021 (Sigma-Aldrich), 10 mM in DMSO, storage at −20 °C.
10. Repsox (Selleckchem), 10 mM in DMSO, storage at 2–8 °C.
11. 6-well ultralow adherent plates (Stemcell Technologies).
12. Accutase (Stemcell Technologies).
13. Paraformaldehyde (PFA) (Sigma-Aldrich).
14. Matrigel (BD).
15. Antifade mounting medium with DAPI (ThermoFisher).

2.3 Mice Strains

1. Nestin-GFP transgenic mice.

3 Methods

3.1 Isolation of Nestin-GFP–Negative Mouse Embryonic Fibroblasts

1. Female Nestin-GFP transgenic mice at day 13.5 postcoitum are sacrificed by cervical dislocation then rinsed in 70% EtOH outside of the tissue culture hood.
2. Mouse is placed on its back and cut in the abdomen skin with scissors.
3. Uterus is cut out by new scissors and transferred by forceps into petri dish with DPBS on ice.
4. Embryos are released from uterus and the head, limbs, visceral tissues, gonads, and vertebral column are carefully torn out.
5. The remaining tissues are placed in a new Petri dish with PBS on ice and chopped into small pieces with ~1 mm by scissors.
6. Minced tissues are digested in 15 mL conical centrifuge tube with trypsin then placed in 37 °C incubator for 15 min. Pipet up and down every 5 min.
7. Digested cell suspension is transferred into 50 mL conical centrifuge tube. 1:1 DMEM medium with 10% FBS is added to inactivate the trypsin.
8. Spin down at 500 × g for 5 min to let the cells sink in the bottom of the tube and the upper supernatant was discarded.
9. The remaining cells and cell clusters are suspended with DMEM with 10% FBS, GlutaMAX-I, NEAA, and Penicillin-Streptomycin, then transferred into 10 cm petri dish. One embryo to one dish with 10 mL medium.

10. Medium with suspended cells is changed 24 h later.
11. Once the cell density reaches ~90% confluency. Cells are harvested and sorted by flow cytometry to remove GFP-positive cells. GFP-negative mouse embryonic fibroblasts are kept or frozen (10% DMSO in FBS) for following assays.

3.2 Chemical Induction of Nestin-GFP–Positive Cells

1. Nestin-GFP–negative mouse embryonic fibroblasts (less than 3 passages) are cultured in DMEM with 10% FBS to reach 80–90% confluency.
2. Medium is changed into M5300 containing VPA (0.5 mM), CHIR99021 (3 μM), Repsox (1 μM), and IL-6 (50 ng/mL) (*see* **Note 1**).
3. 4 days later, the conditional medium is changed into M5300 containing VPA, CHIR99021, Repsox, FGF-5 (50 ng/mL), and LIF (100 ng/mL).
4. Nestin-GFP–positive cells should be observed at this stage and cells with bipolar shape with or without GFP signal should be also monitored at the same time (Fig. 1) (*see* **Note 2**). GFP-positive cell number will gradually increase later with the medium changed every 4 days.
5. GFP-positive cells are sorted out by flow cytometry and cultured in 6-well ultralow adherent plates with neural basal medium containing 20 ng/mL EGF and 20 ng/mL bFGF.
6. About 1 week later, the cells aggregate together and form neurospheres. Medium is half-changed every 2 days.
7. Once the center of the neurospheres appear dark (neurosphere diameter is about 150–200 μm), they should be digested with Accutase and passaged (*see* **Note 3**).

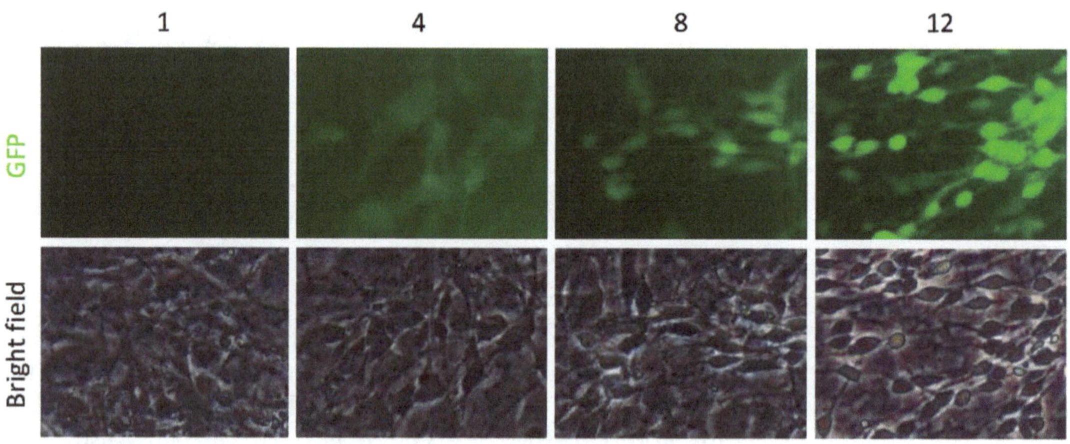

Fig. 1 Nestin-GFP–negative fibroblasts are treated with the chemical cocktail in a conditional medium for different days (1, 4, 8, and 12)

3.3 Neural Stem Cell Marker Characterization of Induced Nestin-GFP–Positive Cells

1. Neurospheres are digested into single cells and cultured on coverslips coated with Matrigel.
2. Once the cells become attached 12 h later. They are fixed by 4% PFA for 15 min at room temperature (RT), then rinsed by PBS and blocked with 0.5% BSA and 0.1% Triton X-100 for 30 min.
3. Samples are incubated with primary antibodies, Sox2 or Pax6, at 4 °C overnight, followed by secondary fluorescently labeled antibodies at RT for 1 h.
4. Coverslips are fixed on slide by mounting medium containing DAPI and protected from light at RT for 30 min. Slides can be stored at 4 °C or directly taken images by fluorescence microscope.

4 Notes

1. Without addition of IL-6, FGF-5, and LIF, Nestin-GFP–positive cells can still be induced from Nestin-GFP–negative cells only by chemical cocktail treatment in the conditional medium. Inducing efficiency is reduced under this condition and much more dead cells will be observed during the inducing procession.
2. For the fibroblasts from wild-type mouse, cells terminally becoming NSCs firstly acquire phenotype change mostly with bipolar shape, which can be considered as an obvious indicator for successful reprogramming or not and be easily checked often under microscope .
3. Passage of chemically induced NSCs using trypsin should be avoided. Pipet the neurospheres up and down gently first, for the cell clusters cannot be dissociated, then digest with Accutase and passage into new plate with high density with increase cell survival and proliferation.

Acknowledgments

This work has been supported by the National Natural Science Foundation of China (31871498), the Medical-Engineering Crossover Fund of Shanghai Jiao Tong University (YG2017MS59), Shanghai Collaborative Innovation Program on Regenerative Medicine and Stem Cell Research (2019CXJQ01).

References

1. Tang Y, Yu P, Cheng L (2017) Current progress in the derivation and therapeutic application of neural stem cells. Cell Death Dis 8(10):e3108. https://doi.org/10.1038/cddis.2017.504
2. Han DW, Tapia N, Hermann A, Hemmer K, Hoing S, Arauzo-Bravo MJ, Zaehres H, Wu G, Frank S, Moritz S, Greber B, Yang JH, Lee HT, Schwamborn JC, Storch A, Scholer HR (2012) Direct reprogramming of fibroblasts into neural stem cells by defined factors. Cell Stem Cell 10 (4):465–472. https://doi.org/10.1016/j.stem.2012.02.021
3. Ring KL, Tong LM, Balestra ME, Javier R, Andrews-Zwilling Y, Li G, Walker D, Zhang WR, Kreitzer AC, Huang Y (2012) Direct reprogramming of mouse and human fibroblasts into multipotent neural stem cells with a single factor. Cell Stem Cell 11(1):100–109. https://doi.org/10.1016/j.stem.2012.05.018
4. Thier M, Worsdorfer P, Lakes YB, Gorris R, Herms S, Opitz T, Seiferling D, Quandel T, Hoffmann P, Nothen MM, Brustle O, Edenhofer F (2012) Direct conversion of fibroblasts into stably expandable neural stem cells. Cell Stem Cell 10(4):473–479. https://doi.org/10.1016/j.stem.2012.03.003
5. Xu A, Cheng L (2016) Chemical transdifferentiation: closer to regenerative medicine. Front Med 10(2):152–165. https://doi.org/10.1007/s11684-016-0445-z
6. Cheng L, Hu W, Qiu B, Zhao J, Yu Y, Guan W, Wang M, Yang W, Pei G (2014) Generation of neural progenitor cells by chemical cocktails and hypoxia. Cell Res 24(6):665–679. https://doi.org/10.1038/cr.2014.32
7. Zhang M, Lin YH, Sun YJ, Zhu S, Zheng J, Liu K, Cao N, Li K, Huang Y, Ding S (2016) Pharmacological reprogramming of fibroblasts into neural stem cells by signaling-directed transcriptional activation. Cell Stem Cell 18 (5):653–667. https://doi.org/10.1016/j.stem.2016.03.020
8. Tang Y, Xiong S, Yu P, Liu F, Cheng L (2018) Direct conversion of mouse fibroblasts into neural stem cells by chemical cocktail requires stepwise activation of growth factors and Nup210. Cell Rep 24(5):1355–1362.e1353. https://doi.org/10.1016/j.celrep.2018.06.116

Chapter 17

Efficient RNA-Based Reprogramming of Disease-Associated Primary Human Fibroblasts into Induced Pluripotent Stem Cells

Patrick S. McGrath, Shennea S. McGarvey, Igor Kogut, and Ganna Bilousova

Abstract

Reprogramming a patient's somatic cells into induced pluripotent stem cells (iPSCs) holds great promise for disease modeling and the development of autologous cellular therapeutics. However, it remains challenging to consistently reprogram primary human cells, as they are frequently aged, diseased, or in low abundance. Here we present a modified highly efficient and clinically relevant RNA-based method for reprogramming disease-associated and other difficult-to-reprogram human primary fibroblast lines into iPSCs. We also describe optimizations that can be employed for consistent reprogramming of these difficult-to-reprogram cells. With the provided protocol, integration-free iPSC lines can be successfully generated from a small number of primary human fibroblasts in approximately 5–7 weeks.

Key words Induced pluripotent stem cells, iPSCs, Reprogramming, Modified mRNA, microRNAs, Primary cells, Human, Fibroblasts, Clinically relevant

Abbreviations

bFGF	basic fibroblast growth factor
DPBS	Dulbecco's phosphate buffered saline
FBS	fetal bovine serum
FEM	fibroblasts expansion media
iPSC	induced pluripotent stem cell
MEM	minimum essential media
mod-mRNA	modified mRNA
ReproM	reprogramming media

Benjamin L. Kidder (ed.), *Stem Cell Transcriptional Networks: Methods and Protocols*, Methods in Molecular Biology, vol. 2117, https://doi.org/10.1007/978-1-0716-0301-7_17, © Springer Science+Business Media, LLC, part of Springer Nature 2020

1 Introduction

It was first shown in 2007 that forced expression of a core set of transcription factors that regulate pluripotency could directly reprogram human somatic cells into induced pluripotent stem cells (iPSCs) [1]. IPSCs have the unique ability to differentiate into any cell type of the body and hold great promise for clinical application. It is therefore critical that the reprogramming factors are only expressed transiently and are nonintegrating to avoid safety risks associated with insertional mutagenesis [2]. The most widely adopted nonintegrating approaches utilize episomal plasmids [3] or Sendai virus [4], but these require lengthy dilution passaging steps to eliminate the vectors, which results in a higher mutational load and increased cost. Transiently expressing reprogramming factors from modified messenger RNAs (mRNAs) has also been used to generate iPSCs from somatic cells [5] although mRNAs alone can be very inefficient for reprogramming difficult-to-transfect primary fibroblasts.

We recently described a highly efficient reprogramming procedure that overcomes many of the inconsistencies associated with mRNA-based reprogramming by optimizing various parameters [6]. In this method, the reprogramming modified mRNA cocktail includes SOX2, KLF4, cMYC, NANOG, LIN28A, and M_3O (OCT4 fused to the MyoD transactivation domain). Mature miRNA-367/302 mimics, previously shown to induce pluripotency [7], are also included in the cocktail, which synergizes with the reprogramming factors to enhance reprogramming efficiency. Lastly, the transfection reagents and culture conditions were adjusted such that human dermal fibroblasts could be reproducibly transfected with extremely high efficiency. Collectively, the improved method can consistently yield thousands of reprogrammed iPSC colonies from just a single well of a 6-well plate seeded with as few as 500 human fibroblasts [6].

Despite the improved protocol, we still encounter significant line-to-line variability in reprogramming amongst patient-derived fibroblasts. This can be due to age and disease status of the patient or sometimes from excessive handling and extended culture of the primary fibroblasts. We have observed that cell density throughout the reprogramming procedure critically impacts efficiency. Plating the fibroblasts sparsely is important to induce adequate cell cycling, which facilitates reprogramming [8], but this must be balanced with plating cells at a sufficiently high density to maintain overall health of the primary cells without overgrowth during the procedure.

It is usually difficult to predict how any given fibroblast line will behave during reprogramming. For this reason, we now initiate fibroblast reprogramming with a range of plating densities

(3×10^3, 10×10^3, and 30×10^3 cells per well of a 6-well plate), one of which will be the ideal density to maximize efficiency. The addition of this step, together with our highly optimized reprogramming protocol using modified mRNAs and miRNA-mimics, has yielded high-quality iPSC lines from even the most difficult-to-reprogram fibroblasts on the first attempt. The improved reprogramming protocol and plating procedure is described here using an example of somatic fibroblasts derived from a patient with Down syndrome (Trisomy 21).

2 Materials

2.1 Fibroblast Culture

1. Tissue culture treated 10 cm dish.
2. Fibroblast expansion medium (FEM): Minimum essential medium (MEM) supplemented with 10% FBS, 1× MEM non-essential amino acids, 1× GlutaMAX, 55 μM 2-mercaptoethanol, 1× Pen/Strep/Fungizone (GE Healthcare). Store at 4 °C.
3. DPBS (no calcium, no magnesium, Thermo Fisher Scientific).
4. Trypsin–EDTA (0.25%), with Phenol Red (Thermo Fisher Scientific).
5. Primary human dermal fibroblasts isolated using standard procedures.

2.2 Plating Fibroblasts for Reprogramming

1. DPBS (no calcium, no magnesium, Thermo Fisher Scientific).
2. rhLaminin-521 (Thermo Fisher Scientific).
3. FGF Basic (bFGF, Thermo Fisher Scientific).
4. Vaccinia Virus B18R (Carrier-free, Thermo Fisher Scientific).
5. Plating medium: ReproM (*see* Subheading 2.3, **item 1** below) supplemented with 5% FBS. Add bFGF to a final concentration of 100 ng/mL and B18R to a final concentration of 200 ng/mL immediately before use.

2.3 Reprogramming Fibroblasts

1. Reprogramming medium (ReproM): Dulbecco's Modified Eagle Medium/Nutrient Mixture F-12 without HEPES (DMEM/F12, Thermo Fisher Scientific) supplemented with 20% KnockOut Serum Replacement (Thermo Fisher Scientific), 0.5× MEM nonessential amino acids, 0.5× GlutaMAX, 55 μM 2-mercaptoethanol, 50 μg/mL ascorbic acid, 1× Pen/-Strep/Fungizone (GE Healthcare). Store at 4 °C. Add bFGF to a final concentration of 100 ng/mL and B18R to a final concentration of 200 ng/mL immediately before each use to an aliquot sized for that day's use.

	RNA Component	Source	Catalog #	Transcript Size (bp)	Reprogramming Mix (6x) Concentration	Units	Amount Transfected per 6-Well	Total
Modified mRNAs	M3O	In vitro transcribed using template DNA PCR-amplified from plasmids (Addgene)	112638	1529	26.5	ng/μL	371 ng	1000 ng modified mRNA
	SOX2		26817	1214	7.0		98 ng	
	KLF4		26815	1673	9.7		136 ng	
	cMYC		26818	1580	9.1		128 ng	
	NANOG		112639	1178	6.8		95 ng	
	LIN28A		26819	890	5.1		72 ng	
	mWasabi		112640	971	7.14		100 ng	
miRNAs	Syn-hsa-miR-302a-3p	Synthesized (Qiagen)	MSY0000684	--	0.29	pmol/μL	4 pmol	20 pmol miRNA
	Syn-hsa-miR-302b-3p		MSY0000715	--	0.29		4 pmol	
	Syn-hsa-miR-302c-3p		MSY0000717	--	0.29		4 pmol	
	Syn-hsa-miR-302d-3p		MSY0000718	--	0.29		4 pmol	
	Syn-hsa-miR-367-3p		MSY0000719	--	0.29		4 pmol	
					14 μL of reprogramming mix transfected per 6-well (ie 84 μL for full plate)			

Fig. 1 Comprehensive list of factors in the reprogramming mix

2. Lipofectamine RNAiMax Transfection Reagent (Thermo Fisher Scientific).
3. Transfection buffer (*see* **Note 1**): Adjust the pH of Opti-MEM (Thermo Fisher Scientific) to 8.2 with 1 M NaOH as previously described [9]. Filter-sterilize and aliquot the adjusted Opti-MEM into 2– mL tubes with minimal air space. Aliquots can be stored for up to 3 months at 4 °C or up to 1 year at −80 °C. Limit aliquot usage to two transfections since the exposure of Opti-MEM to atmospheric air changes the pH of the buffer.
4. Reprogramming RNA cocktail: a comprehensive procedure for the in vitro transcription of modified mRNA is beyond the scope of this protocol but has been described in depth elsewhere [5, 6, 10]. Prepare the mixture of in vitro transcribed modified mRNAs and micro-RNA (miRNA) mimics as described in Fig. 1. The mRNAs are mixed in a molar ratio of 3:1:1:1:1:1 (M_3O–SOX2–KLF4–cMYC–NANOG–LIN28A) with an additional 10% mWasabi modified mRNA to control for transfection efficiency. For each transfection, each well of a 6-well plate is transfected with a total of 1,000 ng of modified mRNA and 20 pmol of miRNA (4 pmol of each miRNA) diluted in 14 μL of nuclease-free water supplemented with 1 U/μL of Riboguard RNase inhibitor (Lucigen). Prepare seven 88 μL aliquots and store at −80 °C. 88 μL is sufficient to reprogram all wells of a 6-well format plate and includes 4 μL excess for pipetting error.

2.4 Picking iPSC Colonies

1. 6-well plate, tissue culture treated.
2. iPSC passaging reagent: Add 500 μL of 0.5 M EDTA (pH 8.0) to 500 mL of DPBS (no calcium, no magnesium). Filter-sterilize using a 0.22 μm vacuum filtration system. Store at room temperature.

3. mTeSR1 complete medium (mTeSR, STEMCELL Technologies).
4. Matrigel hESC-Qualified Matrix (LDEV-free, Corning).

2.5 Equipment

1. Biological Safety Cabinet.
2. 37 °C water bath or bead bath.
3. 37 °C/5% CO_2 humidified tissue culture incubator.
4. 37 °C/5% CO_2/5% O_2 humidified trigas tissue culture incubator.
5. Inverted microscope configured to visualize EGFP.
6. Hemocytometer.
7. Centrifuge.

3 Methods

Work under RNase-free conditions and use aseptic techniques when possible. Perform all cell culture-related manipulations in a biological safety cabinet using aseptic techniques. Follow institutional biosafety standards for work with human cells. Once the reprogramming procedure is initiated, daily maintenance is required for approximately 30–40 days. Be sure to plan accordingly.

3.1 Fibroblast Culture

1. Prewarm 15 mL of FEM and 3 mL of trypsin–EDTA to 37 °C.
2. Carefully aspirate the spent cell culture media from the fibroblasts. Rinse the cells once with 5 mL of DPBS to remove residual serum. Add 3 mL of trypsin–EDTA. Gently rock the plate to ensure complete and even coverage.
3. Aspirate excess trypsin–EDTA off the monolayer, leaving approximately 500 μL. Do not overaspirate, since this can dry the fibroblasts and kill them. Incubate the fibroblasts with trypsin–EDTA for 3 min at 37 °C.
4. During incubation, add 10 mL of prewarmed FEM to a tissue-culture treated 10 cm dish. Set aside.
5. Remove the plate from the incubator and firmly but gently tap the side of the plate to dislodge cells. Check the cells under the microscope. If the cells are detached and floating, proceed. If the cells are still attached, place them back in the 37 °C incubator and check every 2 min until detached.
6. Quickly rinse and collect the detached cells using 5 mL of prewarmed FEM to neutralize the trypsin–EDTA. Move the fibroblast suspension into a 15 mL conical tube. Ensure that the cells are well mixed, and then count the cells on a hemocytometer. DO NOT centrifuge the fibroblasts to pellet.

7. After counting, gently but thoroughly resuspend the fibroblasts. Transfer 2.5×10^5 cells into the 10 cm dish with FEM. Incubate the cells overnight in the 37 °C/5% CO_2 humidified regular tissue culture incubator (*see* **Note 2**). The remaining fibroblasts can be frozen for long-term storage or discarded.
8. Replace the medium with 10 mL of fresh FEM the following day (day −1).

3.2 Plating Fibroblasts For Reprogramming

1. Verify that the fibroblasts to be reprogrammed are at a confluency of 40–60%. If the cells are over- or underconfluent, passage the fibroblasts as previously described (Subheading 3.1) and adjust the plating density accordingly. Culture for another 2 days.
2. To prepare a coated 6-well plate for reprogramming, transfer 6 mL of DPBS into a 15 mL conical tube. Add 150 μL of rhLaminin-521. Pipette up and down to mix thoroughly. Add 1 mL of diluted rhLaminin-521 into each well of a 6-well plate. Incubate the coated plate at 37 °C for 2 h.
3. Prewarm 3 mL of trypsin–EDTA and 20 mL of plating media supplemented with B18R (200 ng/mL) and bFGF (100 ng/mL) to 37 °C.
4. After 2 h of incubation, remove the coated plate from the incubator and aspirate diluted rhLaminin-521 from the wells. Do not allow the surface of the wells to dry. Add 1.9 mL, 1.7 mL, or 1 mL of plating media into the appropriate wells as shown in Fig. 2a and set aside.
5. Carefully aspirate the spent medium from the fibroblasts. Rinse the cells with 5 mL of DPBS and add 3 mL of trypsin–EDTA. Gently rock the plate to cover the cells with trypsin–EDTA. Aspirate excess trypsin–EDTA, leaving approximately 500 μL. Take care not to allow the cells to dry. Incubate the fibroblasts with trypsin–EDTA for 3 min at 37 °C. Remove the plate from the incubator and firmly but gently tap the side of the plate to dislodge cells. Check cells under the microscope. If the cells are detached and floating, proceed. If the cells are still attached, place them back in the 37 °C incubator and check every 2 min until the cells are detached.
6. Rinse and collect the detached cells using 5 mL of plating media to neutralize the trypsin–EDTA. Move the fibroblast suspension into a 15 mL conical tube. Ensure that the cells are well mixed and then count on a hemocytometer (*see* **Note 3**). Do not centrifuge the fibroblasts at any point.
7. Resuspend 90,000 cells to a final volume of 3 mL using prewarmed plating media and mix gently but thoroughly mix. The final cell concentration is 30,000 cells/mL.

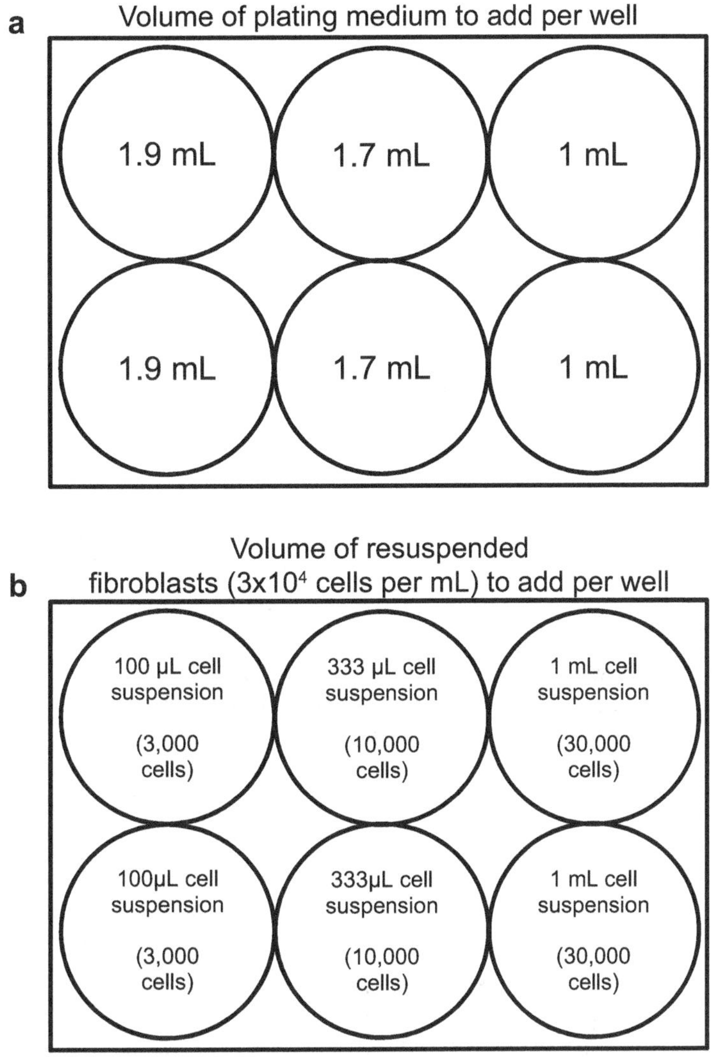

Fig. 2 A procedure for fibroblast plating. To optimize the efficiency of reprogramming, fibroblasts are plated at multiple densities ranging from 3,000 to 30,000 cells per well. The total plating volume after addition of plating medium (**a**) and resuspended fibroblasts (**b**) equals 2 mL

8. As shown in Fig. 2b, pipette 100 μL of resuspended fibroblasts into the two wells prepared with 1.9 mL plating media (from Subheading 3.2, **step 4**) to seed 3,000 cells per well. Pipette 333 μL of resuspended fibroblasts into the two wells prepared with 1.7 mL plating media to seed 10,000 cells per well. Pipette 1 mL of resuspended fibroblasts into the two wells prepared with 1 mL plating media to seed 30,000 cells per well. The final volume in each well is 2 mL.
9. Place the plated cells into a trigas tissue culture incubator with O_2 set to 5% (low-O_2). Once the plate is set down, gently but

thoroughly disperse the cells by alternating between an up/down then left/right motion. Repeat the motions two more times. Do not swirl the plate to mix. Incubate the cells overnight.

10. Pipette 7 mL of reprogramming media (ReproM) into a 15 mL conical tube and place in the low-O_2 incubator with a loosened cap to equilibrate overnight. Do not add bFGF and B18R until the following day.

3.3 Reprogramming Fibroblasts

Fibroblasts undergoing reprogramming are transfected seven times, every-other-day (days 1, 3, 5, 7, 9, 11, and 13). Reprogramming media is replaced prior to the first transfection (day 1) and each day the cells are not transfected (days 2, 4, 6, 8, 10, 12, 14, 15, 16, 17, 18) (Fig. 3a).

1. Before the day 1 transfection, the plating media must be replaced with ReproM. One hour prior to transfecting the fibroblasts, supplement the preequilibrated ReproM with bFGF to a final concentration of 100 ng/mL and B18R to a final concentration of 200 ng/mL. Proceeding 1 well at a time, use a 1 mL pipette to remove the spent 2 mL of plating medium and replace with 1 mL of ReproM supplemented with bFGF and B18R. Repeat for each well (*see* **Note 4**). Move the plate with cells back into the low-O_2 incubator. A media change is not performed on any other days of transfection (i.e., days 3, 5, 7, 9, 11, 13).
2. Equilibrate an aliquot of the transfection buffer for approximately 1 h at RT. Do not use a 37 °C water bath or incubator to warm up the transfection buffer.
3. Remove a single aliquot of reprogramming RNA (88 μL) from −80 °C and warm to RT for approximately 3–5 min, until thawed. Do not thaw the aliquot at 37 °C. Spin down briefly in a microfuge.
4. Warm the Lipofectamine RNAiMAX to room temperature for approximately 3–5 min. Do not use a 37 °C water bath or incubator. Invert the closed tube 2–3 times to mix the reagent. Spin down in a microfuge.
5. To prepare the transfection mix (*see* Fig. 4), transfer 486 μL of the room temperature transfection buffer into an RNase-free 1.5 mL tube. Add 54 μL of the RNAiMAX to achieve a final volume of 540 μL (includes an excess 12 μL to allow for pipetting error). Mix thoroughly by pipetting, do not vortex. Incubate the tube at RT for 1 min.
6. Add 440 μL of the room temperature transfection buffer to the 88 μL aliquot of reprogramming RNA. Gently pipette to mix: final volume is 528 μL (*see* **Note 5**).

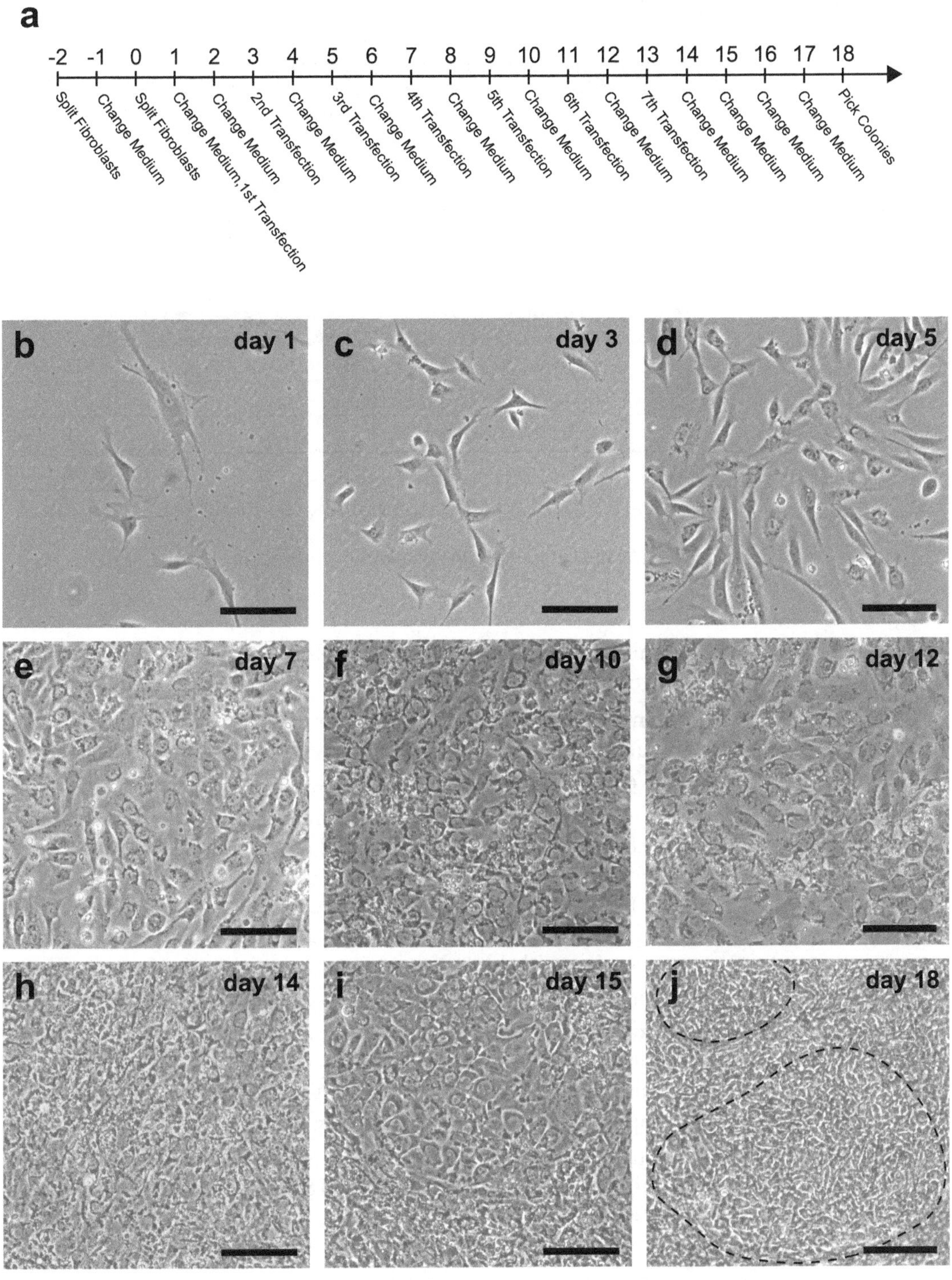

Fig. 3 Visual representation of the reprogramming protocol. The full reprogramming procedure takes approximately 20 days to pick iPSC colonies for expansion (**a**). 10,000 fibroblasts plated into one well of a 6-well plate appear relatively sparse on the first day of transfection (**b**). Following the first transfection, fibroblasts adopt a more compact morphology with fewer cellular projections (**c**, **d**). By days 6–7, cells begin to rapidly proliferate (**e**) and should become confluent (**f**). Colonies may first be observed on day 12 following the sixth transfection (**g**) and will grow larger and more clearly distinguishable by day 18 (**h**–**j**). Dashed line indicates an iPSC colony. Scale bar = 50 μm

Tube 1 - RNAiMax dilution (1st mix)		
Reagent	Concentration	For 6 wells
Transfection Buffer		486 µL
RNAiMax (add 2nd)	10x	54 µL
		540 µL

(incubate 1 min at room temperature)

Tube 2 - modified mRNA mix (2nd mix)		
Reagent	Concentraton	For 6 wells
RNA reprogramming mix	6x	88 µL
Transfection Buffer		440 µL
		528 µL

(add equal volume of diluted RNAiMax from Tube 1)

Fig. 4 An outline of the RNA complexation procedure

7. Add 528 µL of the diluted RNAiMAX to the diluted reprogramming RNA. Pipette to mix: final volume is 1056 µL.
8. Incubate for 15 min at room temperature to allow the Lipofectamine RNAiMAX to complex with the reprogramming RNA.
9. After the 15 min incubation, remove the plate with fibroblasts from the low-O_2 incubator. Gently pipet the complexed transfection cocktail to mix. Add 168 µL of the complexed transfection mix to each well, dropwise throughout the well (*see* **Note 6**). Disperse transfection complexes by gently but thoroughly agitating the plate with an up/down then left/right motion. Do not swirl the plate to mix.
10. Place the transfected cells into the trigas incubator with O_2 set to 5% (low-O_2). Once the plate is set down, disperse transfection complexes again by gently but thoroughly agitating the plate with an up/down then left/right motion.
11. Pipet 7 mL of ReproM into a 15 mL conical tube and place in the low-O_2 incubator with loosened cap to equilibrate overnight. Do not add bFGF and B18R until the following day (*see* **Note 7**).
12. Monitor expression of mWasabi daily to confirm transfection quality. mWasabi expression should be minimally apparent on day 2 and increase in brightness with each additional transfection.

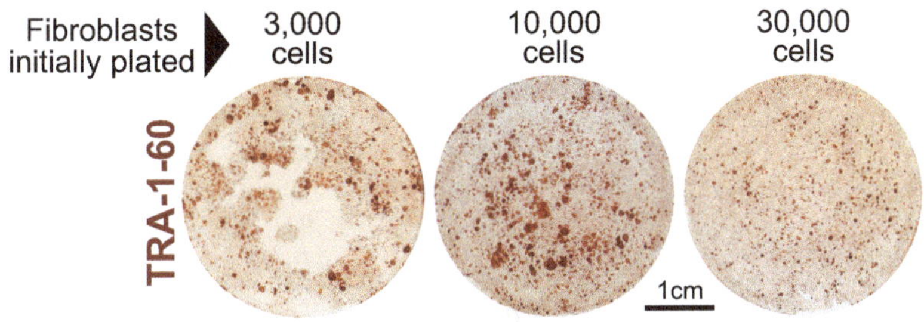

Fig. 5 Representative TRA-1-60 immunostaining. 3×10^3, 10×10^3, and 30×10^3 dermal fibroblasts derived from a patient with Down syndrome were reprogrammed and stained for the pluripotency marker TRA-1-60. For this cell line, the most robust reprogramming was achieved in the 10×10^3 cell-seeding condition. Plating 3×10^3 fibroblasts was subconfluent with observable barren patches throughout the well by the end of reprogramming. Seeding 30×10^3 fibroblasts became overconfluent, which also inhibited reprogramming efficiency

3.4 Picking Reprogrammed iPSC Colonies

During the reprogramming procedure, fibroblasts rapidly change their morphology (Fig. 3b–i), and the formation of iPSC colonies usually occurs by day 18 but may take as long as day 22 (Figs. 3j and 5).

1. Using an inverted microscope, identify the best plating condition by confirming the presence of well-formed reprogrammed iPSC colonies (Fig. 3j).
2. Prewarm 16 mL of mTeSR to 37 °C.
3. Coat all wells of a single 6-well plate with hESC-qualified Matrigel following the manufacturer's instructions. Seal plates with Parafilm and incubate for 1 h at RT (*see* **Note 8**).
4. Aspirate the cell culture medium from the well being used to collect iPSCs. Rinse once with 1 mL of iPSC passaging reagent and aspirate. Add 1 mL of iPSC passaging reagent per well and incubate for 4 min at 37 °C.
5. While the cells are incubating, aspirate the Matrigel from the coated plate and replace with 2 mL of mTeSR medium per well. Set the plates aside.
6. Gently remove the plate with iPSCs to be picked from the incubator and place in the biosafety cabinet (*see* **Note 9**).
7. Carefully aspirate the iPSC passaging reagent. Very gently add 3 mL of prewarmed PSC medium, taking care not to dislodge iPSC colonies.
8. Move the plate to an inverted or dissecting scope to better visualize colonies. Prepare a 200 μL pipette with a sterile tip. Depress the plunger fully, then use the pipette tip to gently scrape a colony while slowly drawing liquid into the tip to

collect the colony. Draw as little medium as possible while picking the iPSC colony.

9. To transfer the colony, pipette up and down 3–4 times in a single well of the prepared plate from Subheading 3.4, **step 5**. Repeat until six colonies have been picked and transferred into individual wells (*see* **Note 10**).
10. Use standard iPSC culture procedures to expand the picked iPSC lines.

4 Notes

1. The pH of the transfection buffer is critical. Ensure the pH meter is calibrated. pH fluctuates with temperature, so it is important to allow the Opti-MEM to fully equilibrate to room temperature before adjusting the pH. Reprogramming will be successful if the pH of Opti-MEM is slightly below 8.2 (between 8.15 and 8.2) but may fail if the pH is higher than 8.2.
2. The cell density is critical to achieve high reprogramming efficiency, as it is important that the fibroblasts are healthy and rapidly dividing. Plating 2.5×10^5 cells yields a 40–60% confluency 2 days later, for most fibroblast samples. Adjust accordingly for diseased or senescent cells to attain the desired 40–60% confluency 48 h later.
3. A 10 cm dish of fibroblasts at the correct density should yield approximately $1 \times 10^6 \pm 20\%$ fibroblasts.
4. Do not use vacuum aspiration, which excessively dries the cells, causes stress, and reduces reprogramming efficiency.
5. Each tube of transfection buffer can be used for two transfections. If this is the first use, tightly close the tube and store at 4 °C. Use for the next transfection.
6. 168 μL of complexed mix contains 1,000 ng of modified mRNAs and 20 pmol of miRNAs.
7. After the final transfection (day 13), it is not necessary to equilibrate media overnight.
8. Picking six colonies for expansion is usually sufficient to establish 2–3 high-quality iPSC lines (i.e., karyotypically normal, expresses all pluripotency markers, good morphology, small number of differentiating cells).
9. At this point, cells may be very loosely adhered and can be easily dislodged. Handle the plate gently.
10. Once the ideal plating density is identified, the reprogramming efficiency is sufficiently high that all six clonal lines can be

established from a single reprogrammed well. The second well can be left in culture as backup in case more colonies need to be picked. Alternatively, the second well can be processed for immunohistochemistry for a pluripotency marker such as TRA-1-60 to assess reprogramming efficiency (Fig. 5).

Acknowledgments

We are grateful for funding support from the National Institutes of Health (P30 AR057212 and R21 AR074642). We also thank Epidermolysis Bullosa (EB) Research Partnership, the EB Medical Research Foundation, the Cure EB Charity, Dystrophic Epidermolysis Bullosa Research Association (DEBRA) International and the Gates Frontiers Fund.

References

1. Takahashi K, Tanabe K, Ohnuki M, Narita M, Ichisaka T, Tomoda K, Yamanaka S (2007) Induction of pluripotent stem cells from adult human fibroblasts by defined factors. Cell 131 (5):861–872. https://doi.org/10.1016/j.cell.2007.11.019
2. Hacein-Bey-Abina S, Von Kalle C, Schmidt M, McCormack MP, Wulffraat N, Leboulch P, Lim A, Osborne CS, Pawliuk R, Morillon E, Sorensen R, Forster A, Fraser P, Cohen JI, de Saint Basile G, Alexander I, Wintergerst U, Frebourg T, Aurias A, Stoppa-Lyonnet D, Romana S, Radford-Weiss I, Gross F, Valensi F, Delabesse E, Macintyre E, Sigaux F, Soulier J, Leiva LE, Wissler M, Prinz C, Rabbitts TH, Le Deist F, Fischer A, Cavazzana-Calvo M (2003) LMO2-associated clonal T cell proliferation in two patients after gene therapy for SCID-X1. Science 302 (5644):415–419. https://doi.org/10.1126/science.1088547
3. Junying Y, Kejin H, Kim SO, Shulan T, Stewart R, Slukvin II, Thomson JA (2009) Human induced pluripotent stem cells free of vector and transgene sequences. Science 324:797–802. https://doi.org/10.1126/science.1172482
4. Fusaki N, Ban H, Nishiyama A, Saeki K, Hasegawa M (2009) Efficient induction of transgene-free human pluripotent stem cells using a vector based on Sendai virus, an RNA virus that does not integrate into the host genome. Proc Jpn Acad Ser B Phys Biol Sci 85(8):348–362. https://doi.org/10.2183/pjab.85.348
5. Warren L, Manos PD, Ahfeldt T, Loh YH, Li H, Lau F, Ebina W, Mandal PK, Smith ZD, Meissner A, Daley GQ, Brack AS, Collins JJ, Cowan C, Schlaeger TM, Rossi DJ (2010) Highly efficient reprogramming to pluripotency and directed differentiation of human cells with synthetic modified mRNA. Cell Stem Cell 7:618–630. https://doi.org/10.1016/j.stem.2010.08.012
6. Kogut I, McCarthy SM, Pavlova M, Astling DP, Chen X, Jakimenko A, Jones KL, Getahun A, Cambier JC, Pasmooij AMG, Jonkman MF, Roop DR, Bilousova G (2018) High-efficiency RNA-based reprogramming of human primary fibroblasts. Nat Commun 9:745. https://doi.org/10.1038/s41467-018-03190-3
7. Anokye-Danso F, Trivedi CM, Juhr D, Gupta M, Cui Z, Tian Y, Zhang Y, Yang W, Gruber PJ, Epstein JA, Morrisey EE (2011) Highly efficient miRNA-mediated reprogramming of mouse and human somatic cells to pluripotency. Cell Stem Cell 8:376–388. https://doi.org/10.1016/j.stem.2011.03.001
8. Guo S, Zi X, Schulz VP, Cheng J, Zhong M, Koochaki SH, Megyola CM, Pan X, Heydari K, Weissman SM, Gallagher PG, Krause DS, Fan R, Lu J (2014) Nonstochastic reprogramming from a privileged somatic cell state. Cell 156(4):649–662. https://doi.org/10.1016/j.cell.2014.01.020
9. McGrath PS, Diette N, Kogut I, Bilousova G (2018) RNA-based reprogramming of human primary fibroblasts into induced pluripotent

stem cells. J Vis Exp (141). https://doi.org/10.3791/58687

10. Mandal PK, Rossi DJ (2013) Reprogramming human fibroblasts to pluripotency using modified mRNA. Nat Protoc 8:568–582. https://doi.org/10.1038/nprot.2013.019

Chapter 18

Direct Reprogramming of Mouse Embryonic Fibroblasts to Induced Trophoblast Stem Cells

Benjamin L. Kidder

Abstract

Trophoblast cells are the first committed lineage to emerge during mammalian preimplantation embryo development. Trophoblast stem (TS) cells can be derived from the trophectoderm (TE) of blastocyst-stage embryos and differentiate into extraembryonic trophoblast cells of the placenta. While mouse TS cells are an indispensable tool to study placental development, and reproductive diseases such as implantation failure and recurrent miscarriage, human TS cells have not been isolated. To model human trophoblast development and to investigate trophoblast-specific causes of reproductive diseases, it will be important to derive human induced trophoblast stem (iTS) cells. Recent studies have shown that fibroblasts can be reprogrammed to iTS cells by overexpressing four transcription factors (TFs) including TFAP2C, GATA3, EOMES, and ETS2. Here, we describe a protocol to directly convert mouse embryonic fibroblasts (MEFs) to iTS cells following overexpression of 10 TFs. iTS cells are capable of self-renewing using conventional TS cell culture media supplemented with the external signal FGF4 and heparin. iTS cells are also able to differentiate into trophoblast lineages.

Key words Trophoblast stem cells, iTS cells, Trophectoderm, FGF4, Derivation, Blastocyst, Differentiation

1 Introduction

The formation of the trophectoderm (TE), which forms a layer of cells surrounding the inner cells mass (IMC) of preimplantation blastocyst stage embryos, represents the first lineage commitment during mammalian development, and facilitates uterine implantation. While pluripotent cells residing in the ICM of preimplantation stage embryos are capable of giving rise to all cell types represented in the embryo proper and adult, trophoblast stem (TS) cells derived from polar TE of blastocyst-stage embryos are multipotent and are restricted in their potential to differentiate into trophoblast cells of extraembryonic tissue. The TE is an essential lineage during mammalian development, where it facilitates implantation, and serves as a source for trophoblast progenitors, which differentiate into

Benjamin L. Kidder (ed.), *Stem Cell Transcriptional Networks: Methods and Protocols*, Methods in Molecular Biology, vol. 2117, https://doi.org/10.1007/978-1-0716-0301-7_18,

epithelial trophoblast lineages of the placenta such as giant cells, glycogen trophoblast cells, spongiotrophoblasts, and syncytiotrophoblasts [1]. Specialized trophoblast cells in the placenta function to exchange nutrients and waste between the fetus and mother. TS cells derived from the TE of blastocysts are capable of self-renewing indefinitely in the presence of FGF4/heparin [2], while ES cells can be derived from the ICM in the presence of LIF [3]. TS cell self-renewal has also been shown to be supported by additional external signals such as INHBA, NODAL, and TGFB1 [2]. Mouse TS cells are an ideal model to interrogate placental development, recurrent miscarriage, and implantation failure. However, bona fide human TS cells have not been isolated. Two studies have shown that mouse embryonic fibroblasts (MEFs) can be converted to induced trophoblast stem (iTS) cells following overexpression of four TFs including TFAP2C, GATA3, EOMES, and ETS2 [4, 5]. Moreover, additional studies have shown that expression of several TFs is enriched in TS cells such as Cdx2 [6], Eomes [7], Esrrb [8], and Tead4 [9]. Additional combinations of TFs, including GATA3, EOMES, TFAP2C, MYC, and ESRRB, have recently been shown to facilitate conversion of MEFs to iTS cells [10]. Moreover, we previously found that TFAP2C and EOMES comprise a core transcriptional regulatory circuit that maintains self-renewal in TS cells [11]. Combined, these results suggest that alternate combinations of TFs are capable of reprogramming fibroblasts to iTS cells. Here, we describe a protocol to derive iTS cells from mouse fibroblasts using a combination of 10 TFs including CDX2, ELF5, EOMES, ESRRB, ETS2, GATA3, KLF5, TFAP2C, TEAD4, and YAP1.

2 Materials

2.1 Reagents for Cell culture

1. DMEM (Dulbecco's Modified Eagle's Medium, High Glucose.
2. Penicillin–streptomycin (100×).
3. L-glutamine (200 mM).
4. Fetal bovine serum (FBS).
5. Nonessential amino acids (NEAA, 100×).
6. PBS (Phosphate buffered saline without calcium or magnesium, 1×).
7. 0.25% Trypsin–EDTA (Invitrogen)
8. 0.05% Trypsin–EDTA (Invitrogen)
9. RPMI (Roswell Park Memorial Institute) 1640 medium.
10. Sodium pyruvate (100 mM).
11. EmbryoMax 2-mercaptoethanol (100×).

12. FGF4 (25 μg, R&D Systems). Resuspend lyophilized FGF4 in PBS with 0.1% BSA. Aliquot tubes and freeze at −80 °C until use (*see* **Note 1**). To make a solution of 0.1% BSA in PBS, dissolve BSA in PBS, filter using a 0.45 μM syringe, aliquot, and store at −80 °C until use.
13. Heparin (10,000 units, Sigma H3149). Prepare a 1000× stock tube of heparin by resuspending in PBS to 1 mg/mL (1000×). Store at −80 °C until use.
14. Cell culture dishes (6-well, 12-well, and 24-well).
15. Conical tubes (polystyrene, 15 mL and 50 mL).
16. 0.1% gelatin solution in water
17. Water bath.
18. 70% ethanol.

2.2 Equipment

1. Pulled capillary tube or drawn Pasteur pipette.
2. Aspirator tube assembly: aspirator mouth piece, tubing (Sigma A5177).
3. Flushing needle (1 cc syringe and 30–32 gauge needle).
4. Dissecting instruments (fine forceps, fine scissors).
5. Sterile petri dishes (10 cm, 35 mm).
6. Stereomicroscope.
7. Clinical centrifuge.
8. Cell culture incubator.

2.3 Lentivirus Production

1. HEK 293T cells.
2. Opti-MEM medium (Life Technologies).
3. Lipofectamine 2000 (Life Technologies).
4. Lentiviral vectors: packaging plasmid (e.g., psPAX2), envelope plasmid (e.g., plpVSVG), cDNA lentiviral expression vector (e.g., pLEX).
5. 10 cm culture dish.
6. 0.45 μM cell strainer (BD Biosciences).
7. Polybrene.

2.4 Culture Media

1. MEF (Mouse embryonic fibroblast) media for culture of MEFs for reprogramming and for culture of mitotically inactivated MEFs (iMEFs). MEFs and iMEFs should be stored at −150 °C or lower. DMEM high glucose with L-glutamine, 10% FBS, penicillin–streptomycin (1×), at 37 °C with 5% CO_2. Mitotically inactivated MEFs (iMEFs) derived from E13.5 to E14.5 mouse embryos [12] should be purchased from a commercial

vendor. It is recommended that reprogramming should be performed with the use of an iMEF feeder layer (*see* **Note 1**).

2. iTS cell reprogramming media: RPMI 1640 with L-glutamine, 20% FBS, EmbryoMax 2-mercaptoethanol (1×), penicillin–streptomycin (1×), 25 ng/mL FGF4, 1 μg/mL heparin. iTS cell media should be stored at 4 °C, and FGF4 should be added fresh.
3. iMEF-conditioned iTS cell media: Prepare conditioned media (CM) by culturing iMEFs in iTS cell media in the absence of FGF4/heparin. Collect the CM after 2–3 days. Conditioned media can be collected and stored at −20 °C or lower, or at 4 °C for 1–2 weeks (*see* **Note 2**).

3 Methods

3.1 Prepare Mouse Embryonic Fibroblasts

1. Prewarm MEF media (10% FBS, DMEM, pen/strep) for 10–15 min. in a water bath at 37 °C.
2. Thaw a vial of MEFs for 1–2 min in a water bath at 37 °C.
3. Add MEFs to 3 mL of prewarmed MEF media in a 15 mL conical tube, and spin at 1500 rpm (500 × *g*) for 5 min. Resuspend cells in MEF media and plate on a 6-well cell culture plate. For 6-well plates, cells should be resuspended in 2 mL of media.
4. Once the cells are ~50–60% confluent, they are ready to be transduced.

3.2 Lentiviral Particle Production

1. Thaw a vial of HEK 293T cells in water bath. Next, add cells to 3 mL of MEF media in a 15 mL conical tube. Spin at 1500 rpm (500 × *g*) for 5 min.
2. Resuspend HEK 293T cells in MEF media and plate on a 10 cm culture dish. When HEK 293T cells are 60–70% confluent, passage to 10 cm dishes.
3. After one week of culture, passage HEK 293T cells 24 h before producing lentiviral particles (3–4 × 10^6 cells per 10 cm dish).
4. The next day, prewarm Opti-MEM media for 15–20 min. at room temperature.
5. Using a 15 mL conical tube, add 1.5 mL of Opti-MEM media and 36 μL Lipofectamine 2000. Invert the tube gently 3–4 times to mix. Incubate at room temperature for 5 min.
6. Using another 15 mL conical tube, add 1.5 mL of Opti-MEM media and 10 μg packaging vector (e.g., psPAX2), 4 μg of envelope plasmid (plpVSVG), and 10 μg of the cDNA lentiviral expression vector (e.g., pLEX). Invert the tube gently 3–4 times to mix.

7. Next, add 1.5 mL of solution containing the plasmids to the tube containing Lipofectamine 2000 (1.5 mL). Invert the tube gently 3–4 times to mix. Incubate at room temperature for 20 min.
8. In the meantime, change MEFs media by aspirating media, and adding 10 mL of MEF media without penicillin–streptomycin to the HEK 293T cells.
9. After the 20 min. Incubation, slowly add 3 mL of the plasmid–Lipofectamine mixture (transfection mix) dropwise to the HEK 293T cells and very gently mix (*see* **Note 3**).
10. Incubate for at least 4–6 h at 37 °C.
11. After incubation for at least 4–6 h, aspirate media and add fresh 10 mL of iTS cell media without FGF4/heparin and incubate at 37 °C.
12. Collect the supernatant containing the viral particles after 24–48 h. Filter the collected media through a 0.45 μM cell strainer (*see* **Note 4**).

3.3 Reprogramming of MEFs to iTS Cells

The scheme for reprogramming is shown in Fig. 1a.

1. Prepare MEFs as described in Subheading 3.1 so they are ~50% confluent in a 6-well or 12-well cell culture plate. Avoid dense cultures of MEFs, which will result in low transduction efficiency and reprogramming rate. Cultures that are too sparse may lead to increased cell death following transduction. It has been previously suggested that an MOI of 2–5 is sufficient to generate iPS cells.
2. Transduce MEFs by adding media containing Polybrene (4–8 μg/mL) and lentiviral particles encoding the 10 TFs to the 6-well or 12-well plate containing MEFs. The next day, aspirate the media and add fresh iTS cell media containing FGF4/heparin.
3. The next day, evaluate the viability of transduced MEFs using an inverted microscope.
4. In the meantime, prepare a layer of feeder (iMEF) cells by plating on 6-well plates.
5. After several days (3–4 days), or when MEFs reach confluency, passage the transduced MEFs onto a 6-well plate containing iMEFs in iTS cell media supplemented with FGF4/heparin.
6. Change the media every 48 h by removing the spent iTS cell media and adding fresh iTS cell media supplemented with FGF4/heparin. Evaluate cellular reprogramming using standard microscopy. Reprogrammed cells will undergo a mesenchymal-to-epithelial (MET) transition, become more

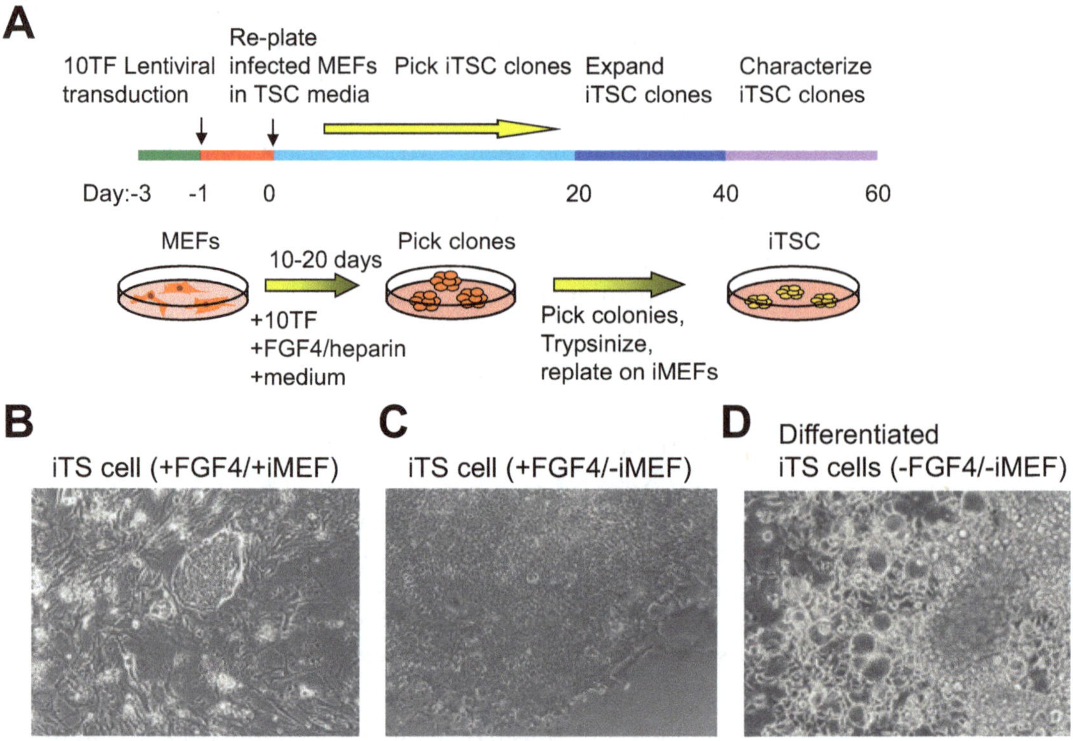

Fig. 1 Derivation of iTS cells from mouse fibroblasts. (**a**) Schematic of experimental design for generating iTS cells from mouse embryonic fibroblasts (MEFs). (**b**) Bright field microscopy of a representative iTS cell colony on a layer of iMEF feeder cells. Note the epithelial sheet-like morphology of iTS cell colonies. (**c**) Feeder-free culture of iTS cells. Bright-field microscopy of a representative iTS cell colony. (**d**) Bright-field microscopy of differentiated iTS cells cultured in the absence of FGF4 in feeder-free conditions

compact and display an increased nuclear–cytoplasmic ratio in an analogous manner as TS cells.

7. Day 14–20. Reprogrammed cells resembling iTS cell-like colonies should begin to be visible.
8. Twenty-four hours before picking iTS cell colonies, prepare 24-well dishes with an iMEF feeder layer to passage iTS cell colonies. Coat 24-well dishes with 0.1% gelatin solution and incubate at 37 °C for 15–20 min. In the meantime, thaw a vial of frozen iMEFs in a water bath at 37 °C, resuspend in 3 mL of MEF media, and spin at 1500 rpm (500 × *g*) for 5 min. Remove gelatin solution from the culture dish, and add 1 mL MEF media, and incubate at 37 °C with 5% CO_2.
9. Pick iTS cell colonies when they become visible using a 10–200 μL pipette tip. Briefly wash in a 50 μL drop of PBS on a lid of a petri dish, trypsinize briefly (0.05% trypsin), and plate on a layer of iMEFs in a 24-well plate.
10. Evaluate iTS cell colony viability 24 h later.

11. Change iTS cell media every 48 h by aspirating the old media, and adding 1 mL of fresh iTS cell media supplemented with FGF4/heparin.
12. Once iTS cells reach 60–70% confluency, passage using 0.05% trypsin and replate on a 12-well dish, and subsequently passage to a 6-well plate (Fig. 1b). iTS cells should be frozen and stored at −150 °C or lower.

3.4 Culture of iTS Cells in Feeder-Free Conditions

1. iTS cells can be cultured in feeder-free conditions on cell culture dishes using iTS cell conditioned media.
2. iTS cells can be propagated in media containing: 70% iTS cell conditioned media and 30% iTS cell media supplemented with FGF4/heparin. An evaluation of iTS cell colony morphology should resemble iTS cells cultured on a layer of iMEF feeder cells (Fig. 1c).
3. Passaging of iTS cells grown in feeder-free conditions should be performed in a similar manner as iTS cells cultured on a layer of iMEF feeder cells. Maintaining small clumps of cells without dissociating into a single-cell suspension aids in the viability of passaged iTS cells.

3.5 Differentiation of iTS Cells

1. iTS cells can be differentiated by culturing in iTS cell media in the absence of FGF4 in feeder-free conditions. iTS cells can be passaged as described above, and replated in 6-well plates in iTS cell media without FGF4. Media should be changed every 48 h by aspirating spent media and adding 2 mL of fresh iTS cell media without FGF4.
2. Differentiated iTS cells should become visible after several days. iTS cell differentiation can be evaluated using an inverted microscope and standard bright-field microscopy. The cytoplasmic to nuclear ratio should increase during iTS cell differentiation. Trophoblast cells should become visible, such as giant cells (Fig. 1d).

4 Notes

1. Culture of transduced MEFs on a feeder-layer of iMEFs increases the rate of reprogramming.
2. It is important to avoid multiple freeze–thaw cycles.
3. When changing media for HEK 293T, it is important to add media slowly to avoid detachment of the cells from the dish, which will lead to decreased virus production.
4. It is important to filter supernatant containing viral particles to avoid contamination of HEK 293T cells.

References

1. Cross JC, Baczyk D, Dobric N, Hemberger M, Hughes M, Simmons DG, Yamamoto H, Kingdom JC (2003) Genes, development and evolution of the placenta. Placenta 24 (2-3):123–130
2. Tanaka S, Kunath T, Hadjantonakis AK, Nagy A, Rossant J (1998) Promotion of trophoblast stem cell proliferation by FGF4. Science 282(5396):2072–2075
3. Martin GR (1981) Isolation of a pluripotent cell line from early mouse embryos cultured in medium conditioned by teratocarcinoma stem cells. Proc Natl Acad Sci U S A 78 (12):7634–7638
4. Benchetrit H, Herman S, van Wietmarschen N, Wu T, Makedonski K, Maoz N, Yom Tov N, Stave D, Lasry R, Zayat V, Xiao A, Lansdorp PM, Sebban S, Buganim Y (2015) Extensive nuclear reprogramming underlies lineage conversion into functional trophoblast stem-like cells. Cell Stem Cell 17(5):543–556. https://doi.org/10.1016/j.stem.2015.08.006
5. Kubaczka C, Senner CE, Cierlitza M, Arauzo-Bravo MJ, Kuckenberg P, Peitz M, Hemberger M, Schorle H (2015) Direct induction of trophoblast stem cells from murine fibroblasts. Cell Stem Cell 17 (5):557–568. https://doi.org/10.1016/j.stem.2015.08.005
6. Strumpf D, Mao CA, Yamanaka Y, Ralston A, Chawengsaksophak K, Beck F, Rossant J (2005) Cdx2 is required for correct cell fate specification and differentiation of trophectoderm in the mouse blastocyst. Development 132(9):2093–2102. https://doi.org/10.1242/dev.01801
7. Russ AP, Wattler S, Colledge WH, Aparicio SA, Carlton MB, Pearce JJ, Barton SC, Surani MA, Ryan K, Nehls MC, Wilson V, Evans MJ (2000) Eomesodermin is required for mouse trophoblast development and mesoderm formation. Nature 404(6773):95–99
8. Luo J, Sladek R, Bader JA, Matthyssen A, Rossant J, Giguere V (1997) Placental abnormalities in mouse embryos lacking the orphan nuclear receptor ERR-beta. Nature 388(6644):778–782. https://doi.org/10.1038/42022
9. Nishioka N, Yamamoto S, Kiyonari H, Sato H, Sawada A, Ota M, Nakao K, Sasaki H (2008) Tead4 is required for specification of trophectoderm in pre-implantation mouse embryos. Mech Dev 125(3-4):270–283
10. Benchetrit H, Jaber M, Zayat V, Sebban S, Pushett A, Makedonski K, Zakheim Z, Radwan A, Maoz N, Lasry R, Renous N, Inbar M, Ram O, Kaplan T, Buganim Y (2019) Direct induction of the three pre-implantation blastocyst cell types from fibroblasts. Cell Stem Cell 24(6):983–994 e987. https://doi.org/10.1016/j.stem.2019.03.018
11. Kidder BL, Palmer S (2010) Examination of transcriptional networks reveals an important role for TCFAP2C, SMARCA4, and EOMES in trophoblast stem cell maintenance. Genome Res 20(4):458–472. https://doi.org/10.1101/gr.101469.109
12. Kidder BL, Oseth L, Miller S, Hirsch B, Verfaillie C, Coucouvanis E (2008) Embryonic stem cells contribute to mouse chimeras in the absence of detectable cell fusion. Cloning Stem Cells 10(2):231–248

Chapter 19

Characterization of Arsenic-Induced Cancer Stem-Like Cells

Qingshan Chang, Zhuoyue Bi, Yao Fu, M' Kya Anique Rice, Qian Zhang, Priya Wadgaonkar, Bandar Almutairy, Wenxian Zhang, Yongju Lu, Liping Xu, Chitra Thukar, and Fei Chen

Abstract

Arsenic is a well-known human carcinogen. However, the mechanisms underlying arsenic-induced carcinogenesis remain elusive. Here we show that chronic and low level of arsenic stress induces transformation of the human bronchial epithelial cells, BEAS-2B, and that some of the transformed cells show characteristics of cancer stem-like cells (CSCs). Meanwhile, we demonstrate that arsenic stress dedifferentiates $CD61^+$ BEAS-2B cells into CSC-like $CD61^-$ cells featured with noncanonical epithelial–mesenchymal transition (EMT), enhanced chemoresistance, and metastasis. Finally, we show that oncogene c-Myc expression is associated with arsenic induced tumor initiation and progression. Altogether, our findings highlight a unique mechanism of arsenic-induced transformation of human bronchial epithelial cells and provide a novel therapeutic target for arsenic-initiated lung cancer.

Key words Arsenic, Transformation, Cancer, Cancer stem cells, Environment

1 Introduction

Arsenic is a widely distributed toxic and carcinogenic metalloid. Arsenic contamination via drinking water is a major worldwide public health threat [1]. It was estimated that over 137 million people in more than 70 countries are chronically exposed to the arsenic-contaminated drinking water, which exceeds the WHO standard of 10 μg/L [2]. Long-term exposure to arsenic impairs development and has been associated with skin lesions, cardiovascular disease, diabetes, and cancers in skin, lung, urinary bladder, kidney, and liver [3].

Lung cancer is the leading cause of cancer-related death worldwide, with more than one million deaths annually [4]. Although smoking is the major reason of lung cancer, approximately 25% of lung cancer cases worldwide are never smokers, accounting for over 300,000 deaths each year [5]. The differences in the epidemiological and molecular characteristics of lung cancers between never

Benjamin L. Kidder (ed.), *Stem Cell Transcriptional Networks: Methods and Protocols*, Methods in Molecular Biology, vol. 2117, https://doi.org/10.1007/978-1-0716-0301-7_19,

smokers and smokers suggest that they are separate entities. Arsenic is one of the major nontobacco carcinogens strongly associated with lung cancer of never-smokers [6]. Relative to other histological subtypes of lung cancer, lung squamous cell carcinomas (*SqCCs*) occur at higher frequency following *arsenic* exposure [7]. However, the underlying mechanisms by which arsenic induces lung cancer remain to be fully elucidated.

Cancer stem cells (CSCs) refer to a subset of tumor cells that has the ability of self-renewal and differentiate into heterogeneous lineages of cancer cells that form the bulk of the tumor [8, 9]. Although the first evidence for CSCs was documented in hematological malignancies, in which only a small subset of cancer cells were capable of forming new tumors when transplanted into immunodeficient mice [10, 11], increasing evidence suggests that many solid tumors are driven by a subpopulation of CSCs [12–15]. However, the origin of the CSCs is still a matter of debate. In some cancers, critical gene mutations could transform normal stem cells or progenitor cells into CSCs [16, 17]. Indeed, recent cell lineage tracing analysis further conforms CSCs exist in tumor's native environment and can be derived from the mutation of somatic stem cells [18, 19]. Other recent studies, however, demonstrate that differentiated cells could dedifferentiate into CSC-like cells [20–22]. In the present chapter, we demonstrate that long-term, low-level arsenic stress dedifferentiates human bronchial epithelial cells into CSC-like cells via c-Myc activation.

2 Materials

2.1 BEAS-2B Cells

The human bronchial epithelial cell line (BEAS-2B) was purchased from the American Type Culture Collection (ATCC, Manassas, VA) and cultured in DMEM medium supplemented with 10% FBS (*see* **Note 1**). All cell culture mediums were purchased from Sigma (Ronkonkoma, NY). The BEAS-2B cell line was derived from human bronchial epithelial cells through immortalization by SV-40 large T antigen. The cells retain the ability to undergo squamous differentiation in response to serum as well stress signals. BEAS-2B cells is the only human lung cell line that had been widely used in malignant transformation studies by heavy metals or metalloids. Because of SV-40 large T immortalization, it was assumed that p53 was inactivated in this cell line. However, genetic test demonstrated functional p53 protein in BEAS-2B cells, although the p53 gene showed missense mutation from CCG to TCG at codon 47 that results in Pro to Ser substitution, and the codon 72 polymorphism from CGC to CCC, causing Arg to Pro conversion [23, 24]. Indeed, our previous p53 reporter gene activity assay suggested a normal p53 response to arsenic of the BEAS-2B cells (data not shown). Other studies also indicated Ser15 and Ser392 phosphorylation of p53 induced by toxic metals [24].

2.2 Sodium Arsenic

Arsenic was commonly viewed as a carcinogenic metal for human, although element arsenic is classed as a metalloid chemically. Arsenic has been known since antiquity in the form of its sulfide compounds. In carcinogenic experiments, the commonly used arsenic include sodium arsenic ($NaAsO_2$), arsenic trioxide (As_2O_3), and arsenic chloride ($AsCl_3$). All of these forms of arsenic exhibited similar absorption, distribution, metabolism, excretion, and the stress responses in animal tests. In the present study, we used sodium arsenic purchased from Sigma in cell transformation and CSC experiments (*see* **Note 2**).

2.3 Antibodies

All antibodies were purchased from Santa Cruz Biotechnology (Santa Cruz, CA), Cell Signaling Technology (Danvers, MA), BD Biosciences (San Jose, CA), or Abcam (Cambridge, MA) (*see* **Note 3**).

3 Methods

3.1 Consecutive Arsenic Treatment of the BEAS-2B Cells (see Note 1)

For arsenic treatment, BEAS-2B cells were treated with 0.25 μM $NaAsO_2$ continuously for six months (*see* **Note 1**). Cells cultured without As^{3+} treatment for six months were used as parental cell control in all experiments. Cell transformation was evaluated by using in vitro soft agar colony-forming assay and in vivo tumorigenicity experiment in athymic nude mice. For the soft agar colony-forming assay, 1% agar was melted and mixed with 2× DMEM medium in a 1:1 ratio to produce a supporting layer (0.5%) in a 6-well tissue culture plate (1 mL/well). The bottom agar layer was allowed to solidify at room temperature for 20 min. The top layer containing 0.3% agar (1 mL/well) was prepared by mixing stock agar solutions with 5000 cells in 2× DMEM medium and was laid on the top of the bottom agar layer. Fresh DMEM medium was added every 3 days. Three weeks later, colonies were stained with a 0.05% crystal violet solution, counted, and photographed under a microscope.

3.2 Flow Cytometry Analysis and Fluorescence-Activated Cell Sorting (FACS) (see Note 5)

The cells were dissociated, and one million individual cells were resuspended in 100 μL of sorting buffer (PBS containing 0.5% FBS, 2 mM EDTA) and then stained with 20 μL of the indicated antibodies conjugated with PE or FITC purchased from BD Biosciences (San Jose, CA) for 30 min at 4 °C (*see* **Note 5**). Antibodies against human immunoglobulin conjugated to PE or FITC (BD) were used as antibody isotype controls. The cells were washed in PBS and centrifuged at 800 × *g* for 2 min twice and then resuspended in 500 μL sorting buffer. FACS analysis was performed using CellQuest software (BD Biosciences). The cells were sorted on the FACS Vantage SE/SORP with a Diva flow cytometer (BD Biosciences).

3.3 In Vitro Tumor Sphere Formation (see Note 4)

The self-renewal capability of the transformed cells was determined by an in vitro tumor sphere formation assays. Briefly, 5000 cells were seeded onto ultralow-attachment 6-well plates (Corning life sciences, Tewksbury MA) in serum-free DMEM/F12 (Invitrogen, Grand Island, NY) medium containing 1 × B27 supplement (Invitrogen), 20 ng/mL EGF (Invitrogen), and 20 ng/mL bFGF (Invitrogen). Fresh medium was added every 3 days for 2 weeks. Tumor spheres were counted and photographed using a microscope. For tumor sphere passage assays, the primary spheres were dissociated and reseeded in ultralow-attachment 6-well plates for the second round of tumor sphere formation. All experiments were performed in triplicate for each condition and were repeated at least three times.

3.4 Western Blot Analysis (see Notes 3 and 6)

Cells were lysed in 1 × RIPA lysis buffer containing protease and phosphatase inhibitors on ice. Each sample, containing 40 μg of protein, was resolved on a 4–12% Bis-Tris Gel with MOPs Running Buffer and transferred to PVDF membranes. The blots were then probed with various antibodies. Horseradish peroxidase (HRP)-conjugated secondary antibodies were used to detect the specific immunoreactions. Blots were developed with an enhanced chemiluminescence regent (Pierce) and exposed to X-ray film. The following primary antibodies were used: Oct4, Sox2, Klf4, c-Myc, Nanog, TCF4, PBX1, EBF3, CD44, CD49f, E-Cad, KRT5, Fibronectin, p63, Vimentin, β-catenin, EZH2, pEZH2, JNK, pJNK, Stat3, pStat3, Akt, pAkt, Smad3, pSmad3^{s208}, Actin, and GAPDH. All antibodies were purchased from Cell Signaling Technology (Danvers, MA), Abcam (Cambridge, MA), or Santa Cruz Biotechnology, INC. (Santa Cruz, CA).

3.5 Gene Expression Profiling and Real-Time PCR (see Note 7)

Total RNA was extracted from 1×10^7 parental cells, As^{3+}-induced transformed cells and tumor sphere-forming cells using TRIzol reagent (Invitrogen) and dissolved in diethyl pyrocarbonate-treated H_2O to a final concentration of 500 ng/μL. RNA integrity was determined by electrophoresis. Gene expression profiling was performed using the Human Gene 1.0 ST Array (Affymetrix). To identify differentially expressed genes between the parental cells, transformed cells and tumor sphere-forming cells, data were uploaded to Partek® software (version 6.6 Copyright © 2012 Partek Inc., St. Louis, MO) using the Exon workflow. Quantile normalization, log2-transformation, probeset summarization (median polish), and background subtraction (RMA) were performed on the data upon import. Exons were summarized to genes by averaging exon expression across each gene. Differential expression was calculated for each comparison by performing an analysis of variance (ANOVA). Genes with a change of at least ±1.5-fold and an FDR-corrected p-value <0.05 are reported. Hierarchical clustering (Euclidean Distance and Average linkage clustering) was performed using the TM4 Microarray Software Suite's MultiExperiment Viewer V4.8. The gene expression profiling data were validated by quantitative real-time PCR, as reported previously.

3.6 Tumorigenicity Assay in Nude Mice (see Note 2)

For the evaluation of tumorigenicity, 1×10^6 parental or transformed cells were injected subcutaneously into the flank of 6-week-old male BALB/c nude mice purchased from the Jackson Laboratory (Bar Harbor, ME). Tumor size was measured on the indicated days with calipers, and tumor volume was calculated based on the formula $(\text{width}^2 \times \text{length})/2$ (*see* **Note 2**). All animal procedures were preapproved by the Institutional Animal and Care Use Committee (IACUC) of the Wayne State University.

3.7 Transplantation Assays Using CSCs in NOD/SCID Il2rγ$^{-/-}$ Mice (see Note 2)

To evaluate the tumorigenicity of the putative CSCs, aliquots of 100 or 500 parental cells, transformed cells, CD61$^+$ cells, or CD61$^-$ cells were injected subcutaneously into the flank of 6-week-old male NOD/SCID Il2rγ$^{-/-}$ mice purchased from Jackson Laboratory (Bar Harbor, ME). The mice were euthanized on the indicated days, followed by the removal of tumors. Tumor size was measured by volume, as calculated based on $(\text{width}^2 \times \text{length})/2$, or by direct weighting. For the serial xenotransplantation of the CSCs, the primary tumors or the secondary tumors that were removed from NOD/SCID mice were dissociated mechanically and enzymatically to generate single cell suspensions for a subsequent injection into the new NOD/SCID mice.

3.8 Cell Migration and Invasion (see Note 7)

In vitro migration assay was carried out in insert of Transwell plate (Corning, #3422). Cells (5×10^3) in 100 μL serum-free DMEM medium were placed in the upper chambers and 500 μL DMEM containing 10% FBS was placed in the lower chamber. After 24 h incubation, the cells were fixed in 4% formaldehyde and stained with 0.5% crystal violet. Nonmigrating cells on the upper side of the chamber were removed using cotton swab. Migrating cells were photographed and counted in three randomly selected fields.

In vitro invasion assay was performed in BD BioCoat™ Matrigel™ Invasion Chamber (BD Biosciences, #354480). After rehydration of the chambers, 1×10^5 cells in 500 μL serum-free DMEM medium were added into the upper chambers and 750 μL DMEM containing 10% FBS was placed in the lower chamber. After 16 h incubation, the cells were fixed in 4% formaldehyde and stained with 0.5% crystal violet. Noninvading cells on the upper side of the chamber were removed using cotton swab. Invading cells were photographed and counted in three randomly selected fields. All experiments were performed in triplicate and repeated three times.

3.9 Apoptosis Analysis (see Note 7)

The apoptotic cells were measured by FACS analysis with annexin-V plus PI staining according to the protocol of FITC Annexin V apoptosis Detection Kit I (BD Biosciences, #556547).

3.10 Statistical Analysis

In general, all data are shown as the mean ± standard deviation. Statistical significance in differences was determined by either SigmaPlot 11 software or unpaired two-tailed Student's *t*-tests. In all analyses, a $p < 0.05$ was considered statistically significance.

4 Notes

1. To address the carcinogenic effect of arsenic, we treated the BEAS-2B (B-2B) cells with 0.25 μM $NaAsO_2$ for 6 month, and found morphologic changes of the cells from epithelial-like to fibroblast-like cells as compared to the control counterpart cells (Fig. 1a). Moreover, the colony formation capacity of arsenic-treated BEAS-2B cells (Tas-B-2B) in soft agar increased significantly compared to BEAS-2B cells (Fig. 1b).
2. To further test the tumorigenic capacity of Tas-B-2B cells in vivo, 1×10^6 B-2B and 1×10^4 Tas-B-2B cells were injected into the nude mice, respectively. As early as 4 days inoculation, tumor formation was observed in 100% of mice inoculated with Tas-B-2B cells (4/4), but no tumors were found in the mice inoculated with B-2B cells (0/4). After 17 days injection, the average diameter of tumors of Tas-B-2B cells was up to 1 cm (Fig. 2). Again, no tumors were detected in the mice inoculated with B-2B cells. Collectively, our results show that low concentration of arsenic could induce malignant transformation of the human bronchial epithelial cells.
3. Increasing evidences suggest that some aggressive cancer cells share the similar gene expression profiles and surface markers with embryonic stem cells and induced pluripotent stem cells [25]. Therefore, we first determined whether the embryonic stem cell markers were expressed in Tas-B-2B cells. Western blot analyses showed that the embryonic stem cell markers Oct3/4, Sox2, Klf4, and c-Myc were all overexpressed in Tas-B-2B cells as compared with B-2B cells (Fig. 3a). To further determine the self-renewal ability of Tas-B-2B cells in vitro, tumorsphere formation assay was performed. Tas-B-2B cells remained viable and formed tumorspheres after 3 days of culture in serum-free medium. In contrast, B-2B cells could not survive in the same culture condition (Fig. 3b).
4. In colony formation assay when the cells were cultured in soft agar, we found the sphere-forming cells retain the capacity of colony formation as compared to the original Tas-B-2B cells (Fig. 4a, b). All of these data clearly indicated to us that some of the $NaAsO_2$-induced Tas-B-2B cells have characteristics of the CSCs.

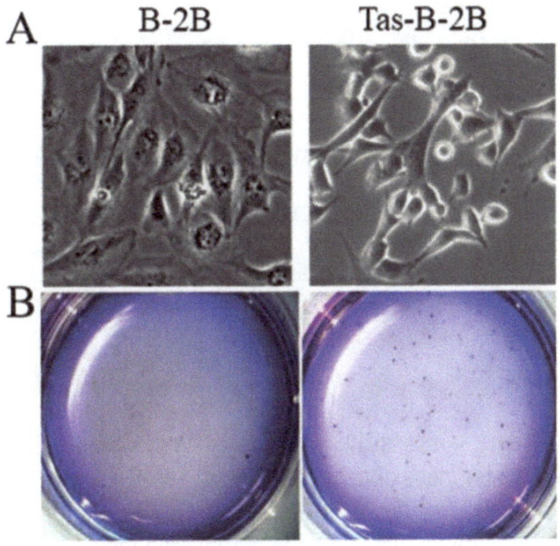

Fig. 1 Long-term and low concentration of arsenic treatment induces malignant transformation of the BEAS-2B cells. (**a**) Representative phase contrast images of BEAS-2B (B-2B) and transformed BEAS-2B (Tas-B-2B) cells. The cells were treated with 0.25 μM $NaAsO_2$ for 6 months. (**b**) Colony formation in soft agar. Both B-2B and Tas-B-2B (5×10^3 cells) were plated in 0.3% soft agar medium. After 4 weeks growth, the colonies were stained and photographed

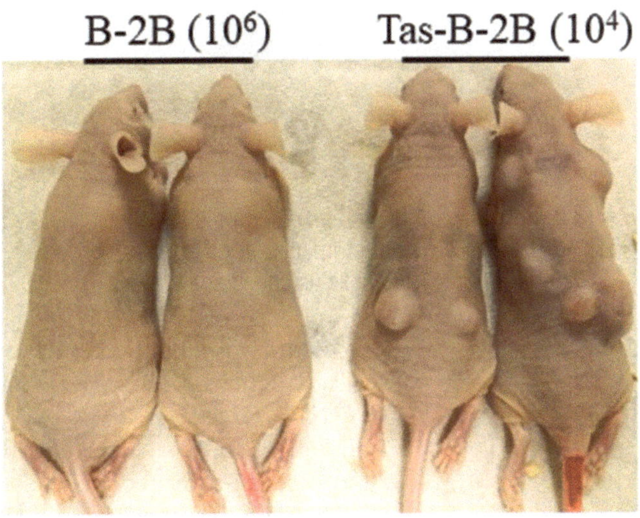

Fig. 2 The arsenic-transformed cells are tumorigenic. B-2B (1×10^6) or Tas-B-2B cells (1×10^4) were injected subcutaneously into nude mice. Tumors were assessed by day 17

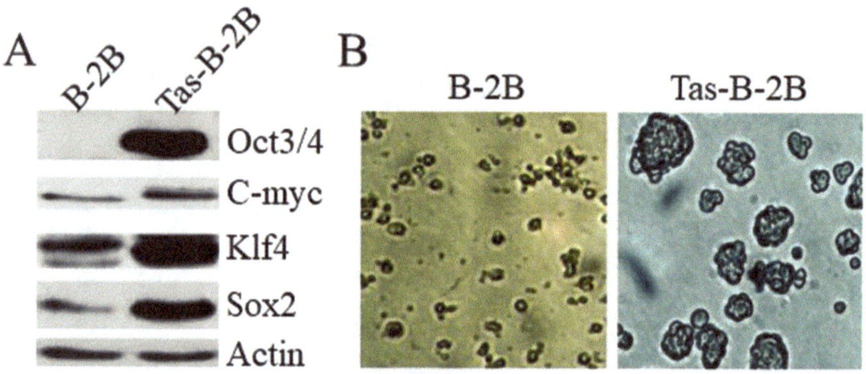

Fig. 3 Tas-B-2B cells possess the characteristics of CSC-like cells. (**a**) Western Blot analysis of embryonic stem cell markers (stemness genes) in B-2B and Tas-B-2B cells. (**b**) Representative phase-contrast images of tumor sphere formation of the B-2B (left) and Tas-B-2B (right) cells. Both cells (5×10^3 cells/well) were seeded in six-well plate in suspension in serum-free medium for 5 days

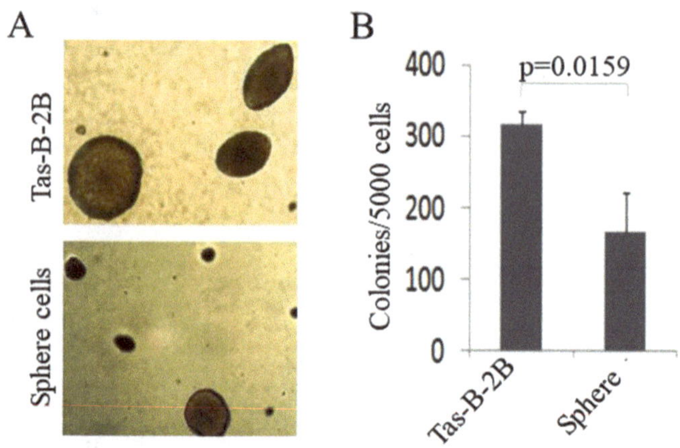

Fig. 4 Tas-B-2B forms large colonies in soft agar. (**a**) Representative phase-contrast images of colonies in soft agar. (**b**) Quantification of colonies formed from cells as shown in (**a**)

5. To identify potential cell surface markers of CSCs in Tas-B-2B cells, a panel of 15 stem/progenitor cell markers, including CD133, CD44, c-kit, CD24, EpCAM and CXCR4, and CD166, were first screened in B-2B and Tas-B-2B cells by fluorescence-activated cell sorting (FACS) analysis. Among the markers screened, only CD61 showed differential expression between B-2B and Tas-B-2B cells. In B-2B cells, almost all the cells (99.54%) were CD61 positive cells, whereas only 13.01% of the long-term $NaAsO_2$-induced Tas-B-2B were CD61 positive cells, the rest of the Tas-B-2B cells were CD61 negative (Fig. 5a). Morphologically, the $CD61^+$ Tas-B-2B cells remained epithelial-like, whereas $CD61^-$ Tas-B-2B cells showed fibroblast-like morphology (Fig. 5b).

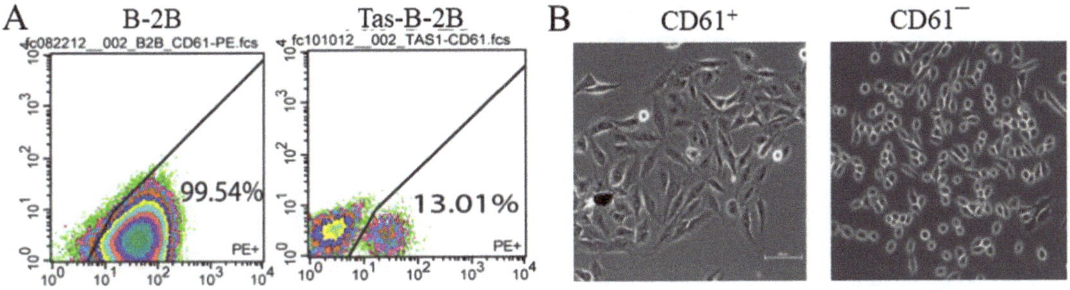

Fig. 5 The arsenic-induced CSC-like cells are CD61−. (**a**) FACS analysis of B-2B and Tas-B-2B cells with CD61 antibody staining. (**b**) Phase-contrast images of CD61+ Tas-B-2B and CD61− Tas-B-2B cells at low confluency

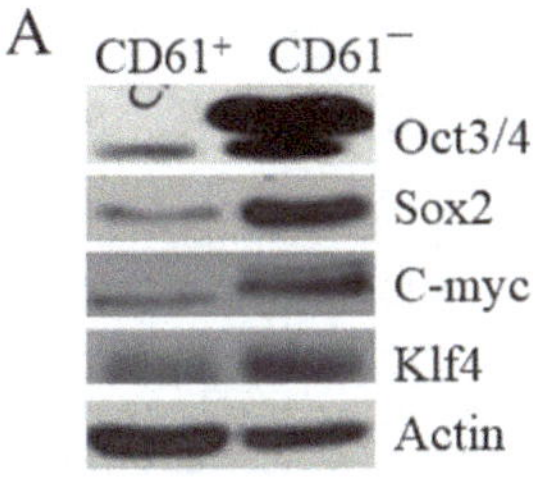

B

Tumorigenesis in NOD/SCID mice

Cell Number	100,000	1000	100
CD61+ Cells	3/3	1/4	0/4
CD61- Cells	3/3	4/4	4/4

Fig. 6 CD61− Tas-B-2B cells express higher levels of the stemness genes and tumorigenic. (**a**) Western Blot analysis of stem cell markers in CD61+ Tas-B-2B and CD61− Tas-B-2B cells. (**b**) Tumor formation of CD61+ Tas-B-2B or CD61− Tas-B-2B cells in NOD/SCID mice. Left flanks were injected with CD61+ Tas-B-2B cells while right flanks were injected with CD61− Tas-B-2B cells. Tumor formation were evaluated and determined at day 27 after inoculation

6. Next, the key stemness genes were screened between $CD61^+$ Tas-B-2B and $CD61^-$ Tas-B-2B cells. We found that Oct3/4, Sox2, Klf4, and c-Myc were all upregulated in $CD61^-$ Tas-B-2B cells compared to $CD61^+$ Tas-B-2B cells (Fig. 6a). Furthermore, tumorigenic capacity was determined in NOD/SCID mice. When 1×10^5 cells were injected, 100% (3/3) of both $CD61^+$ Tas-B-2B and $CD61^-$ Tas-B-2B cells could form tumors (Fig. 6b). When 1000 cells injected, only 25% of mice generated tumors (1/4) with $CD61^+$ cell inoculation, compared to 100% of mice generated tumors (4/4) with $CD61^-$ Tas-B2B cell inoculation. Subcutaneous injection of as few as 100 $CD61^-$ cells conferred 100% tumor formation in NOD-SCID mice (4/4). In contrast, no tumors were detected in $CD61^+$ cells injection when only 100 cells were inoculated (0/4) (Fig. 6b). Together, these results indicated that $CD61^-$ Tas-B-2B cells possess much higher tumor-initiating capacity compared to the counterpart $CD61^+$ Tas-B-2B cells.
7. In addition to the features described above, we also noted resistance of the $CD61^-$ Tas-B-2B cells to arsenic- and typical chemodrug-induced cell apoptosis, differentiation of the $CD61^-$ Tas-B-2B cells to the $CD61^+$ cells, and these cells

showed atypical epithelial–mesenchymal transition (EMT). In conclusion, we found that consecutive lower concentration of arsenic (0.25 μM) treatment for 6 months can induce transformation of the BEAS-2B cells, and some of the transformed cells clearly have characteristics of the CSCs.

Acknowledgments

This work was supported by NIH grants R01 ES028263, R01 ES028335, and P30 ES020957 to FC.

References

1. Lubin JH, Beane Freeman LE, Cantor KP (2007) Inorganic arsenic in drinking water: an evolving public health concern. J Natl Cancer Inst 99(12):906–907
2. Li L et al (2019) Metabolic and epigenetic reprogramming in the arsenic-induced cancer stem cells. Semin Cancer Biol 57:10–18
3. Liu J, Waalkes MP (2008) Liver is a target of arsenic carcinogenesis. Toxicol Sci 105 (1):24–32
4. Jemal A et al (2011) Global cancer statistics. CA Cancer J Clin 61(2):69–90
5. Sun S, Schiller JH, Gazdar AF (2007) Lung cancer in never smokers-a different disease. Nat Rev Cancer 7(10):778–790
6. Mead MN (2005) Arsenic: in search of an antidote to a global poison. Environ Health Perspect 113(6):A378–A386
7. Guo HR (2004) Arsenic level in drinking water and mortality of lung cancer (Taiwan). Cancer Causes Control 15(2):171–177
8. Clarke MF et al (2006) Cancer stem cells-perspectives on current status and future directions: AACR Workshop on cancer stem cells. Cancer Res 66(19):9339–9344
9. Chang Q, Chen B, Thakur C, Lu Y, Chen F (2014) Arsenic-induced sub-lethal stress reprograms human bronchial epithelial cells to CD61 cancer stem cells. Oncotarget 5 (5):1290–1303
10. Lapidot T et al (1994) A cell initiating human acute myeloid leukaemia after transplantation into SCID mice. Nature 367(6464):645–648
11. Bonnet D, Dick JE (1997) Human acute myeloid leukemia is organized as a hierarchy that originates from a primitive hematopoietic cell. Nat Med 3(7):730–737
12. Al-Hajj M, Wicha MS, Benito-Hernandez A, Morrison SJ, Clarke MF (2003) Prospective identification of tumorigenic breast cancer cells. Proc Natl Acad Sci U S A 100 (7):3983–3988
13. O'Brien CA, Pollett A, Gallinger S, Dick JE (2007) A human colon cancer cell capable of initiating tumour growth in immunodeficient mice. Nature 445(7123):106–110
14. Singh SK et al (2004) Identification of human brain tumour initiating cells. Nature 432 (7015):396–401
15. Eramo A et al (2008) Identification and expansion of the tumorigenic lung cancer stem cell population. Cell Death Differ 15(3):504–514
16. Barker N et al (2009) Crypt stem cells as the cells-of-origin of intestinal cancer. Nature 457 (7229):608–611
17. Krivtsov AV et al (2006) Transformation from committed progenitor to leukaemia stem cell initiated by MLL-AF9. Nature 442 (7104):818–822
18. Driessens G, Beck B, Caauwe A, Simons BD, Blanpain C (2012) Defining the mode of tumour growth by clonal analysis. Nature 488 (7412):527–530
19. Schepers AG et al (2012) Lineage tracing reveals Lgr5+ stem cell activity in mouse intestinal adenomas. Science 337(6095):730–735
20. Friedmann-Morvinski D et al (2012) Dedifferentiation of neurons and astrocytes by oncogenes can induce gliomas in mice. Science 338 (6110):1080–1084
21. Landsberg J et al (2012) Melanomas resist T-cell therapy through inflammation-induced reversible dedifferentiation. Nature 490 (7420):412–416
22. Schwitalla S et al (2013) Intestinal tumorigenesis initiated by dedifferentiation and acquisition of stem-cell-like properties. Cell 152 (1–2):25–38

23. Matlashewski GJ et al (1987) Primary structure polymorphism at amino acid residue 72 of human p53. Mol Cell Biol 7(2):961–963
24. Park YH, Kim D, Dai J, Zhang Z (2015) Human bronchial epithelial BEAS-2B cells, an appropriate in vitro model to study heavy metals induced carcinogenesis. Toxicol Appl Pharmacol 287(3):240–245
25. Wong DJ et al (2008) Module map of stem cell genes guides creation of epithelial cancer stem cells. Cell Stem Cell 2(4):333–344

INDEX

Benjamin L. Kidder (ed.), *Stem Cell Transcriptional Networks: Methods and Protocols*, Methods in Molecular Biology, vol. 2117, https://doi.org/10.1007/978-1-0716-0301-7, © Springer Science+Business Media, LLC, part of Springer Nature 2020

J

L

M

N

O

P

Q

R

Kolibri 360 GmbH
Ferdinand-Jühlke-Straße 7
99095 Erfurt, Deutschland
produktsicherheit@kolibri360.de